Computational Materials Science

AN INTRODUCTION

SECOND EDITION

Computational Materials Science

AN INTRODUCTION

SECOND EDITION

June Gunn Lee

CRC Press is an imprint of the
Taylor & Francis Group, an informa business

CRC Press
Taylor & Francis Group
6000 Broken Sound Parkway NW, Suite 300
Boca Raton, FL 33487-2742

Printed on acid-free paper
Version Date: 20160830

International Standard Book Number-13: 978-1-4987-4973-2 (Hardback)

Library of Congress Cataloging-in-Publication Data

Names: Lee, June Gunn, author.
Title: Computational materials science : an introduction / June Gunn Lee.
Description: Second edition. | Boca Raton : CRC Press, Taylor & Francis, 2017. | Includes bibliographical references and index.
Identifiers: LCCN 2016025825 | ISBN 9781498749732 (alk. paper)
Subjects: LCSH: Materials--Mathematical models. | Materials--Data processing. | Molecular dynamics--Mathematics.
Classification: LCC TA404.23 .L44 2017 | DDC 620.1/10113--dc23
LC record available at https://lccn.loc.gov/2016025825

Visit the Taylor & Francis Web site at
http://www.taylorandfrancis.com

and the CRC Press Web site at
http://www.crcpress.com

Dedication

to Hak Nam, Susan and Ray, Sandra and Dongho, Gia, Joseph, James, and Gina

Contents

Preface

No longer underestimated, computational science has emerged as a powerful partner to experimental and theoretical studies. Accelerated by the ever-growing power of computers and new computational methods, it is one of the fastest growing fields in science these days. Its predictive power in atomic and subatomic scales benefits all disciplines of science, and materials science is definitely one of them. Note that, for example, materials under extreme conditions such as high temperature or pressure, high radiation, on a very small scale, can be rather easily examined via the keyboard in computational materials science.

Computational science has been a familiar subject in physics and chemistry, but in the materials field it was considered of secondary importance. It is now in the mainstream, and we have to catch up with the knowledge accumulated in the subject, which strongly involves physics and mathematics. Here, we are forced to deal with an obvious question: how much catch-up will be enough to cover the major topics and to perform computational works as materials researchers? Dealing with the entire field might be most desirable, but many certainly prefer to cover only the essential and necessary parts and would rather be involved in actual computational works. That is what this book is all about.

As listed in the "Further Readings" sections in several chapters, a number of excellent and successful books are already available in this field. However, they are largely physics- or chemistry-oriented, full of theories, algorisms, and equations. It is quite difficult, if not impossible, for materials students to follow all those topics in detail. This book intends to do some sorting and trimming of the subject, so that many are able to venture into the field without too much difficulty. Thus, this book is exclusively devoted to students who would like to venture into computational science and to use it as a tool for solving materials problems. Once this book is completed, one may go further to the more advanced books as listed in "Further Readings."

A materials system may be described at three different levels: the electronic structure level of nuclei and electrons, the atomistic or molecular level, and the finite element level of coupled structural elements. In this book, we will deal with only the first two levels of the materials system and their computational treatments with molecular dynamics (MD) and first-principles methods, which are most relevant and essential in materials science and engineering. With these tools, we simply try to bring a very small part of nature on computer as a system, and apply the known rules of nature to solve a certain problem at hands, especially on materials.

This book is organized into nine chapters, starting with Chapter 1, which gives a general overview of computational science. Chapter 2 introduces MD methods based on classical mechanics: Its implementation into actual calculations follows in Chapter 3 with run examples of XMD and LAMMPS, respectively. Chapter 4 introduces first-principles methods based on quantum mechanics on a brief introductory level. Here, various illustrations and appropriate analogies will be presented to assist students to understand this tough subject. Chapter 5 is dedicated solely to the density functional theory (DFT) in detail, because this is the very first-principles method that can handle materials practically.

Chapter 6 exclusively deals with solids and reveals how bulk materials can be represented with a handful of *k*-points. The chapter also provides how each orbital of electron leads to particular properties of solids such as total energy, band structure, and band gap. Finally, Chapters 7 through 9 implement the DFT into actual calculations with various codes such as Quantum Espresso, VASP, and MedeA-VASP, respectively. They cover from an atom to solids, and from simple GGA to GGA+U and hybrid methods. Chapter 9 specifically deals with advanced topics in DFT counting dispersion, +U, DFT with hybrid XC potentials, and *ab initio* MD by using a convenient GUI program, MedeA-VASP.

Note that methods once considered as "too expensive" are now practical enough to treat materials, owing to the ever-increasing power of computers. Various postprocessing programs such as VESTA, VMD, and VTST will be exercised through the runs.

For using this book as a textbook in the classroom, here are some tips and suggestions for a course outline:

- The contents are so arranged that one semester will be enough to cover this book.
- Lectures and run exercises may be conducted simultaneously, for example, Chapter 2 with Chapter 3, and Chapters 5 and 6 with Chapters 7 through 9.
- Most exercises with XMD (Chapter 3), LAMMPS (Chapter 3), and Quantum Espresso (Chapter 7) are so arranged that they can be carried out on student's notebook computers or PCs in a reasonable time.
- Most exercises with VASP (Chapter 8) can be carried out on any mini-supercom with more than eight CPUs via remote access from

the classroom. However, only starting the run during class and providing the results at the next class will be an excellent option.
- The exercises with MedeA-VASP (Chapter 9) can be carried out on any mini-supercom with more than eight CPUs via remote access from the classroom. Again, only starting the run during class and providing the results at the next class will be an excellent option.

Writing a book is a variational process. One minimizes mistakes iteratively, but it never goes down to the ground-state (zero mistake). Erratum will be posted under http://www.amazon.com/June-Gunn-Lee/e/B005N2XON0/ref=ntt_dp_epwbk_0.

Some final remarks concerning the preparation of this book:

- Analogies and illustrations used in this book may have been exaggerated to emphasize certain points.
- Figures used in this book often show the general features, neglecting the details or exact numbers.

During the course of writing this book, I have been privileged to have much support. I am particularly grateful to three of my colleagues at the Computational Science Center, Korea Institute of Science and Technology (KIST): Dr. Kwang-Ryoel Lee, Dr. Seung-Cheol Lee, and Dr. Jung-Hae Choi. Without their support and advice, this book would not have been possible. I am indebted to all my friends who kindly provided examples and scripts for this book: Professor Ho-Seok Nam (Kookmin University, Seoul), Professor Aloysius Soon (Yonsei University, Seoul), Dr. Sang-Pil Kim (Samsung Electronics, Suwon), Dr. Jinwoo Park (Sejong University, Seoul), Na-Young Park (KIST, Seoul), Dr. Joo-Whi Lee (Kyoto University, Kyoto), Byung-Hyun Kim (Samsung Electronics, Suwon), Dr. Jung-Ho Shin (Humboldt-Universität zu Berlin, Berlin), Professor Yeong-Cheol Kim (KoreaTech, Cheonan, Korea), and Dr. Ji-Su Kim (KoreaTech, Cheonan, Korea).

Author

June Gunn Lee is an emeritus research fellow at the Computational Science Center, Korea Institute of Science and Technology (KIST), Seoul, where he served for 28 years. He has also lectured at various universities in Korea for over 20 years. He has published about 70 papers both on engineering ceramics and computational materials science. Dr. Lee is a graduate of Hanyang University, Seoul, and acquired his PhD in materials science and engineering from the University of Utah. He has been involved in computational materials science ever since he was a visiting professor at Rutgers University, New Jersey, in 1993. Currently, he is lecturing at University of Seoul, Seoul.

chapter one

Introduction

It is amazing how much computing power has progressed since the invention of the Chinese abacus: from slide rule to mechanical calculator, vacuum-tube computer, punch-card computer, personal computer, supercomputer, and cluster-supercomputer (Figure 1.1). Calculation speed has increased roughly 10^{10} times in the span of about 50 years and is still growing. Its immense impact on every sector of society is astonishing. In this chapter, the significance of computational science in materials is addressed, and a brief description of the various methods is presented. The last section will provide remarks on the development of computers.

1.1 Computational materials science

1.1.1 Human beings versus matter

We may say that we are all alike since all of us talk, work, play, and eat in a very similar manner. While this is true, our individual behaviors are so versatile as a result of different feelings, emotions, ideologies, philosophies, religions, and cultures that no individuals actually behave in exactly the same way. Despite scientific studies on human behaviors and mental processes, it is very difficult to predict events that involve humans. The numerous upsets in battle and sports aside, we cannot even tell whether the person next to us will stand up or remain seated in the very next moment.

On the contrary, atoms and their constituting electrons and nuclei (the main characters at the arena of computational materials science) always follow specific rules without exceptions. If an exception is observed, it is more likely that it is a mistake due to a human error. Unlike humans, an electron never kills itself or other electrons because of love. The same goes for nuclei, atoms, molecules, and materials. They simply follow the known laws of nature, which are mainly classical and quantum mechanics in terms of electromagnetic force (Figure 1.2). Therefore, we can foresee practically all forms of phenomena encountered in materials by tracing down the interactions between them.

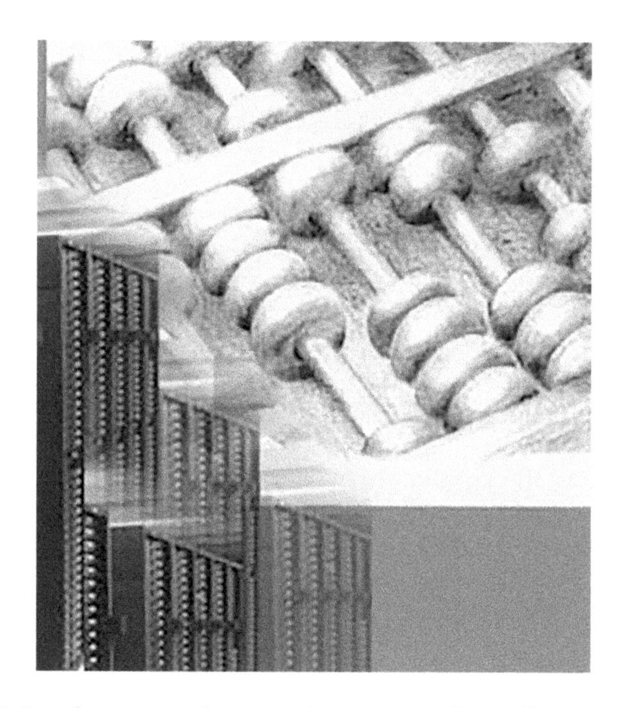

Figure 1.1 Developments of computing power: from Chinese abacus to cluster-supercomputer.

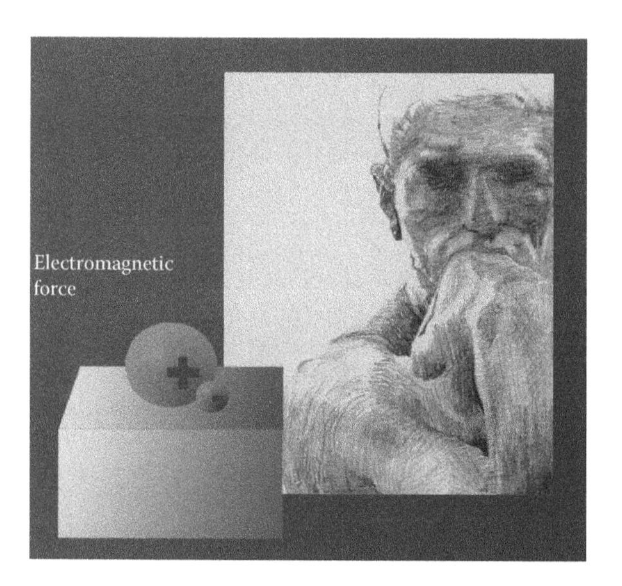

Figure 1.2 Complexity involved in human events and simplicity associated with matters.

1.1.2 Computational materials science

1.1.2.1 Goals

In computational materials science, we aim to understand various properties and phenomena of materials and achieve designing and making better materials for society. This goal is realized by modeling materials with computers that are programmed with theories and algorithms based on physics, mathematics, chemistry, material science, and computer science. For example, the sintering behavior of a metal or a ceramic can be normally studied with the usual sintering furnace in a laboratory. However, it can be done on a computer by using molecular dynamics (MD) on atomic scale. By changing various input conditions, the entire spectra of data can be generated efficiently and accurately if the runs are set up properly.

In many cases, a computational approach may become the only way to handle materials under extreme and hostile conditions that can never be reached in a laboratory: under high pressures, at high temperatures, and in the presence of toxic substances or nuclear radiation. For example, the materials under nuclear fusion environments are of great concern these days. The various damages occurring in fusion materials by neutron irradiation can be simulated without worrying about expensive equipment and danger of radiation.

Let us look at another example that has great impact on our daily lives. Every day we use a cell phone, smart phone, smart pad, TV, computer, and so on, which employ IC chips usually made of silicon. Using computational materials science, we can design better materials and develop faster, smaller, and lighter IC chips. To summarize, there is no doubt that computational materials science will change the paradigm of materials research. It will change "lab experiments" with heavy equipment to "keyboard science" on computers. Currently, computational materials science is no longer a specialized topic. It has become familiar and routine such as analyzing XRD curves and examining SEM or TEM images. Furthermore, most people recognize that computational materials science is not just an optional topic but an essential one. It is not surprising to hear scientists say "computation first, then experiment" or "material design by computation."

1.1.2.2 Our approach

I wrote the following comparison as an illustration in my book (Lee 2003) published in 2003, but allow me to restate it. For a materials scientist, computational science is a convenient tool such as a car (Figure 1.3). In order to operate a car, it is not necessary to understand how the engine block is cast with molten metal or how the combustion energy transfers from engine to wheels as the mechanical energy of rotation. We only need to

Figure 1.3 Driving on the roads of computational materials science.

know how to use the accelerator, brakes, and steering wheel, as well as be aware of a few traffic rules.

Then, we check the car for maintenance if we notice anything wrong. Similarly, we use computational methods with well-proven codes, run it, obtain data, and check the results with experts if we find something amiss. It is not necessary to understand all the theories, algorithms, and equations in detail. It is also not necessary to understand the codes or programs down to the bits and bytes. This knowledge is reserved for other professionals such as physicists and chemists.

However, it is undeniable that there are still certain basics to be fully comprehended even for materials people. Otherwise, there will be limitations not only to our understanding but also to our performance on actual simulations. This book intends to cover the essential basics and to provide assistance to materials people with many useful illustrations, figures, quotes, and, most of all, step-by-step examples. This will help readers to be on track during a simulation process without losing the connection between the particles involved and the underlying concepts. Then this will lead to a more meaningful interpretation of the results.

1.2 Methods in computational materials science

> Do you want to calculate it, or do you want it to be accurate?
>
> **John C. Slater (1900–1976)**

In this section, the basic procedures and methods in computational materials science are briefly introduced.

1.2.1 Basic procedures of computational materials science

If matter is affected in any way in this world, we can safely say that it is a result of one of the four known fundamental interactions: electromagnetic, strong nuclear, weak nuclear, and gravitational. Fortunately, as far as materials science is concerned, we need to consider only the electromagnetic interaction; we rarely encounter cases with the other three. Thus, what happens in any materials in any circumstance is narrowed down to the electromagnetic interactions between nuclei, electrons, and atoms. Based on this simple fact, the basic procedures of computational materials science may be stated as follows:

- Define what to calculate.
- Make a model system that represents the real system properly.
- Select the relevant rules (classical mechanics, quantum mechanics, theories, algorithms, etc.).
- Select a code/module/program/package to do the job for the system.
- Run simulation, analyze the results, and refine the run under better-defined conditions.
- Produce data and compare them with reported data by other relevant studies and experiments.

In short, we are recreating a part of nature in the frame of our simulation system in a simplified and well-controlled manner. Among these steps, the last should not be underestimated. The relevance of simulation results is derived from somewhat idealized situations and therefore must be critically examined using experimental data.

In the following subsections, the four typical methods for performing the above-mentioned processes are briefly outlined. As shown in Figure 1.4, all four methods have advantages and limitations in terms of system size and simulation time. Multiscale methods, which combine two or more methods, are also included in the figure. Note also that, as time or size scale increases, the discipline changes from physics to chemistry, material science, and engineering.

1.2.2 Finite element analysis

Finite element analysis (FEA) involves dividing the system into many small elements and calculating variables such as stress, strain, temperature, and pressure. Sets of algebraic, differential, and integral equations are solved by a computer-based numerical technique. This method provides solutions to a wide variety of complex engineering problems including materials properties and phenomena (elastic and plastic deformation, fracture, heat transfer, etc.) for the scale of real components.

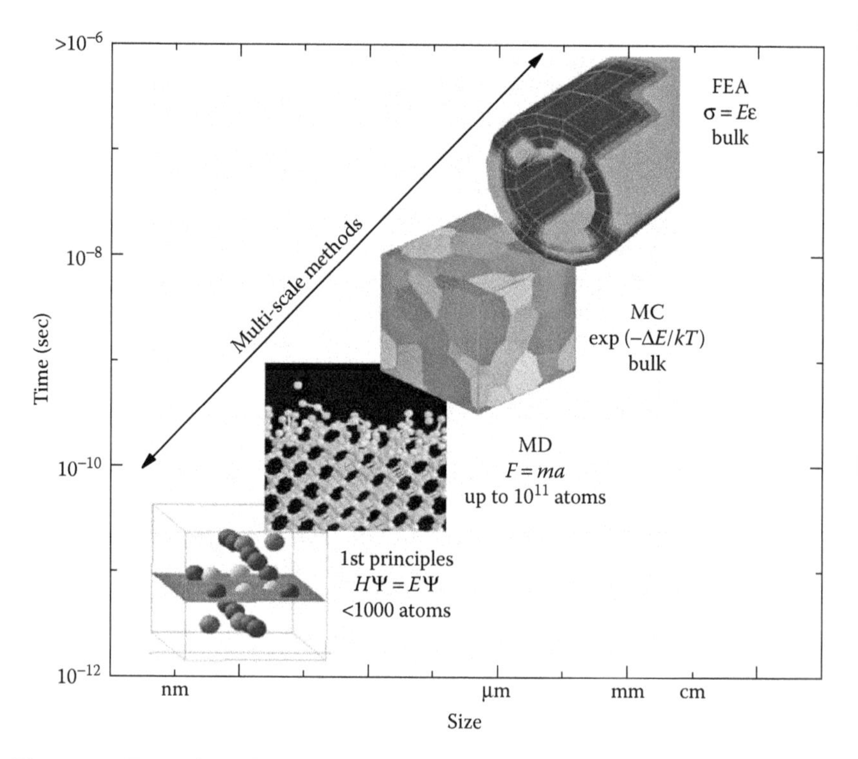

Figure 1.4 Typical methods in computational materials science in terms of size and time.

1.2.3 Monte Carlo method

The Monte Carlo (MC) method is a statistical technique that involves using discrete and random walks for sampling and the Boltzmann factor of $\exp(-U/k_BT)$ for probability to solve problems in materials and many other systems. Here, U is the energy of a state, k_B is the Boltzmann constant, and T is temperature. The method is conceptually very simple and easy to implement since it simply lets the atoms jump around randomly and picks the lowest energy state. It can equilibrate any degree of freedom, and no dynamics is needed since it is based on statistical mechanics. The so-called Metropolis MC normally simulates a thermodynamic ensemble and is used for energy minimization of the system. The Kinetic Monte Carlo (KMC) method generally works for activated processes such as atomic migration. The scale of the MC method is in microns and can thus treat microstructures of materials.

1.2.4 *Molecular dynamics*

MD considers atoms as the basic particles and disregards nuclei or electrons. Therefore, the system can be well described classically by using Newton's equations of motion that involve relatively easy differential equations for only atom–atom interactions. In the 1960s, only several hundreds of atoms were simulated by MD, but the number of atoms has increased to more than 100 billion these days. The catch is that we have to empirically generate the interacting potential between atoms since the origins of potentials (nuclei and electrons) are completely excluded in this method. For the same reason, electronic and magnetic properties cannot be obtained. Once the potential is known and the initial positions and velocities are given, the time-evolutions of the atoms can be revealed in a rather straightforward fashion. Conceptually, MD is not much different from playing screen golf.

1.2.5 *First-principles methods* (ab initio *methods)*

The first-principles methods consider nuclei and electrons as the basic particles and describe events in a subatomic world. The system, therefore, can be expressed only by quantum mechanics that involves relatively difficult partial differential equations. Thus, first-principles calculations do not depend on any external parameters except the atomic numbers of the constituent atoms to be simulated. In the 1930s, only simple hydrogen-like atoms were calculated by quantum mechanical methods, but now the number of atoms has increased to over several thousands by the advent of density functional theory (DFT). All properties including electronic and magnetic ones can be obtained given that nuclei and electrons are considered in the method.

1.2.6 *Remarks*

Computation is a continuous compromise between speed and accuracy because an all-encompassing tool of computation does not exist and never will. Therefore, we have to accept that the system is either too small or the simulation time is too short for our purpose, and the following questions should be thoroughly answered before doing any computational work:

- What is the most suitable method for what I need?
- What is the most suitable software for what I need?
- If MD methods are used, is the potential available or generated?
- Is the computer power at my disposal big enough to do the job?
- If computer resources are limited, what will be an affordable system size?

Table 1.1 Various computational methods and their scaling with system size and the maximum number of atoms, N_{max}, that can be handled in a reasonable time

Methods	Scaling	N_{max} (atoms)
Schrödinger	$O(e^N)$	1
Hartree–Fock	$O(N^4)$	~50
DFT	$O(N) \sim O(N^3)$	~200
MD	$O(N)$	10^{7-11}

Table 1.1 summarizes the various methods introduced in this book and shows how the computation load scales with system size and the maximum number of atoms that can be handled within a reasonable time. With Figure 1.4, this table will provide a general picture of computational materials science. In fact, there is no good or bad method since each approach represents a different balance among computational efficiency, tractable system size, accuracy, and predictive power. For materials science, however, it is apparent that we have no choice but to go for the MD methods or DFT approach, considering that the maximum number of atoms can be handled.

1.3 Computers

Thomas Watson, the former chairman of IBM, surely would like to take back his original statement and restate it as "I think there is a world market for everyone except for maybe five people." Indeed, computers are everywhere, and they have changed our lives in many different aspects. In this age of information technology, even if one does not own a personal computer, his or her daily life still depends on hundreds of other running computers. The sociotechnical impact generated by computers is simply beyond measure. Father may take care of his investment portfolio on the computer within seconds and enjoy screen golf with his friends. Mother may go shopping in a 3D mall. Kids may become legendary knights and save the captured princess in 3D fantasy epics.

Computers have transformed computational science in the same drastic way. Simple problems can be run on a PC. Some complex tasks can be solved on a mini-supercomputer that consists of only tens of CPUs. Complex tasks of big systems can be dealt with via remote access to supercomputers anywhere in the world that consist of hundreds and thousands of CPUs. In fact, a supercomputer with over three million cores, now under operation, recorded a performance speed of 34 petaflop/s (10^{15} floating-point operations per second). Note that Moore's law is still valid. Also, take note of the comment once made by Isaac Asimov (1920–1992): "Part of the inhumanity of the computer is that, once it is competently programmed and working smoothly, it is completely honest." In other words, computers are here to serve us, and the limit could be our imagination and our imagination only.

Reference

Lee, J. G. 2003. *Computational Materials Science: Introduction* (in Korean). Seoul, South Korea: Young Publishing.

chapter two

Molecular dynamics

Whether the famous apple actually fell on him or not, we will never know. Nevertheless something opened Isaac Newton's ingenious mind in 1686 to a new perception of nature and to conceive the equations of motion (Figure 2.1). Mankind seemed finally in a position to understand the rules of the game that nature played. The main equation describing this Newtonian world is as simple as

$$\boldsymbol{F} = m\boldsymbol{a} \tag{2.1}$$

where:

$\boldsymbol{F}$ is force vector
m is atomic mass
$\boldsymbol{a}$ is acceleration vector

Note that, in this chapter and throughout this book, vector quantities are written as bold letters, which signify both magnitude and direction.

The simple Equation 2.1 of three letters and a single mathematical symbol accurately predicted the motion of any particle-like objects: Newton's legendary apple, atoms, a flying baseball over the stadium, planets around the sun, and so on. The method using this classical mechanics for atomic movements is called *molecular dynamics* (MD). As the name indicates, it initially simulated a handful of molecules but soon was extended to liquids, solids, and materials in parallel with the growth in computer power.

In this chapter, by using atoms as the lowest level of information, MD will be treated for the predictions of static and dynamic properties of materials at the introductory level. Readers may find more detail and advanced subjects in the books listed in "Further Reading." For beginners, dealing with MD first will provide a smoother introduction to computational materials science because the subject is relatively easy and straightforward. With this knowledge on MD, we will be in a comfortable position to deal with first-principles methods in later chapters. Note that, however, MD has all the relevant basics for atomic-level resolutions of materials by computation.

Figure 2.1 Sir Isaac Newton and the apple.

2.1 Introduction

The primary goal of materials science is to improve existing materials and to design new materials. In computational materials science, we aim to achieve this goal by modeling and simulation. MD, the subject of this chapter, first started simulating just a bucketful of hard spheres to see the liquid–solid phase transition (Alder and Wainwright 1957). Since then, efficient algorithms and powerful codes along with the ever-upgrading computing power have greatly accelerated the progress. A drastic demonstration is the simulation of 320 billion atoms (Kadau et al. 2006). Although first-principles methods are becoming more and more popular these days, MD is still here to serve us where quantum effects are negligible.

Recall that quantum effects become significant only when the particle wavelength, λ, is comparable with the interatomic distance (1–3 Å). Otherwise, the use of the easier Newton's equations of motion is well justified. For most elements and at higher temperatures, therefore, atomic dynamics can be predicted using Newton's equations of motion. Practically, MD is the only option for big systems of more than thousands of atoms such as melting nanospheres, sintering particles, deforming nanowires, and crack-propagating blocks. In this section, several essential topics for MD will be introduced before we solve Newton's equations of motion.

2.1.1 Atomic model in MD

When we talk about gravity, matter as complex as Earth is often represented as a simple sphere even though we could describe it much better than that. We often present only the abstract instead of the whole paper,

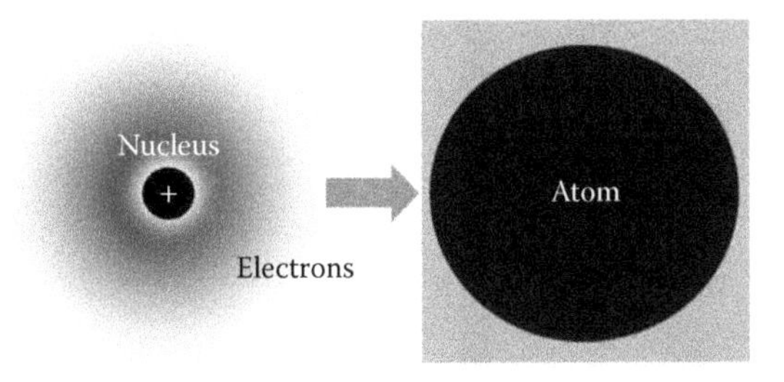

Figure 2.2 Schematic representation of atomic model in MD methods.

and it could still carry enough information to be useful. Approximation works the same way in computations, but only if we do it right. In MD, an atom is basically approximated as a sphere with point mass at the center as schematically drawn in Figure 2.2. This implies that the electron's role is totally neglected, and the electronic wave function is, as will be discussed in Chapters 4 and 5, assumed to instantaneously adapt to the current configuration of atoms. Thus, the computation becomes drastically simple because there is no need to do anything about electrons. There is, however, a negative consequence. Because of the presence of electrons, which is the origin of interatomic potential, is totally neglected, potential between atoms should be generated empirically to carry out MD.

2.1.2 Classical mechanics

In this subsection, some key features of classical mechanics will be reviewed, and this will provide readers some necessary basics to carry out MD. Figure 2.3a represents the characteristics of a classical-mechanics system with a jumping atom with a given energy in a potential well. One may recognize several obvious points here:

- Given the present position and velocity of an atom, its future and past trajectories can be calculated precisely by using Newton's equations of motion.
- Energy changes during a jump are continuous from the very bottom of the well (zero-energy point) to the maximum.
- If an atom's energy is lower than the potential walls, the atom can never jump over the walls.

It all sounds normal and, in fact, nothing is special at all. When we discuss the quantum world in Chapter 4, however, we will realize that this familiar world is a rather exceptional part of the quantum world. Thus, MD is an extension of our everyday experiences such as bouncing and throwing a basketball.

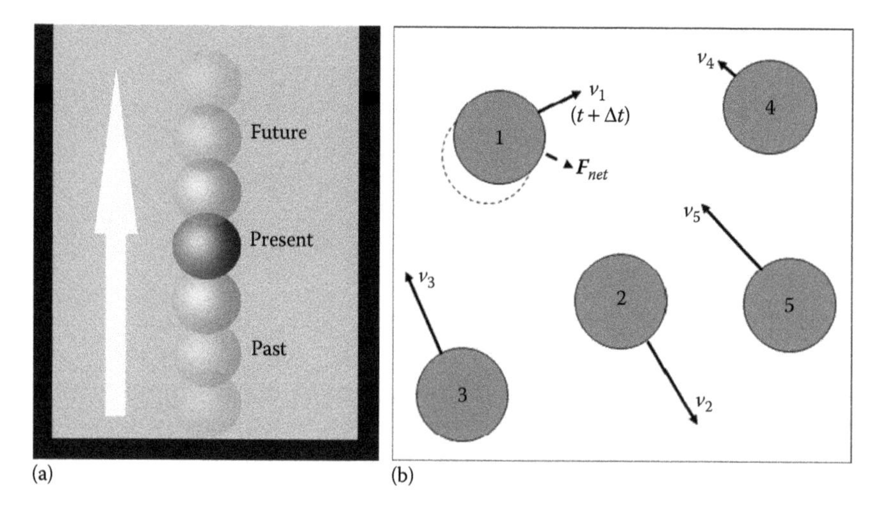

Figure 2.3 (a) A jumping atom (black ball) in a potential well of classical world, (b) A typical procedures of MD in a five-atom system (#1 atom moved from the dotted- to the solid-ball position in Δt by the net force generated with other four atoms).

2.1.3 *Molecular dynamics*

MD is an integration of Newton's equations of motion over time to obtain the time evolution of the system and thus the properties of our interests. It generally proceeds as Figure 2.3b:

- Given the initial positions and velocities of every atom, and using the provided interatomic potential, the forces on each atom are calculated.
- Using this information, the initial positions are advanced toward lower energy states through a small time interval (called a timestep, Δt), resulting in new positions, velocities, and so on.
- With these new data as inputs, the above steps are repeated, typically for more than thousands of such timesteps until an equilibrium is reached, and the system properties do not change with time.

During and after equilibration, various raw data are stored for each or some timesteps that include atomic positions and momenta, energies, forces, and so on. Properties that can be calculated directly or via statistical analysis from these data are as follows:

- Basic energetics, structural and mechanical properties. (note that some of these data were used to fit the potentials empirically.)
- Thermal expansion coefficient, melting point, and phase diagram in terms of pressure and volume.

- Defect structure and diffusion, grain boundary structure, and sliding.
- Heat capacity, free energy differences between phases, and thermal conductivity.
- Radial distribution function and diffusion coefficient for liquids.
- Descriptions for processes and phenomena such as sputtering, vapor deposition, fast plastic flow, crack growth and fast fracture, nanoindentation, propagation of shock wave, detonation, irradiation, ion bombardment, cluster impact, and operation of nanogear.

Compared with the first-principles methods, MD is extremely fast and thus can handle much bigger systems. However, some limitations of MD include

- Availability of potential is limited especially for multicomponent systems, and the accuracy of a potential is always under question.
- Length scale is still not macroscopic, and time scale is also limited to nanoseconds.
- No electromagnetic properties can be obtained.

2.2 *Potentials*

Unlike human beings, who normally try to have more potential, more money, and higher positions, matter always tries to reduce its potential and to be at the bottom of the potential well. Note that rivers run always to the sea and raindrops always form in sphere-like shapes. In this section, we will discuss this potential acting on atoms and how they are generated empirically and yet are accurate enough to be able to describe atomic systems.

When atoms are near enough to feel each other, the balance between attraction and repulsion takes place, and it is determined by so-called interatomic potential. Atoms will eventually settle down at the minimum-potential states at the equilibrium distances following Newton's equations of motion. The sum of all forces acting on an atom, $\boldsymbol{F}$, is

$$\boldsymbol{F} = m\boldsymbol{a} = m\frac{d\boldsymbol{v}}{dt} = m\frac{d^2\boldsymbol{r}}{dt^2} = \frac{d\boldsymbol{p}}{dt} \tag{2.2}$$

where:

$\boldsymbol{a}$ is acceleration
$\boldsymbol{v}$ is velocity
t is time
$\boldsymbol{r}$ is position
$\boldsymbol{p}$ is momentum

Since position $\boldsymbol{r}$ is a vector, its first and second derivatives, $\boldsymbol{v}$ and $\boldsymbol{a}$, and corresponding $\boldsymbol{p}$ and $\boldsymbol{F}$, are also vectors.

If the total energy E is constant in time ($dE/dt = 0$), which is the case of an isolated system for MD simulations, $\boldsymbol{F}$ is related to the negative gradient of potential with respect to position

$$\boldsymbol{F} = -\Delta U \rightarrow \boldsymbol{F} = -\Delta U(\boldsymbol{r}_1, \boldsymbol{r}_2, \ldots, \boldsymbol{r}_N) \tag{2.3}$$

where U is the potential. Therefore, if we know the potential of a system as a function of interatomic distance, then we can have forces on atoms and thus can solve the above equation for time evolution of the system. Once again, unlike human beings who choose money instead of love or love instead of money, atoms always follow the line to the minimum energy according to this potential.

Potentials for specific systems are generated by fitting certain functions with parameters of experimental data or with calculated data from first-principles methods. The experimental data used for fitting includes equilibrium lattice parameter, cohesive energy, bulk modulus, elastic modulus, vacancy formation energies, thermal expansion coefficient, dielectric constants, vibration spectrum, and surface energy. Note that these empirical potentials are system specific. One should ensure that it is *transferable* and be careful about using one specific potential for other systems or conditions. Even under the same system and conditions, some potentials may result in different dynamics since most of the potentials are usually constructed by fitting to static properties.

Significant progress has been made in recent years in generating accurate and reliable potentials for many metallic systems, semiconductors, and some ceramics, and they are generally available in tabulated forms for free download such as:

http://xmd.sourceforge.net/eam.html
http://cst-www.nrl.navy.mil/ccm6/ap/eam/index.html
http://riodb.ibase.aist.go.jp/apot/toppage_e.html
http://enpub.fulton.asu.edu/cms/potentials/main/main.htm
http://lammps.sandia.gov/
http://www.ctcms.nist.gov/potentials
http://www.dfrl.ucl.ac.uk/Potentials/O/index.html

In this section, four popular types of potentials will be reviewed in some detail: pair potentials, potentials by EAM, Tersoff potentials, and potentials for ionic solids. Force field potentials designed for organics, polymers, and bioorganic systems (for example, proteins) are not included here, but the underlying concept and construction are similar. The hybrid-potential approach adopts different types of potentials into a system, and the subject will only be demonstrated with an example runs in Sections 3.7 and 3.10.

2.2.1 *Pair potentials*

In an N-atom system where N is the number of atoms, an atom i interacts with all other atoms at the same time (i.e., there are interactions of two atoms, three atoms, and so forth at any moment):

$$U = \sum_{i<j}^{N} U_2(\mathbf{r}_i, \mathbf{r}_j) + \sum_{i<j<k}^{N} U_3(\mathbf{r}_i, \mathbf{r}_j, \mathbf{r}_k) + \cdots \tag{2.4}$$

where the terms are potentials for two atoms, three atoms, and so forth, respectively. Here, the summation notations indicate sums over all distinct pairs, triplets, and so forth without counting any of them twice. Let us leave this $3N$-dimensional potential until the next section and, to get started, consider the two-atom pair interactions only. There are $(N - 1)$ interactions per atom, and thus the number of pairs, N_{pair}, is on the order of N^2:

$$N_{\text{pair}} = \frac{(N-1)N}{2} \propto O(N^2) \tag{2.5}$$

Figure 2.4 shows 10 pair interactions in a system of five atoms as an example.

Pair potentials are the simplest form of a potential that considers only these two-atom interactions and neglects all others. A typical example

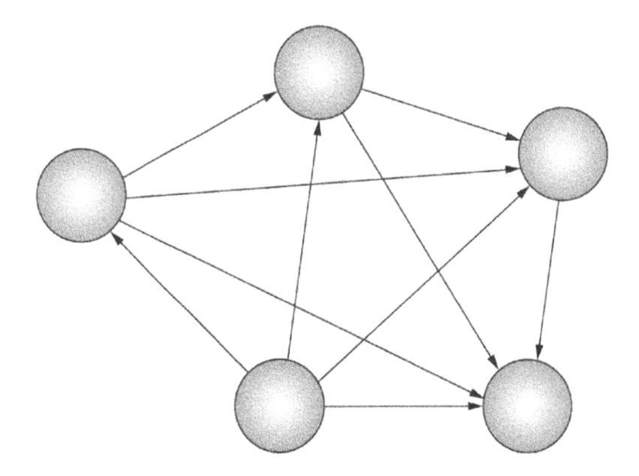

Figure 2.4 Pair interactions (arrows) in a five-atom system.

is the Lennard–Jones potential (Lennard–Jones 1924), $U_{LJ}(r)$, expressed in terms of interatomic distance, r, with the following two parameters:

$$U_{LJ}(r) = 4\varepsilon\left[\left(\frac{\sigma}{r}\right)^{12} - \left(\frac{\sigma}{r}\right)^{6}\right] \tag{2.6}$$

where:

parameter ε is the lowest energy of the potential curve (≡ well depth, cohesive energy)
parameter σ is the interatomic distance at which the potential is zero as shown in Figure 2.5.

Then its force form of repulsive and attractive term is

$$F_{LJ} = \frac{24\varepsilon}{\sigma}\left[2\left(\frac{\sigma}{r}\right)^{13} - \left(\frac{\sigma}{r}\right)^{7}\right] \tag{2.7}$$

Several features are noted here when two Lennard–Jones atoms (such as inert element Ar) approach each other from a long distance:

At $r = \infty$, U_{LJ} and F_{LJ} are zero.
When they get closer, dipole–dipole attraction takes place, and the r^{-6} term best describes this van der Waals interaction.
When they get even closer, the Pauli repulsion due to overlapping electron clouds takes place, and the arbitrary r^{-12} term describes this steep increase in repulsion.

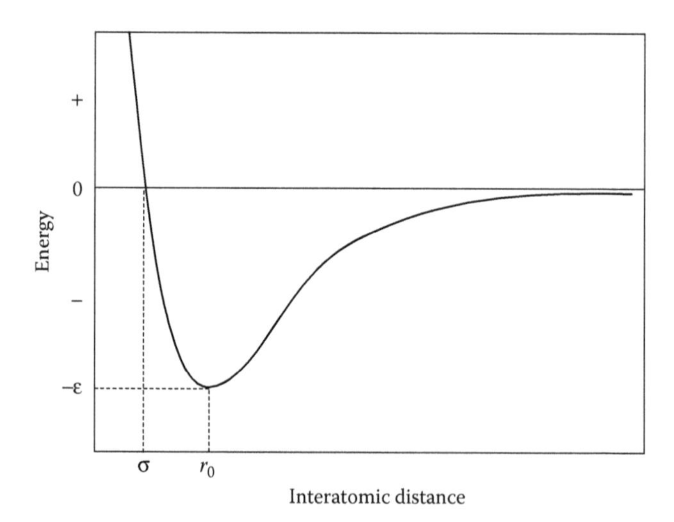

Figure 2.5 Schematic representation of the Lennard–Jones pair potential.

At equilibrium interatomic distance, r_0 (= $2^{1/6}\sigma = 1.122\sigma$), two forces are balanced to a net force of zero, and the corresponding energy becomes minimum ($dU_{LJ}/dr = 0$).
The energy passes through 0 at $r = \sigma$ and increases steeply as r decreases further due to the Pauli repulsion.

This simple pair potential can express the atomic interactions of noble gases (Ne, Ar, Kr, etc.) as shown in Section 3.4, spherical molecules, and secondary bonds very well. However, since many-atom effects (~10% of the total potential in most materials) of three atoms and so forth are completely neglected, this potential cannot be applied to metals, semiconductors, and other solids.

2.2.2 Embedded atom method potentials

In metals and transition metals, atoms (or ions) are in the middle of the electron sea and mostly coordinated with 8–12 other atoms. The coulomb interactions between them are long-ranged ones, covering easily tens of atoms. Significant efforts have been devoted to the inclusion of these many-atom effects within the framework of empirical potentials. The embedded atom method (EAM) considers the effective electron density at a given atomic site as one of the parameters, thereby capturing some electronic effects while maintaining the simplicity of a potential.

Valence electrons in metals are delocalized as an electron cloud and thus exhibit an additional attraction toward atom cores (nuclei plus core electrons). It is obvious that the near-sighted pair potentials cannot describe this circumstance properly. The EAM considers both contributions from the pair potential and the embedding energy that approximates the missing N-atom effect in a pair potential (Daw and Baskes 1984, Daw et al. 1993). This approach is conceptually the same as the mean-field approximation adopted to deal with n-electron effect in first-principles methods (see Chapter 4).

As schematically shown in Figure 2.6, the embedding energy is the approximated energy that is required to embed positively charged atom cores into the electron cloud (electron densities). This energy will be negative since the interaction is attractive. Note that the pair interactions are primarily repulsive. Thus, the mathematical form of an EAM potential, U_{EAM}, has two terms, a term for pair interaction and another term for embedding energy as a function of electron density ρ_i at atom i:

$$U_{EAM} = \sum_{i<j} U_{ij}(r_{ij}) + \sum_{i} F_i(\rho_i) \tag{2.8}$$

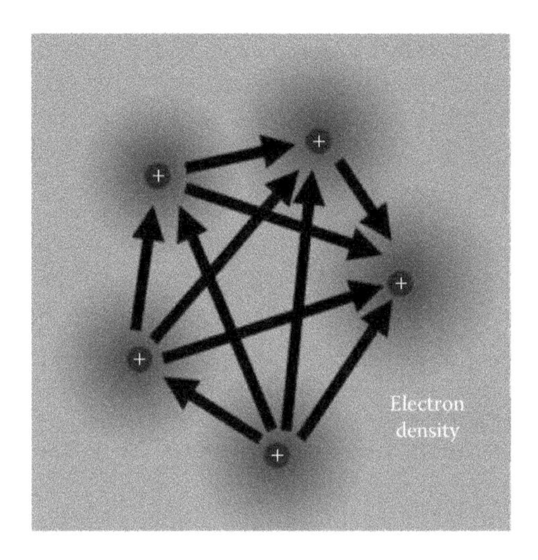

Figure 2.6 Model of a five-atom system for EAM potential showing pair interactions (arrows) and embedding energy (grey area) that represents the N-atom effects.

where:

$F_i(\rho_i)$ is the embedding energy function

r_{ij} is the scalar distance between atom i and j

and represents atom–atom distances in the x-, y-, and z-axis as

$$r_{ij} = |\boldsymbol{r}_i - \boldsymbol{r}_j| = \sqrt{(x_i - x_j)^2 + (y_i - y_j)^2 + (z_i - z_j)^2} \tag{2.9}$$

In Equation 2.8, the first summation notation represents a sum of all unique pair interactions excluding any double counting:

$$U = \sum_{i<j}^{N} U_{ij}(r_{ij}) = U_{12} + U_{13} + \cdots + U_{23} + U_{24} + \cdots + U_{34} + U_{35} + \cdots \tag{2.10}$$

The electron density at site i is simply the linear superposition of valence-electron clouds from all other atoms:

$$\rho_i = \frac{1}{2} \sum_{j(\neq i)} \rho_j(r_{ij}) \tag{2.11}$$

The following is, for example, an EAM potential for the Ni–Al system tabulated by Voter and Chen (1987), and it shows seven data sets (three for pair potentials, two for electron densities, and two for embedding energies).

```
##############################################################
.....
# A.F. Voter and S.P. Chen, Mat. Res. Soc. Symp. Proc. {\bf 82},
175 (1987).
.....
potential set eam 2
*
* Read in individual potential functions
*
POTENTIAL PAIR 1 1 3000 1.000000E+00 4.789502E+00
0.000000E+00 9.146377E+05 9.097009E+05 9.047881E+05
.....
-5.145341E-02 - 2.813742E-02 - 1.177822E-02 - 2.393252E-03
POTENTIAL PAIR 1 2 3000 1.100000E+00 5.463932E+00
6.850880E+05 5.270333E+05 5.237156E+05 5.204156E+05
.....
-3.723284E-02 - 2.035746E-02 - 8.518450E-03 - 1.728402E-03
POTENTIAL PAIR 2 2 3000 1.000000E+00 5.554982E+00
0.000000E+00 7.524923E+05 7.480864E+05 7.437026E+05
.....
-6.492704E-02 - 3.556257E-02 - 1.493719E-02 - 3.074944E-03
POTENTIAL DENS 1 3000 1.000000E+00 4.789501E+00
0.000000E+00 3.775296E-25 3.770337E-25 3.765355E-25
.....
5.570677E-32 3.030678E-32 1.254542E-32 2.438011E-33
POTENTIAL DENS 2 3000 1.000000E+00 5.554981E+00
4.567510E-11 4.321021E-25 4.317819E-25 4.314566E-25
.....
3.627624E-32 1.968903E-32 8.107946E-33 1.541933E-33
POTENTIAL EMBED 1 2362 0.000000E+00 1.705184E-24
-1.137440E-13 - 7.041029E+02 - 9.087178E+02 - 1.037061E+03
.....
3.308477E+05 3.309134E+05
POTENTIAL EMBED 2 2213 0.000000E+00 2.054483E-24
-8.057838E-14 - 5.634532E+02 - 5.423222E+02 - 4.451216E+02
.....
6.848900E+05
.....
echo on
##############################################################
```

For an MD run, the electron densities at various sites are computed first, and the corresponding embedding energies are evaluated and added to pair potentials. The nearby atoms contribute most for the electron density buildup on any site. Normally, embedding energy forms a slowly changing concave curve with electron density: it starts at zero at the electron density of zero and decreases to more negative values as electron density increases up to the point where negative charges of electron density roughly match the positive charges of the ion core. The nearby ion cores and electron densities, of course, form the strongest bonds.

This type of potential works very well for most metals and transition metals, especially for FCC metals. For element metals, any error arising from using the atomic electron density instead of the actual N-atom effect is reduced when the embedding function is fitted. For alloys, however, fitting embedding function for all constituting elements is a very difficult job to do. This is one of the reasons that there are only limited potentials available for alloy systems. This type of potential has extended to the MEAM, which can even treat directional bondings (Baskes 1992).

2.2.3 *Tersoff potential*

In covalent solids, deviations from the scheme of pairwise additives are even larger since these solids are less closely-packed (often with a coordination number of only 4) and have strong directional bonds of equivalent strength. Diamond or zincblende structures belong to this group of solids and are characterized by a bond angle of ~109.47° and bonds of the sp^3 hybridization of orbitals. Therefore, bond angle and bond order are the prime characteristics of these materials, and the strength of a bond between two atoms depends on the local environment such as coordination number and bond angle. Tersoff (1988) recognized this geometrical fact and developed a potential in the following general form:

$$U_{\text{Tersoff}} = \frac{1}{2}\sum_{i \neq j} U_R(r_{ij}) + \frac{1}{2}\sum_{i \neq j} B_{ij} U_A(r_{ij}) \quad (2.12)$$

where:

U_R and U_A represent repulsive and attractive potentials

B_{ij} is the bond order for the bond between atom i and j, which is normally a decreasing function of the coordination number, N_{coor}, as

$$B_{ij} \propto \frac{1}{\sqrt{N_{\text{coor}}}} \quad (2.13)$$

In other words, the bonding B_{ij} of atom i with atom j is reduced by the presence of another bond B_{ik}. The degree of weakening depends on where this other bond is and what is the angle.

Although both terms depend only on distance, this type of potential usually has more than six parameters to be fitted. With a good parameterization, this potential has worked quite well and has been applied to various covalently bonded solids and molecules such as SiC, Si, diamond, graphite, amorphous carbon, and hydrocarbons. However, one must be careful in the case of graphite since its layer-to-layer description is out of the range of the Tersoff potential.

2.2.4 *Potentials for ionic solids*

Most potentials are short-ranged and approach to zero exponentially within the second neighbors. In ionic solids, however, ions can be polarized or charged, and the coulombic component can spread into a very long-range. To describe potentials for these solids, therefore, one must consider two pair-interaction terms; short-range and long-range potentials as follows:

$$U_{\text{ionic}} = U_{\text{short}}(r_{ij}) + \frac{Z_1 Z_2}{r_{ij}} \tag{2.14}$$

where Z_1 and Z_2 are charges of the constituting atoms. The first term consists of typically two contributions—one by the repulsive interactions between electrons in close range and the other by the attractive interactions of van der Waals nature:

$$U_{\text{short}}(r_{ij}) = C_1 \exp\left(-\frac{r_{ij}}{\rho}\right) - \frac{C_2}{r_{ij}^2} \tag{2.15}$$

where C_1 and C_2 are constants.

The second term for coulomb interaction is the long-range interaction that decays as $1/r_{ij}$. For an ionic crystal, this term under periodic boundary conditions cannot be treated with the usual energy-cutoff approach (see Section 2.4.1) because the potential periodically accumulates and imposes a serious divergent problem. However, it can be efficiently dealt with by a technique called the Ewald sum (Ewald 1921). This method generates an artificial charge distribution localized around each ion twice: first with the opposite sign and next with the *same sign* as the ions (the magnitude is the same for both times).

The first charge generation effectively cancels most of the original charge distribution and thus makes the problem a short-ranged one with no divergent problem. The second charge generation restores the original system, which can be done effectively when the charge distributions are expressed in a spherically symmetric Gaussian type, by FFT (fast Fourier transformations). Remember that the FFT technique replaces the continuous-charge distribution into discrete charges on a finite grid in reciprocal space where any long dimension becomes a short one. One should be very careful when using ionic potentials since anions such as oxygen ions have a tendency to change their size and charge states, especially with transition metal ions. It is also known that potentials for the mixed systems such as the Al–Al_2O_3 system do not work properly. The fixed-charge potentials incorrectly predict the cohesive energy of ionic materials, and they cannot be used to simulate oxidation at metal surfaces or metal/oxide interfaces where the local ion charge can be significantly different from that in the bulk oxide.

2.2.5 *Reactive force field potentials*

Reactive force field (ReaxFF) potentials (van Duin et al. 2001) made of tens of parameters that are determined by first-principles calculations. These potentials are designed to respond dynamically to the local environment and to describe bond forming/breaking in situ. Although they are roughly 10 times slower than the above-mentioned potentials, their successful applications for hydrocarbons, nanotube systems, Si systems, and so on, have made the approach very popular and promising (see Homework 3.5).

For systems with high-energy particles (>10 eV), a special potential called ZBL universal potential (Ziegler et al. 1985) is available and is very successful in counting the repulsive potential in a very short-range.

2.3 *Solutions for Newton's equations of motion*

During an MD run or any simulation run, we do not notice any sign of actual calculations except lines of results and flickering of tiny LED lights on our computers or nodes. A computer does all the calculations to solve a set of given differential equations for N atoms interacting via a potential U and to give the evolution of the system in time. Most computing action takes place by the given command lines in the execution file, and it is necessary to understand the background of these actions to carry out MD properly and effectively. In this section, within the framework of classical mechanics, Newton's second law that governs the motion of atoms will be discussed.

2.3.1 N-*atom system*

Let us consider the total force of an N-atom system in terms of a complete set of 3N atomic coordinates:

$$\boldsymbol{F}(\boldsymbol{r}_1, \boldsymbol{r}_2, \ldots \boldsymbol{r}_N) = \sum_i m_i \boldsymbol{a}_i = \sum_i m_i \frac{d^2 \boldsymbol{r}_i}{dt^2} \tag{2.16}$$

The force on atom i can be obtained from the negative derivative of the potential energy with respect to a position if energy is conserved ($dE/dt = 0$), which is the case for most MD simulations. With the empirical potentials described in the previous section, Newton's equations of motion can relate the derivative of the potential energy to the changes in position as a function of time, and we now have a practical equation for MD:

$$\boldsymbol{F}_i = m_i \frac{d^2 \boldsymbol{r}_i}{dt^2} = -\frac{dU(\boldsymbol{r}_i)}{d\boldsymbol{r}_i} \tag{2.17}$$

Note that the force acting on the atom at a given time can be obtained exactly from the interatomic potential, which is a function only of the positions of all atoms. The next step is solving the preceding equation, which is a coupled ordinary differential equation for N atoms. In principle, the first and second integrations of the equation will result in velocities and positions. However, analytical solutions for a $6N$-dimensional system ($3N$ positions and $3N$ momenta) are simply impossible. MD simulation, therefore, uses the finite-difference method in numerical step-by-step fashion. It replaces differentials, such as $d\boldsymbol{r}$ ($\equiv d^3r$) and dt, with finite differences, $\Delta\boldsymbol{r}$ and Δt, and thus transforms differential equations into finite-difference equations. For example, the position at a later time $t+\Delta t$ is projected forward from the position at time t in the Taylor expansion:

$$\boldsymbol{r}(t+\Delta t) = \boldsymbol{r}(t) + \boldsymbol{v}(t) + \frac{1}{2!}\boldsymbol{a}(t)\Delta t^2 + \frac{d^3\boldsymbol{r}(\mathrm{t})}{3!dt^3}\Delta t^3 + \cdots \tag{2.18}$$

where Δt is a small discretized timestep, in which all the rates are assumed to remain constant. Here, the actual function is approximated by a polynomial that is constructed such that the accuracy can be improved by including more terms. Then, the integration of Newton's equations of motion for the N-atom system proceeds as follows:

- Calculate forces on all atoms from the given potential.
- Calculate accelerations, $\boldsymbol{a}_i$ on all atoms from the calculated $\boldsymbol{F}_i$ using $\boldsymbol{a}_i = \boldsymbol{F}_i/m$.
- Calculate $\boldsymbol{r}_i$, $\boldsymbol{v}_i$, and $\boldsymbol{a}_i$ at later time, $t+\Delta t$, numerically using finite-difference methods such as Equation 2.18.
- Using the calculated data as inputs for the next round, repeat the process until equilibrium is reached and stable trajectories of atoms result.

In practical MD runs, terms up to the third-order term in the Taylor expansion are normally considered, and higher-order terms become truncation error expressed as $O(\Delta t^4)$. The solutions are, of course, not exact, but are close to exact. In the following subsections, the three most popular algorithms that are derived from the Taylor expansions are reviewed for the efficient numerical solutions of Newton's second law.

2.3.2 Verlet algorithm

There are several algorithms available for the numerical integration of Newton's equations of motion and calculation of atomic trajectories in practical MD runs. All of them are aiming for good stability of the runs as well as high accuracy of the results. Verlet (1967) first proposed an algorithm obtained by the following procedure:

Writing the Taylor expansion for forward and backward positions in time and taking up to the third-order terms,

$$\boldsymbol{r}(t+\Delta t)=\boldsymbol{r}(t)+\boldsymbol{v}(t)\Delta t+\frac{1}{2!}\boldsymbol{a}(t)\Delta t^2+\frac{d^3\boldsymbol{r}(t)}{3!dt^3}\Delta t^3 \tag{2.19}$$

$$\boldsymbol{r}(t-\Delta t)=\boldsymbol{r}(t)-\boldsymbol{v}(t)\Delta t+\frac{1}{2!}\boldsymbol{a}(t)\Delta t^2-\frac{d^3\boldsymbol{r}(t)}{3!dt^3}\Delta t^3 \tag{2.20}$$

Adding two equations and giving the final expression for position at $t+\Delta t$,

$$\boldsymbol{r}(t+\Delta t)=2\boldsymbol{r}(t)-\boldsymbol{r}(t-\Delta t)+\boldsymbol{a}(t)\Delta t^2 \tag{2.21}$$

Note that the backward step in Equation 2.20 involves the plus term, the minus term, and so forth due to the order of differentials involved in each term.

So we have an algorithm that essentially does predictions of $\boldsymbol{r}(t+\Delta t)$ with $\boldsymbol{r}(t)$, $\boldsymbol{r}(t-\Delta t)$, and $\boldsymbol{a}(t)$. The first timestep needs the position at one step backward, $\boldsymbol{r}(t-\Delta t)$, as shown in Equation 2.21. However, we can avoid its calculation by the use of the normal Taylor expansion for the first timestep:

$$\boldsymbol{r}(t+\Delta t)=\boldsymbol{r}(t)+\boldsymbol{v}(t)\Delta t+\frac{\boldsymbol{a}(t)}{2}\Delta t^2 \tag{2.22}$$

After this first timestep, we use Equation 2.21 for the second timestep, $\boldsymbol{r}(t+2\Delta t)$. Note also that velocities do not appear in the time evolution by the Verlet algorithm. They are, however, required to compute the kinetic energy (see Equation 2.42) and are, thus, calculated indirectly by using position change in $2\Delta t$ as

$$\boldsymbol{v}(t)=\frac{\boldsymbol{r}(t+\Delta t)-\boldsymbol{r}(t-\Delta t)}{2\Delta t} \tag{2.23}$$

The Verlet algorithm is simple and requires only one force evaluation per timestep. It is relatively accurate with a small error of $O(\Delta t^4)$, since the fourth- and higher-order terms are truncated. It is also time reversible (invariant to the transformation $t \rightarrow -t$), although accumulation of computational round-off errors eventually breaks the time-reversibility. However, energy fluctuations during a run normally happen since the error associated to velocity is of $O(\Delta t^2)$, and $\boldsymbol{r}$ and $\boldsymbol{v}$ are obtained at different timesteps.

2.3.3 *Velocity Verlet algorithm*

The velocity Verlet algorithm (Verlet 1968) is one of the most frequently implemented methods in MD. In this scheme, the positions, velocities, and accelerations at time $t + \Delta t$ are obtained from the corresponding quantities at time t in the following way:

Advance $\boldsymbol{v}$ by a half step and $\boldsymbol{r}$ by a full step using the half-step advanced $\boldsymbol{v}$:

$$\boldsymbol{v}\left(t + \frac{\Delta t}{2}\right) = \boldsymbol{v}(t) + \frac{1}{2!}\boldsymbol{a}(t)\Delta t \tag{2.24}$$

$$\boldsymbol{r}(t + \Delta t) = \boldsymbol{r}(t) + \boldsymbol{v}(t)\Delta t + \frac{1}{2!}\boldsymbol{a}(t)\Delta t^2 = \boldsymbol{r}(t) + \boldsymbol{v}\left(t + \frac{\Delta t}{2}\right)\Delta t \tag{2.25}$$

Advance $\boldsymbol{a}$ by a full step from the potential relationship:

$$\boldsymbol{a}(t + \Delta t) = -\left(\frac{1}{m}\right)\frac{dU[\boldsymbol{r}(t + \Delta t)]}{dr} = \frac{\boldsymbol{F}(t + \Delta t)}{m} \tag{2.26}$$

Advance $\boldsymbol{v}$ by a full step from $\boldsymbol{a}$ at the previous and current timesteps and express it using the half-step advanced $\boldsymbol{v}$ and the full-step advanced $\boldsymbol{a}$:

$$\boldsymbol{v}(t + \Delta t) = \boldsymbol{v}(t) + \frac{\boldsymbol{a}(t) + \boldsymbol{a}(t + \Delta t)}{2}\Delta t = \boldsymbol{v}\left(t + \frac{\Delta t}{2}\right) + \frac{1}{2}\boldsymbol{a}(t + \Delta t)\Delta t \tag{2.27}$$

Note that velocities are updated only after the new positions and accelerations (equivalent to new forces) are computed. This algorithm is simple to implement, is time reversible and accurate, works well both for short and long timesteps, and is stable since positions and velocities are calculated for every timestep. It usually shows some energy fluctuations but no long-time energy drifts.

2.3.4 *Predictor–corrector algorithm*

The predictor–corrector algorithm (Rahman 1964) is a higher-order algorithm that uses information of higher-order derivatives of the atom coordinates. It is also one of the most frequently implemented methods in MD.

In this scheme, higher accuracy is expected by using data from the previous steps and using them for the correction of the next step as follows:

Predictor: The positions, velocities, and accelerations at time $t+\Delta t$ are predicted by the Taylor expansions using their current values:

$$\boldsymbol{r}_{\text{pre}}(t+\Delta t)=\boldsymbol{r}(t)+\boldsymbol{v}(t)\Delta t+\frac{1}{2!}\boldsymbol{a}(t)\Delta t^2+\frac{d^3\boldsymbol{r}(t)}{3!dt^3}\Delta t^3+\cdots \tag{2.28}$$

$$\boldsymbol{v}_{\text{pre}}(t+\Delta t)=\boldsymbol{v}(t)+\boldsymbol{a}(t)\Delta t+\frac{1}{2!}\frac{d^3\boldsymbol{r}(t)}{dt^3}\Delta t^2+\cdots \tag{2.29}$$

$$\boldsymbol{a}_{\text{pre}}(t+\Delta t)=\boldsymbol{a}(t)+\frac{d^3\boldsymbol{r}(t)}{dt^3}\Delta t+\frac{1}{2!}\frac{d^4\boldsymbol{r}(t)}{dt^4}\Delta t^2+\cdots \tag{2.30}$$

Error evaluation: The force is computed at time $t+\Delta t$, taking the gradient of the potential at the predicted new positions. The resulting acceleration, $\boldsymbol{a}(t+\Delta t)$, from this force will be in general different from the predicted acceleration, $\boldsymbol{a}_{\text{pre}}(t+\Delta t)$. The small difference between the two values becomes an error range, $\Delta\boldsymbol{a}(t+\Delta t)$:

$$\Delta\boldsymbol{a}(t+\Delta t)=\boldsymbol{a}(t+\Delta t)-\boldsymbol{a}_{\text{pre}}(t+\Delta t) \tag{2.31}$$

Corrector: Assuming the differences for other quantities are also small, they are all considered to be proportional to each other and to $\Delta\boldsymbol{a}$. Thus, the positions and velocities are corrected proportionally to the calculated errors:

$$\boldsymbol{r}_{\text{cor}}(t+\Delta t)=\boldsymbol{r}_{\text{pre}}(t+\Delta t)+c_0\Delta\boldsymbol{a}(t+\Delta t) \tag{2.32}$$

$$\boldsymbol{v}_{\text{cor}}(t+\Delta t)=\boldsymbol{v}_{\text{pre}}(t+\Delta t)+c_1\Delta\boldsymbol{a}(t+\Delta t) \tag{2.33}$$

The constants, c_i depend mainly on how many derivatives are included in the Taylor expansions and vary from 1 to 0. This method is very accurate and stable with almost no fluctuation during the run and is especially good for constant-temperature MD. However, it is not time reversible due to the involvement of error corrections. The method has a tendency to show energy drifts with longer timesteps (>5 fs) and needs much more storage space. This algorithm is further advanced to the Gear algorithm (Gear 1971) for higher accuracy by taking higher-order derivatives of atom positions. For example, the Gear5 method is known to be very accurate and gives minimal oscillations in the total energy.

2.4 Initialization

Normal MD runs proceed through initialization, integration/equilibration, and data production as shown in Figure 2.7. In this section, several topics in initialization are discussed.

2.4.1 Pre-setups

In the computational world, doing unnecessary or unimportant calculations is considered to be a cardinal sin. In this subsection, several tricks designed to minimize these trivial calculations in typical MD are reviewed. All these techniques are important since the computing time is scaled as $O(N)$ instead of $O(N^2)$ by the use of these tricks.

2.4.1.1 Potential cutoff

The most time-consuming part of MD is the computation of forces generated by the potentials acting on atoms as $-dU/dr$. Normally, a potential tails off at long distances and becomes negligible. This allows us to disregard the force calculation, where the interatomic distances are beyond a certain cutoff distance, r_{cut}. The r_{cut} must be smaller than half of the primary box size as indicated as a smaller circle in Figure 2.8. If not, an atom may interact with an atom in the primary box and its image in the image box, which causes unphysical double interaction.

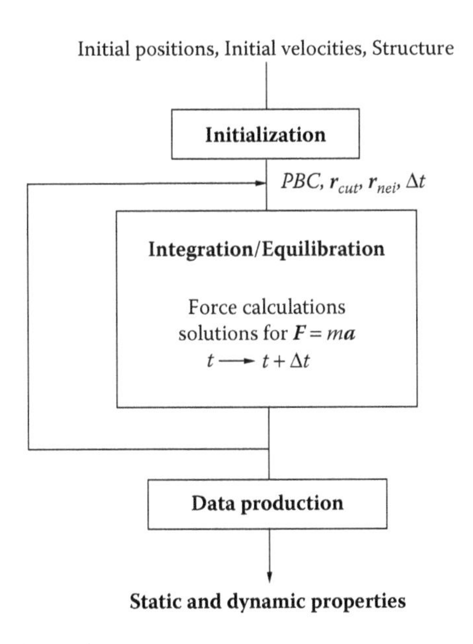

Figure **2.7** Typical flow of an MD run.

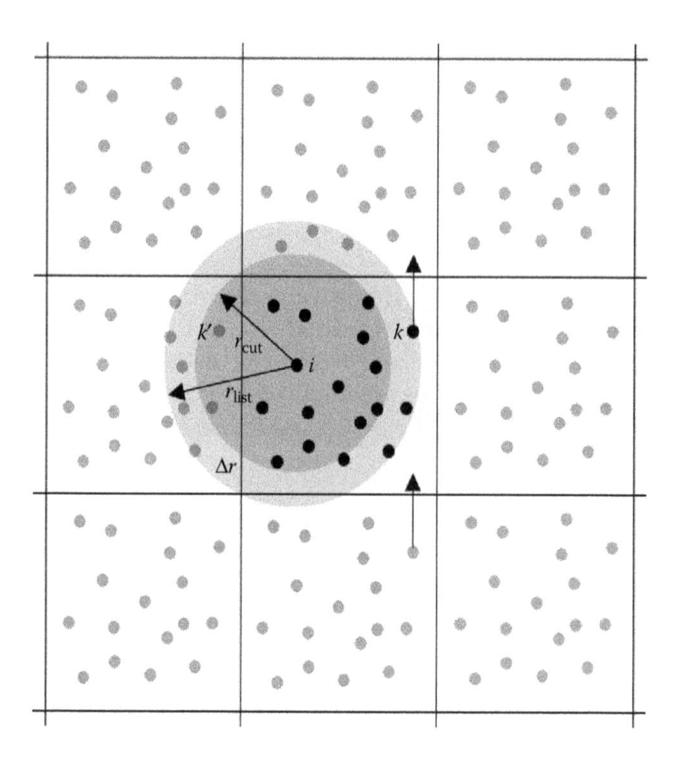

Figure 2.8 Schematic representation of potential cutoff (small circle) and neighbor-list radius (big circle) under periodic boundary conditions in two dimensions (the primary box at the center and its image boxes around it).

After the cutoff, the potential tail is normally tapered such that the tail vanishes at the r_{cut} as shown in Figure 2.9. This will prevent any sudden discontinuity on energy and force. The truncated portion of the potential tail (attractive) is usually less than several percent and may be recovered by the addition of a constant value. Normal cutoff distances for typical potentials are

Lennard–Jones potentials: 2.5–3.2 σ
EAM potentials: ~5 Å
Tersoff potentials: 3–5 Å
Potentials for ionic solids: no cutoff due to the long-range potential that decreases very slowly as $U_{long} \propto 1/r$

With this scheme, interatomic forces depend only on nearby atoms, and overall computational load scales as roughly $O(N)$. For example, the pair interactions may be reduced to ~80N for metals and to ~4N for diamond-structured semiconductors (Si, Ge, GaAs, etc.).

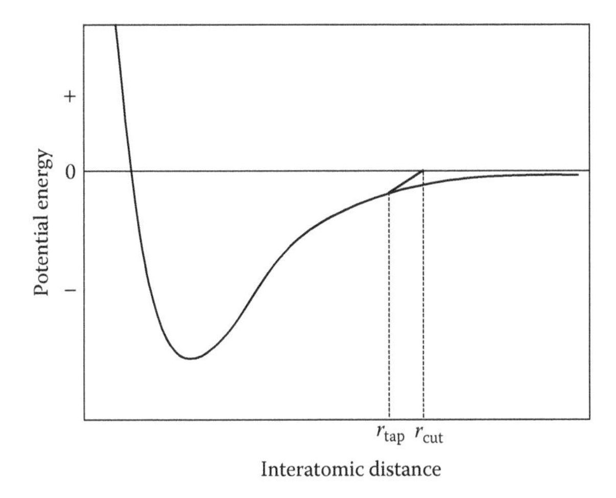

Figure 2.9 Schematic illustration of potential curve with cutoff at r_{cut} and tail – tapering (thicker line) at r_{tap}.

During an MD run, the force calculations of an atom are carried out with all atoms within this potential cutoff. As illustrated in Figure 2.8, the atom *i* has a neighboring atom *k* in the primary box and many *k′* in the surrounding image boxes. Among these atoms, only the atom *k′* in the image box at the left is the closest one to atom *i* and will be considered for force calculation. This is called the minimum image criterion. Thus, two atoms in the image box at the left are included for force calculation instead of their real counter parts. When a potential reaches long-range as in the case of columbic interactions generated by charged or polarized atoms, this criterion is not applicable.

2.4.1.2 *Periodic boundary conditions*

In a simulation box of a 1000-atom face-centered cubic (FCC) lattice, about half of the atoms are exposed to the surface. Certainly, this system cannot be represented as a genuine bulk since the coordination number of 12 is not satisfied for the surface atoms. To be considered as a bulk, atoms of roughly 1 mole (~10^{23} atoms) may be needed, which is impossible to contain in any simulation framework. The so-called periodic boundary conditions (Born and Karman 1912) as shown in Figure 2.8 take care of this problem by surrounding the box with an infinite number of its image boxes all around.

Actual simulation takes place only for the atoms inside the primary box at the center, and all image boxes just copy the primary box. For example, if the atom *k* moves out of the primary box to the upper image box, its image atom in the bottom image box (the atom with the upper arrow) moves in to replace it and keeps the number of atoms constant

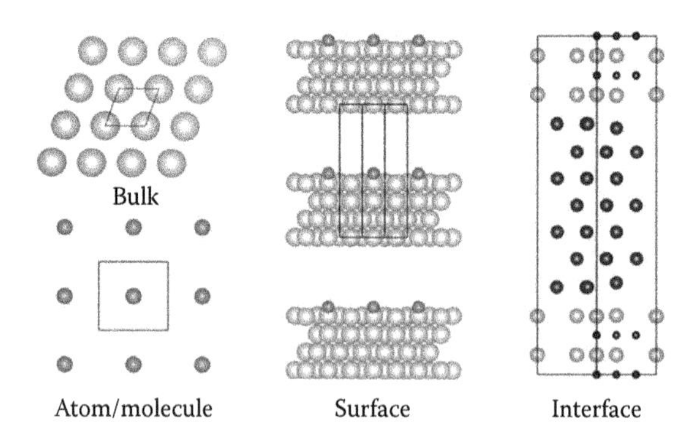

Figure 2.10 Periodic boundary conditions to mimic actual systems with various supercells.

in the primary box. Note that the forces between an atom and its periodic images cancel each other, and simple coordinate transformation can update the positions of atoms in neighboring image boxes if necessary. The same trick can be applied not only to the bulk but also to the single atom, slab, and cluster systems by the use of the supercell approach (see Section 6.2.1) as shown in Figure 2.10.

2.4.1.3 Neighbor lists

During an MD run, each atom moves various distances, and it is obvious that force calculations for the atoms that moved trivial distances are just meaningless. Verlet (1967) proposed a scheme to avoid these unnecessary calculations by listing nearby pairs of atoms and checking their traveled distances. In this technique, a larger circle (sphere in 3D) of radius r_{list} than the potential cutoff is drawn around each atom as shown in Figure 2.8. In most of MD codes, r_{list} is set at 1.1 r_{cut} as the default value, and a list is made of all the neighbors of each atom within the circle. After some fixed number of timesteps for updates, N_{up}, force calculations are carried out only for the atoms moved more than $\Delta r/2$. It is prearranged that the list is updated at every N_{up} before any unlisted atom enters into the circle as (typically after 5–20 MD steps):

$$r_{\text{list}} - r_{\text{cut}} > N_{\text{up}} v \Delta t \tag{2.34}$$

2.4.2 Initialization

To start an MD run, several initial inputs should be prepared. This preparation is important to minimize meaningless data and to have reliable results.

2.4.2.1 Number of atoms (system size)

The number of atoms in the primary box should be as small as possible, as long as it can represent the real system properly. For reliable data production, more than 100 atoms are normally required.

2.4.2.2 Initial positions and velocities

To solve Newton's equations of motion, the initial positions and velocities have to be provided. The initial positions of atoms could be anywhere but are normally specified according to the known lattice positions that are readily available from several websites, including:

http://cst-www.nrl.navy.mil/lattice/spcgrp/index.html
http://rruff.geo.arizona.edu/AMS/amcsd.php

The initial velocities of atoms could be all zero but are normally and randomly chosen from a Maxwell–Boltzmann or Gaussian distribution at a given temperature. For example, the probability that an atom has a velocity v at an absolute temperature T is

$$P(v) = \left(\frac{m}{2\pi k_B T}\right)^{1/2} \exp\left(-\frac{mv^2}{2k_B T}\right) \tag{2.35}$$

where k_B is the Boltzmann constant. The directions of velocities are also chosen randomly to make the total linear momentum become zero.

2.4.2.3 Timestep

A jumping ballerina may travel 2 m and stay in air for 1 second. To follow her trajectory, therefore, we must have a camera that can take at least 5–10 frames in 1 second as shown in Figure 2.11. Similarly, since atoms in a solid lattice vibrate in the range of 10^{-14} s, we have to do a unit MD in the range of 10^{-15} s (= femtosecond, fs). This small time span is called a timestep, Δt, and it is assumed that the velocity, acceleration, and force of each atom are constant in this timestep so that we can perform simple algebraic calculations. Referring to the equations in Section 2.3, atoms will move forward to the new positions by this timestep. To speed up the simulation, therefore, a timestep should be increased as much as possible, as long as energy conservation and stability of a run are maintained. A practical rule is that the atoms should not travel more than 1/30 of the nearest-neighbor distance in this timestep, which leads to typical timesteps from 0.1 to several femtoseconds.

Computational load and accuracy with respect to a timestep is a trade-off problem that often occurs in a computational process. A smaller timestep increases computational load and normally improves accuracy, and a larger timestep works the opposite way. If the total energy becomes

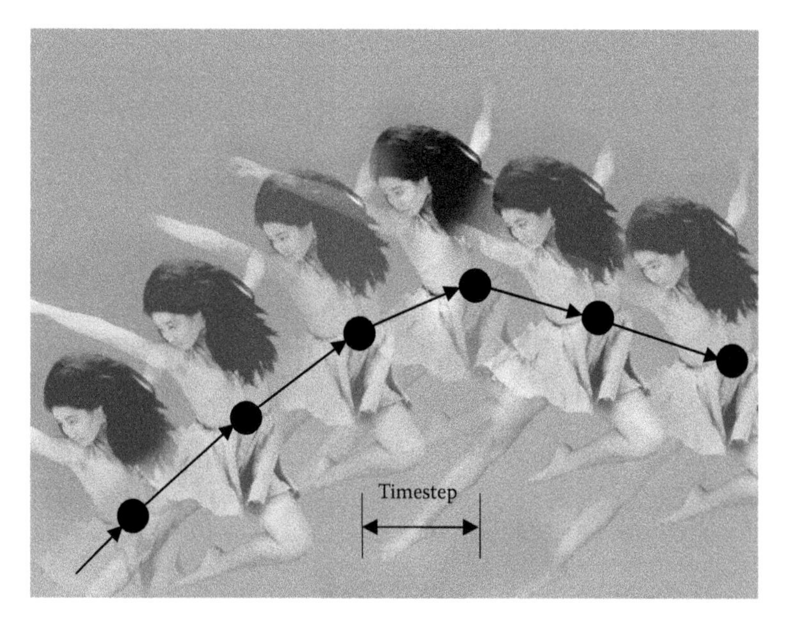

Figure 2.11 Schematic illustration of a trajectory made by a jumping ballerina in five timesteps.

unstable (drifts or fluctuates too much), it is an indication of too large a timestep that overshoots the equilibrium positions. For atoms that move relatively fast due to high temperature, lighter mass, or steep potential curve, a shorter timestep is recommended.

2.4.2.4 *Total simulation time*

A typical MD run may take 10^3–10^6 timesteps that correspond to total simulation time of only a few hundred nanoseconds (10^{-9} s). Normally, this is sufficient to see usual static and dynamic properties in materials. Running an MD too long may cause error accumulation and inefficiency of data production. Total simulation time, however, should be longer than the full relaxation time of the system to have reliable data productions. Especially for phenomena such as phase transition, vapor deposition, crystal growth, and annealing of irradiation damages, the equilibration takes place very slowly, and it is necessary to ensure that the total simulation time was long enough.

2.4.2.5 *Type of ensemble*

In music, an ensemble is a well-harmonized music (or music group) created by instruments of various tones and tunes. Although each instrument has its own characteristics, together the instruments create one piece of unique music. Similarly, in a simulation box of an MD run, each atom

Table 2.1 Various ensembles used in MD and MC

Ensembles	Fixed variables	Remarks
Microcanonical	N, V, E	Isolated system Very common in MD $S = k \ln \Omega_{NVE}$
Canonical	N, V, T	Very common in MC Common in MD $F = -kT \ln \Omega_{NVT}$
Isobaric-isothermal	N, P, T	$G = -kT \ln \Omega_{NPT}$
Grand canonical	μ, V, T	Rarely in MD, more in MC $\mu = -kT \ln \Omega_{NPT} / N$

moves and behaves differently, and a new microstate of the system is generated in every timestep. After a proper simulation and equilibration, however, it becomes an ensemble that is a collection of all possible configurations and yet has the same macroscopic or thermodynamic properties. So, an MD run is simply making a large set of atomic configurations, the ensemble.

Table 2.1 shows various ensembles used in MD and MC with different fixed variables. As with actual experiments in a lab, we impose these external constraints on ensembles to have particular properties out after a run. Especially thermodynamic properties such as entropy S, Helmholtz free energy F, Gibbs free energy G, and the chemical potential μ can be derived from data obtained from these prearranged ensembles. Here Ω is the accessible phase–space volume (or partition function) for the corresponding ensembles. The variables that specify each ensemble in Table 2.1 can be regarded as experimental conditions under which an experiment is carried out.

The microcanonical ensemble fixes the number of atoms (N), box volume (V), and total energy (E) and is an isolated system that can exchange neither matter nor energy with its surroundings. This ensemble is most often used in MD since it represents normal real systems at its best. If we assume this NVE ensemble, the equation of motion for atoms is the usual Newtonian equation.

As shown in Figure 2.12, all other ensembles are artificially surrounded by large external systems to have fixed parameters. The canonical ensemble fixes the number of atoms, box volume, and temperature (T) and is most often used in the Monte Carlo methods. The isothermal–isobaric ensemble fixes the number of atoms, pressure (P), and temperature, whereas the grand canonical ensemble fixes chemical potential (μ), volume, and temperature, and the number of atoms is allowed to change.

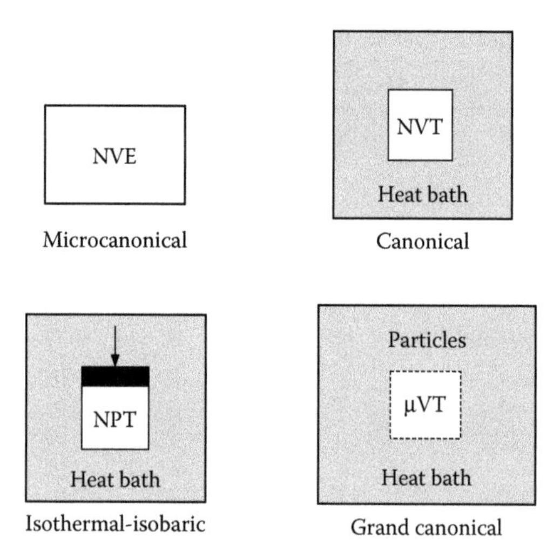

Figure 2.12 Schematics of four ensembles adopted in MD and MC.

2.5 *Integration/equilibration*

Given all pre-setups and initial inputs, an actual MD run starts to bring the unrelaxed initial system to equilibrium under the given conditions, erasing any memory of the initial configuration. In other words, we solve Newton's equations of motion until the properties of the system no longer change with time. Normally, the system is thermally isolated from the external world, and no heat will flow in or out. The result is that the total energy (the sum of potential and kinetic energies) will always remain constant. At the initial part of the equilibration, fluctuations may occur, but eventually, all atoms are driven to minimum values of potential energy where net forces on each atom become zero. The effect of small entropy changes is normally neglected for solid materials, and thus the equilibration here is not the thermodynamic one where the Gibb's free energy ΔG is zero. In this section, we will review how to control temperature and pressure and to establish the intended minimum-energy state.

2.5.1 *Temperature and pressure control*

As in actual experiments, an MD run of NVE ensemble often requires a constant temperature, and it can be achieved by rescaling velocities following the basic knowledge of kinetic energy. Average kinetic energy, $\langle E_{\text{kin}} \rangle$, is related with average velocity and also with the temperature of atoms as

$$\langle E_{\text{kin}} \rangle = \left\langle \frac{1}{2} \sum_i m_i v_i^2 \right\rangle = \frac{3}{2} NkT \tag{2.36}$$

where 3 represents dimensional freedoms in three dimensions. So, temperature is directly related with average velocity as

$$\langle v \rangle = \left(\frac{3kT}{m} \right)^{1/2} \propto T^{1/2} \tag{2.37}$$

Therefore, it is possible to increase or decrease temperature from T to T' by multiplying each velocity component by the same factor:

$$\langle v_{\text{new}} \rangle = \langle v_{\text{old}} \rangle \left(\frac{T'}{T} \right)^{1/2} \tag{2.38}$$

Using this equation, the velocities of all atoms are gradually increased to the desired value in predetermined iteration steps. Although the system is not exactly equivalent to a true NVT ensemble in a strict thermodynamic sense, constant temperature is maintained on average by this practical velocity rescaling.

To have canonical ensemble by MD and obtain the corresponding thermodynamic properties such as Helmholtz free energy F, one needs to add a thermostat interacting with the system by coupling the system to a heat bath (thermostat methods) or by extending the system with artificial coordinates and velocities (extended methods).

It is also often desirable to sample from an isothermal–isobaric ensemble to be relevant with experimental studies under constant pressures. To maintain constant pressure, the use of an extended ensemble (so-called barostats) is often adopted, and the volume of the system is allowed to change. Typically, the simulation box length is coupled to the pressure piston that has its own degree of freedom.

2.5.2 *Minimization in a static MD run*

Finding a minimum-energy configuration in a system is not an easy job. As with a blind man coming down from the hill, the search direction for each atom relies basically on local information of forces. The search may lose its way and may fall into a local minimum instead of global one. Some systematic methods are needed for the efficient search of the global minimum in a static MD run.

2.5.2.1 *Steepest-descent method*

This method starts by calculating U for the initial configuration of atoms, and the atoms move toward lower U until the search line meets the potential gradient perpendicularly. Since any further advance leads only to higher U, the next search starts at that point along the direction of the steepest descent as shown in Figure 2.13. The process repeats iteratively

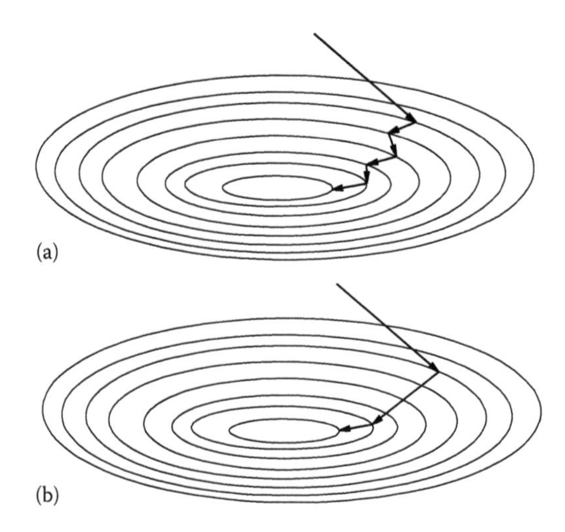

Figure 2.13 Schematics of (a) steepest-descent and (b) conjugate gradient methods.

until the minimum-U configuration of atoms is found. Thus, the search path is in a zigzag fashion with many orthogonal steps that make this method relatively slow.

2.5.2.2 Conjugate gradients method

This method starts by searching along the direction of the steepest descent, stops at a local minimum, and then figures out a new search direction based on both the new energy gradient (negative value of force between atoms) and the previous search direction. Thus, as shown in Figure 2.13, the search path is rather straightforward with fewer steps that make this method relatively fast. An extended scheme is the quasi-Newton method that considers an additional information, the second derivative of U.

2.6 Data production

After an MD run, various data for each atom are saved, which includes $3N$ positions and $3N$ momenta for all or a few timesteps. Some properties are directly obtained from these saved data: energies, forces, atomic stresses, and so on. However, many macroscopic properties of the system are calculated from the trajectories of individual atoms via statistical manipulation. This chapter will explain how to generate static and dynamic properties from this atomic-scale information via time-averaging methods.

2.6.1 Run analysis

2.6.1.1 Conservation of energy

Before we move on to data production after equilibration, it is necessary to question and confirm the validity of the simulation run. First, the conservation of total energy must be confirmed in the case of a microcanonical ensemble. During a proper MD run, the kinetic and potential energies are exchanged, but their sum should remain constant with fluctuation less than 10^{-4}. Instability in energy normally originates from an improper timestep or integration algorithm. A very long run is apt to have slow *drifts* of the total energy, and such drifts can be fixed by setting Δt smaller or by breaking the long run into smaller runs. Increasing the system size can help to reduce fluctuation in many cases.

In addition, the total linear momentum of the cell should also be conserved to zero to prevent the entire cell from moving:

$$p = \sum_i m_i v_i = 0 \tag{2.39}$$

Momentum conservation can be easily achieved, and thus we focus more on energy conservation to make atom trajectories stay on the constant-energy hypersurface in phase space. When using periodic boundary conditions, however, the angular momentum is not conserved.

2.6.1.2 *Confirmation of global minimum*

For an isolated system, equilibrium is reached when the system entropy is at its maximum like the spread of an ink drop into a glass of water. Entropy, however, cannot be directly evaluated from the atom trajectories. Therefore, runs ending up at a local minimum are often a troublesome problem since they show the conserved energy just as in the case of the global minimum. A common way of overcoming this problem is simulated *annealing*. By equilibrating at higher temperatures and slowly cooling down to 0 K, the system has enough chances to look down and identify many different minima on the potential surface and to move over local barriers to arrive at the global minimum.

2.6.1.3 *Time averages under the ergodic hypothesis*

To do the property calculations, the system must become an ergodic system whose accumulated trajectories eventually cover all possible states in the entire phase space. We normally allow the system to evolve over a sufficiently long time (many times longer than the actual relaxation time for any specific property to settle) and assume this hypothesis is closely or generally satisfied. Remember that there are always ample possibilities that one may have hands only on the nose of an elephant instead of on the whole creature and draw a wrong picture.

If the case is ergodic, any observable property, $\langle A \rangle$, can be calculated by taking the temporal average of the trajectories under equilibrium. Let us assume that we have an MD run from timestep $\tau = 1$ to τ_f and have saved τ_f-many trajectories, $A(\Gamma)$. Here, Γ indicates the $6N$-dimensional phase space of $3N$ positions and $3N$ momenta. Then the observable property,

$\langle A \rangle$, is calculated by dividing the total sum of $A(\Gamma)$ by the number of total timesteps:

$$\langle A \rangle = \frac{1}{\tau_f} \sum_{\tau=1}^{\tau_f} A(\Gamma(\tau)) = \frac{1}{\tau_f - 500} \sum_{\tau=501}^{\tau_f} A(\Gamma(\tau)) \tag{2.40}$$

Note that the second expression of the preceding equation considers the appropriate equilibrium established after the first 500 timesteps. The commonly calculated properties by such time averaging will be addressed in the following sections.

2.6.1.4 Errors

The numerical method introduces two kinds of errors: truncation error, which depends on the method itself, and round-off error, which depends on the implementation of the method. The former can be reduced by decreasing the step size or by choosing a better-suited algorithm, and the later by using high-precision arithmetic. The magnitude of these errors will decide the accuracy, while the propagation of these errors reflects the stability of the algorithms. The round-off error, however, is not much of a concern owing to the extended precision available on 64-bit computers currently in use.

2.6.2 Energies

Energies are the most important properties of interest in an MD run. The total energy is a sum of kinetic and potential energies and, during a microcanonical MD run, both terms will balance each other to keep the total energy constant. By averaging its instantaneous value, average kinetic energy is calculated from the velocities of each atom and average potential energy directly from the potential energies of each atom:

$$E_{\text{total}} = \langle E_{\text{kin}} \rangle + \langle U \rangle = \frac{1}{N_{\Delta t}} \sum_{1}^{N_{\Delta t}} \left(\sum_{I} \frac{m_i}{2} v_i^2 \right) + \frac{1}{N_{\Delta t}} \sum_{1}^{N_{\Delta t}} U_i \tag{2.41}$$

where:

$N_{\Delta t}$ is the number of timesteps taken (configurations in the MD trajectory)

U_i is the potential energy of each configuration

Average kinetic energy is proportional to the kinetic temperature according to the equipartition theorem as discussed in Section 2.5.1, and temperature can be calculated from the velocity data as

$$\langle E_{\text{kin}} \rangle = \frac{3}{2} N k_B T = \left\langle \sum_i \frac{1}{2} m v_i^2 \right\rangle \tag{2.42}$$

$$T(t) = \frac{1}{3Nk_B} \sum_i m v_i^2 \tag{2.43}$$

Once the total energy E as a function of T is obtained in different runs, one can construct the E-T curve and predict the first-order transition such as melting. The latent heat of fusion will show up as a jump on the E-T curve. This usually happens at a temperature of 20%–30% higher in simulation than the true melting point since it requires a much longer waiting time for the liquid seed to propagate at liquid phase at the expense of the solid one. On cooling of the liquid, on the contrary, supercooling takes place. The overheating and supercooling problems around the melting point can be solved by artificially setting up the system with 50% solid and 50% liquid and establishing equilibrium for the coexisting two phases across an interface as the melting is defined. Note that the melting temperature also depends on the pressure of the system. Therefore, the system should be allowed to adjust its volume as the system melts or crystallizes.

2.6.3 Structural properties

MD runs with a varying lattice constant can produce a potential curve for the material, and from this curve, various structural properties such as equilibrium lattice constant, cohesive energy, elastic modulus, and bulk modulus can be obtained. Remember, however, that the potential given for the MD run was generated by fitting it with the best available experimental data that normally include these very structural data. Therefore, these calculated results agree very well with experimental data.

2.6.3.1 Equilibrium lattice constant, cohesive energy

Equilibrium lattice constant and cohesive energy are obtained straight from the potential curve as shown in Figure 2.5. Potential energy data of different crystal structures will decide the stable structure for the system, whereas a P-T diagram will give phase stability in terms of pressure.

2.6.3.2 Bulk modulus

Bulk modulus is defined as how much volume is changed with applied pressures:

$$B_V = -V \left(\frac{\partial P}{\partial V} \right)_{\text{NVT}} \tag{2.44}$$

For example, in FCC structures, it can be obtained by taking the second derivative (curvature) of a potential curve at the equilibrium lattice constant, r_0:

$$B_V = \frac{r_0^2}{9\Omega}\left(\frac{d^2U}{dr^2}\right)_{\text{NVT}} \tag{2.45}$$

where Ω is atomic volume (see Section 3.1 for a run example).

2.6.3.3 *Thermal expansion coefficient*

Thermal expansion coefficient is defined as how much volume is changed with temperature under fixed pressures:

$$\alpha_P = \frac{1}{V}\left(\frac{\partial V}{\partial T}\right)_P \tag{2.46}$$

Therefore, it is straightforward to have a thermal expansion coefficient once the simulation box size with temperature is calculated.

2.6.3.4 *Radial distribution function*

As with X-ray diffraction (XRD), the radial distribution function (RDF), $g(r)$, can show how atoms are distributed around a certain atom. It is obtained by summing the number of atoms found at a given distance in all directions from a particular atom:

$$g(r) = \frac{dN/N}{dV/V} = \frac{V}{N}\frac{\langle N(r,\Delta r)\rangle}{4\pi r^2 \Delta r} \tag{2.47}$$

where:

r is the radial distance

$N(r,\Delta r)$ is the number of atoms in a shell volume of $4\pi r^2 \Delta r$ between r and Δr, and the brackets denote the time average

Figure 2.14 shows a typical radial distribution function for Si in various forms: crystalline, amorphous, and liquid.

For crystalline solids, this function consists of sharp peaks at every neighbor distances around a given atom and indicates the crystal structure of the system. The first peak corresponds to an averaged nearest-neighbor distance that is an indication of atomic packing and possibly bonding characteristics. As a solid melts or becomes amorphous, these peaks will broaden according to the changes in neighbor distances so that one can determine structural changes. Coordination numbers, especially for amorphous systems, can also be determined by taking the average number of atoms located between r_1 and r_2 from a given atom:

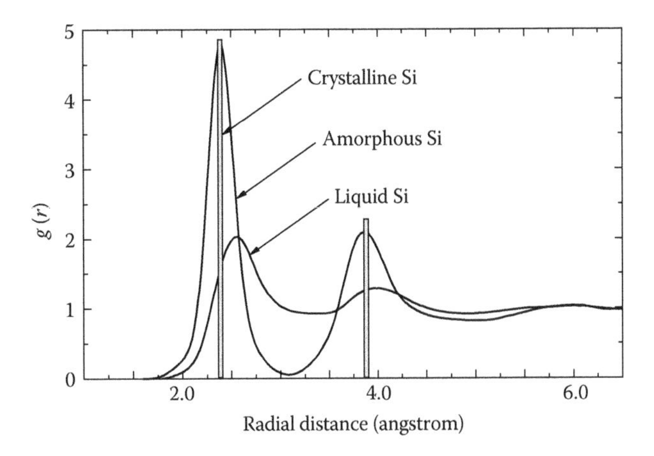

Figure 2.14 Typical radial distribution functions for Si in various forms: amorphous and liquid. (The reference positions for crystalline Si are shown with bars.)

$$N_{\text{coor}} = \rho \int_{r_1}^{r_2} g(r) 4\pi r^2 dr \tag{2.48}$$

2.6.4 *Mean-square displacement*

The vector distance that an atom travels may be close to zero since that atom might take either a forward or a backward step. Nevertheless, if we take the square of these displacements, we will have a nonzero, positive number that is called the mean-square displacement (MSD) over some time interval t. By averaging over all atoms, the MSD will have a linear dependence on t as

$$\text{MSD} \equiv \langle \Delta \boldsymbol{r}^2(t) \rangle = \frac{1}{N} \sum_{i=1}^{N} \left| \boldsymbol{r}_i(t) - \boldsymbol{r}_i(0) \right|^2 = \left\langle \left| \boldsymbol{r}_i(t) - \boldsymbol{r}_i(0) \right|^2 \right\rangle \tag{2.49}$$

Then, sudden changes in the MSD with time are indicative of melting, solidification, phase transition, and so on. In the case of solids, the MSD becomes a finite value. For a liquid, however, atoms will move indefinitely and the MSD continues to increases linearly with time. Thus, the self-diffusion coefficient in liquids, D, can be derived from the slope of the MSD–t relationship:

$$D = \frac{1}{6} \frac{d\left\langle \left| \boldsymbol{r}(t) - \boldsymbol{r}(0) \right|^2 \right\rangle}{dt} = \frac{1}{6} \frac{\left\langle \left| \boldsymbol{r}(t) - \boldsymbol{r}(t) \right|^2 \right\rangle}{t} \tag{2.50}$$

The 6 in the above equation represents the degrees of freedom for atomic jump in 3D (forward and backward jumps per dimension). Note that plotting these self-diffusion coefficients at different temperatures will provide the activation energy via the usual Arrhenius equation.

2.6.5 Energetics, thermodynamic properties, and others

Over the years, people have extended the applicability of MD to a variety of systems and topics that includes

- Calculations for thermodynamic properties.
- Simulations of systems under nonequilibrium states (Hoover and Hoover 2005) that include heat conduction, diffusion, collision, and so on.
- Multitimestep methods that set shorter timesteps for light atoms or interacting atoms, and longer timesteps for other slow-moving atoms in a same system.
- Accelerated MD by adjusting potential surfaces or temperatures (Voter 1997, Sørensen and Voter 2000).
- *First-principles* MD in which ions move classically but under the action of forces obtained by the electronic structure calculations (Car and Parrinello 1985). This subject will be discussed in Chapter 6.

It is evident that, for energetics and thermodynamic calculations, DFT methods are more popular these days since they can provide much more accurate and reliable results with a relatively smaller system (<100 atoms). For bigger systems under drastic dynamic changes, MD is the only alternative we currently have (see Chapter 3).

Homework

2.1 Consider a one-dimensional harmonic oscillator of mass, m, that follows Hooke's law, $F = -kx$, in an isolated system ($E = const.$). Then Newton's second law becomes the second-order ordinary differential equation

$$a(t) = -\frac{k}{m}x(t)$$

Let us assume that the spring is fixed at one side and is an ideal spring with zero mass and spring constant k. Given the oscillator frequency of $\omega = (k/m)^{1/2} = 1$ and initial conditions $x_0 = 5$ and $v_0 = 0$, find and discuss the following:

Write down the potential expression for the system.
Analytical solution for ball positions with time.
Numerical solution for ball positions with time by taking the first two terms in the Taylor series expansion (the Euler method) with $\Delta t = 0.2$ for 10 timesteps.

Compare the two methods above on a position-time plot and discuss what will happen if the numerical calculation proceeds further for more timesteps and how to improve its accuracy.

2.2 A typical potential curve for most materials looks like the one in Figure 2.5. Referring to the curve, explain why most materials expand with increasing temperature.

2.3 If we increase the average velocities of atoms in a system by two times, by how many times will the corresponding system temperature increase in a MD run?

2.4 Predict the positions of the first peak in a diagram of the radial distribution function for a simple cubic solid and an FCC solid.

2.5 Draw the general features of the mean-square displacement (MSD) function with simulation time when a crystalline solid is melted and becomes a liquid and explain its behavior.

References

Alder, B. J. and T. E. Wainwright. 1957. Phase transition for a hard sphere system. *J. Chem. Phys.* 27:1208–1209.

Baskes, M. I. 1992. Modified embedded-atom potentials for cubic materials and impurities. *Phys. Rev. B* 46:2727–2742.

Born, M. and T. Karman. 1912. Über Schwingungen in Raumgittern. *Physik. Z.* 13:297–309.

Car, R. and M. Parrinello. 1985. Unified approach for MD and density-functional theory. *Phys. Rev. Lett.* 55:2471–2474.

Daw, M. S. and M. I. Baskes. 1984. Embedded-atom method: Derivation and application to impurities, surfaces, and other defects in metals. *Phys. Rev. B* 29:6443–6453.

Daw, M. S., S. M. Foiles, and M. I. Baskes. 1993. EAM: A review of theory and application. *Mat. Sci. Rep.* 9:251–310.

Ewald, P. 1921. Die Berechnung optischer und elektrostatischer Gitterpotentiale. *Ann. Phys.* 369:253–287.

Gear, C. W. 1971. *Numerical Initial Value Problems in Ordinary Differential Equations*. Englewood Cliffs, NJ: Prentice Hall.

Hoover, W. G. and C. G. Hoover. 2005. Nonequilibrium molecular dynamics. *Con. Matt. Phys.* 42:247–260.

Kadau, K., T. C. Germann, and P. S. Lomdahl. 2006. Molecular-dynamics comes of age: 320 Billion atom simulation on bluegene/L. *Int. J. Mod. Phys. C* 17:1755–1761.

Lennard-Jones, J. E. 1924. On the determination of molecular fields. II. From the equation of state of a gas. *Proc. Roy. Soc. A* 106:463–477.

Rahman, A. 1964. Correlations in the motion of atoms in liquid argon. *Phys. Rev. A* 136:405–411.

Sørensen, M. R. and A. F. Voter. 2000. Temperature-accelerated dynamics for simulation of infrequent events. *J. Chem. Phys.* 112:9599–9606.

Tersoff, J. 1988. New empirical approach for the structure and energy of covalent systems. *Phys. Rev. B* 37:6991–7000.

van Duin, A. C. T., S. Dasgupta, F. Lorant, and W. A. Goddard. 2001. ReaxFF: A reactive force field for hydrocarbons. *J. Phys. Chem. A* 105:9396–9409.
Verlet, L. 1967. Computer experiments on classical fluids: Thermodynamical properties of Lennard–Jones molecules. *Phys. Rev.* 159:98–103.
Verlet, L. 1968. Computer "experiments" on classical fluids II. Equilibrium correlation functions. *Phys. Rev.* 165:201–214.
Voter, A. F. 1997. A method for accelerating the MD simulation of infrequent events. *J. Chem. Phys.* 106:4665–4677.
Voter, A. F. and S. P. Chen. 1987. Accurate interatomic potentials for Ni, Al, and Ni_3Al. *Mat. Res. Soc. Symp. Proc.* 82:175–180.
Ziegler, J. F., J. P. Biersack, and U. Littmark. 1985. *The Stopping and Range of Ions in Solids*. Vol. 1 of *The Stopping and Ranges of Ions in Matter*. New York: Pergamon Press.

Further reading

Allen, M. P. and D. Tildesley. 1987. *Computer Simulation of Liquids*. Oxford: Clarendon Press.
Delin, A. Simulations and modeling on the atomic scale. http://web.mse.kth.se/~delin/sams.htm.
Ercolessi, F. A MD primer. http://www.fisica.uniud.it/~ercolessi.
Frenkel, D. and B. Smit. 2002. *Understanding Molecular Simulation: From Algorithms to Applications*. San Diego, CA: Academic Press.
Haile, J. M. 1997. *MD Simulation: Elementary Methods*. New York: Wiley.
Hill, J., L. Subramanian, and A. Maiti. 2005. *Molecular Modeling Techniques in Material Science*. Boca Raton, FL: Taylor & Francis Group.
Hinchliffe, A. 2003. *Molecular Modelling for Beginners*. Chichester, England: Wiley.
Leach, A. R. 2001. *Molecular Modelling*, 2nd ed. New York: Pearson Education/Prentice Hall.
Raabe, D. 2005. *Computational Materials Science: The Simulation of Materials Microstructures and Properties*. New York: Wiley.

chapter three

MD exercises with XMD and LAMMPS

In this chapter, we will follow typical examples of MD calculations and see how everything we learned in the previous chapter works in actual runs. We will use two popular programs: XMD for the first five exercises and LAMMPS for the next six.

XMD was originally written by Jon Rifkin and is available for free download.* It is best suited for an MD classroom because it can be readily installed on a notebook computer and run under the MS-DOS environment. It uses the Gear algorithm up to the fifth derivative for the integration of Newton's equations. All types of potentials work with XMD if they are formatted compatibly: pair potential, EAM potential, Tersoff SiC potential (this potential is embedded in XMD), and so on.

LAMMPS was primarily developed by Steve Plimpton, Paul Crozier, and Aidan Thompson, and is also available for free download.† It uses the velocity Verlet algorithm for the integration of Newton's equations. All kinds of potentials can be incorporated with LAMMPS: pairwise potentials, many-body potentials (including EAM, Tersoff, and other potentials), reactive force field potentials, hybrid potentials, and so on. It is best suited for research and development works because it can handle massive systems under the Linux-based parallel mode.

3.1 Potential curve of Al

It is strongly recommended that, before doing any XMD run, you should read the manual briefly. That will make the exercises in this chapter much easier. Then, try to follow the provided examples one by one so you may have a rough idea of what this XMD program is all about.

XMD manual: http://xmd.sourceforge.net/doc/manual/xmd.html
XMD examples: http://xmd.sourceforge.net/examples.html

* XMD. http://xmd.sourceforge.net/.
† LAMMPS. http://lammps.sandia.gov/.

In this exercise, we calculate first the cohesive energies of metallic Al at different lattice parameters and then draw its potential curve at 0 K:

- Program implemented: XMD
- System: bulk Al
- Potential: Voter and Chen's EAM potential for NiAl (Voter and Chen 1987)
- Temperature: 0 K

This is simple and the most often introduced example for newcomers to MD. Remember that an empirical potential such as Voter and Chen's is constructed by fitting a function to experimental data that include cohesive energy, equilibrium lattice parameters, and so on. Thus, we expect a good agreement between our calculated values and experimental values.

3.1.1 *Input files*

We first place three files in a run directory: XMD program file (or installing XMD on the proper path), a run file, and a potential file. The input file for this exercise is a shell script written to calculate cohesive energies with automatically changing lattice parameters by 0.05 Å from 3.80 to 4.30 Å.

3.1.1.1 *Run file*

The following is the run file prepared for this exercise with remarks on some lines with # for brief descriptions of the command. All command lines are self-explanatory, and any inquiry can be referred to in the manual. Note that, to use this file in an actual run, the remarks should be removed or changed to independent lines with # at the start of the line to avoid any error.

```
##############################################################################
# Al-PE-curve.xm                   # XMD ignores any line starting with #. XMD
                                   is not character-sensitive.
echo on                            # display run progress on monitor
read NiAl_EAMpotential.txt         # read given potential file
eunit eV                           # set energy unit in eV
calc Al = 2
calc MassAl = 26.98                # set atomic weight for Al

calc NX = 8                        # set box numbers in x-direction
calc NY = 8
calc NZ = 8
box NX NY NZ                       # make 512 cubic boxes (box size of 1x1x1)
fill particle 4                    # fill each box with 4 Al atoms in FCC
                                   arrangement
2 1/4 1/4 1/4                      # position type 2 atom (Al) at 1/4 1/4 1/4 of
                                   a cubic box
2 3/4 3/4 1/4
2 3/4 1/4 3/4
2 1/4 3/4 3/4
```

```
fill go                          # fill all boxes with 4 Al atoms (total 2,048
                                 Al atoms)

calc A0 = 3.80                   # set A0 = 3.80 Å
calc AX0 = A0                    # set AX0 = A0
calc AY0 = A0
calc AZ0 = A0
calc DEL = 0.05                  # increase lattice size by 0.05 Å for every
                                 repeated run
scale AX0 AY0 AZ0                # make each box 3.80 × 3.80 × 3.80 Å

select all                       # select all Al atoms (total 2,048 Al
                                 atoms)
mass MassAl                      # assign atomic mass for Al atoms
dtime 1.0e-15                    # set timestep

repeat 11                        # repeat all commands until end for
                                 11 times
 calc AX = AX0 + DEL             # set the lattice for 1st run: 3.80 Å, 2nd
                                 run; 3.85 Å,..., 11th run; 4.30 Å

 calc AY = AY0 + DEL
 calc AZ = AZ0 + DEL
 scale AX/AX0 AY/AY0 AZ/AZ0      # adjust each box size accordingly

 calc AX0 = AX                   # set AX0 = AX = AX0 + DEL for the next repeat
 calc AY0 = AY
 calc AZ0 = AZ

 cmd 500                         # solve Newton's equations of motion and
                                 equilibrate 500 timesteps
 write file +Al-lattice.txt AX   # write AX accumulatively (+) into Al-lattice.
                                 txt
 write file +Al-PE.txt energy    # write energy (potential) accumulatively (+)
                                 into Al-PE.txt
 write AX                        # display AX on monitor
 write energy                    # display energy (potential) on monitor
end
##############################################################################
```

3.1.1.2 *Potential file*

The NiAl_EAMpotential.txt will be used for this run:

```
############################################################
** Al EAM potential from Voter and Chen
eunit eV
* Set potential type to EAM
potential set eam 1
* Read in individual potential functions
POTENTIAL PAIR 1 1 3000 1.000000E+00 5.554982E+00
0.0000000E+006.4847665E+016.4467977E+016.4090193E+01
.....
5.8953576E+015.8970665E+015.8987746E+015.9004817E+01
5.9021889E+01
############################################################
```

It is full of numbers as we already reviewed in Section 2.2.2.

3.1.2 Run

We can now start the MD run within the DOS window:

```
########################################################################
Start > Run > cmd >          # open DOS
Microsoft Windows [Version 6.1.7600]
Copyright (c) 2009 Microsoft Corporation. All rights reserved.

>xmd Al-PE-curve.xm          # run Al-PE-curve.xm
XMD (Version 2.5.32 Oct 25 2002)

Not using pthread library.
Not using asm("finit"); patch in thread routines.
# Al-PE-curve.xm
echo on
.....
*** Number new particles 2048
.....
*** Current step is 500      # current timestep number
AX 3.8                       # set lattice parameter at 3.8 Å
EPOT -3.197206702e+000       # potential energy (eV/atom)
.....
AX 4.3
EPOT -3.248231965e+000
.....
Elapsed Time: 0 min 49 sec   # total run time

DYNAMICS STATISTICS
 Number of MD Steps: 5500     # total MD steps = repeat 11 x cmd 500
 Number of Neighbor Searches:  11 # no. of neighbor list updates, N_up
 Time spent on MD steps: 49(secs)
 Time spent on Neighbor Search: 0(secs)

ERROR STATISTICS
 Number of Fatal Errors: 0
 Number of Unknown Command Errors: 0
 Number of Misc. Warnings: 0
########################################################################
```

3.1.3 Results

Let us open two output files, Al-lattice.txt and Al-PE.txt, and make a table:

```
########################################################################
.....
4        -3.36085
4.05     -3.36654
4.1      -3.36094
.....
########################################################################
```

3.1.3.1 *Potential energy curve*

Equilibrium lattice constant, cohesive energy, and bulk modulus can be obtained by first plotting the results with Origin® or MS® Excel. Figure 3.1 shows a plot of energy as a function of lattice parameters for Al, fitted by the fourth-order polynomial. From the fitted curve, we have $a_0 = 4.05$ Å and $E_{coh} = 3.37$ eV/atom, which agrees very well with experimental values of 4.05 Å and 3.39 eV/atom, respectively (Gaudoin et al. 2002).

The next step is calculating bulk modulus, B, which is defined as

$$B = -V \left. \frac{dP}{dV} \right|_{a_0} \tag{3.1}$$

where:

V is volume
P is pressure

Thus, bulk modulus represents the degree of volume change under a given pressure and is one way of expressing material strength. For an FCC unit cell, the total energy per cell is $4E$ with a volume of a^3, and pressure is given by

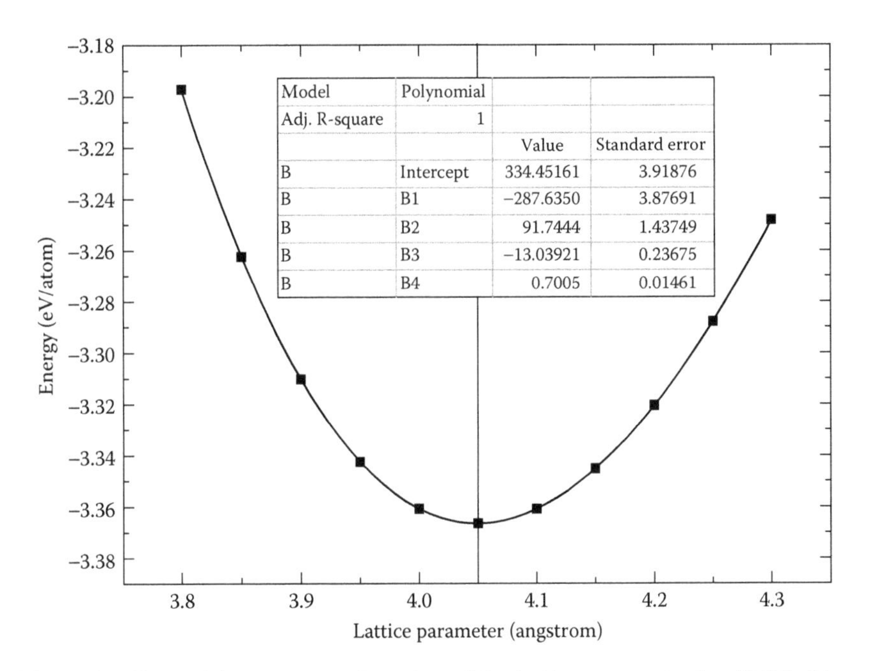

Model	Polynomial		
Adj. R-square	1		
		Value	Standard error
B	Intercept	334.45161	3.91876
B	B1	−287.6350	3.87691
B	B2	91.7444	1.43749
B	B3	−13.03921	0.23675
B	B4	0.7005	0.01461

Figure 3.1 Potential energy as a function of the lattice parameter for Al fitted by the fourth-order polynomial. (The vertical line indicates the equilibrium lattice parameter.)

$$P = -\frac{dE}{dV} = -\frac{4}{3a^2}\frac{dE}{da} \tag{3.2}$$

Then, Equation 3.1 can be rewritten as

$$B = V\frac{\partial^2 E}{\partial V^2} = \frac{4}{9a_0}\frac{d^2 E}{da^2}\bigg|_{a0} \tag{3.3}$$

where a_0 is the equilibrium lattice parameter corresponding to the minimum E.

From the fitted curve in Figure 3.1, one takes the second derivative (curvature of the potential energy curve at a_0) and plugs it into Equation 3.3, and the resulting bulk modulus is 79 GPa. This value compares well with the experimental values of 76 GPa (Gaudoin et al. 2002). The Birch–Murnaghan equation of state (Murnaghan 1944) can be used to fit bulk modulus, but the above polynomial-fitting method is sufficiently accurate and more convenient. Further runs in a similar manner at 400, 600, and 800 K can lead to the plot of the equilibrium lattice parameters versus temperature for the practical calculation of the thermal expansion coefficient.

3.2 Melting of Ni cluster

In this exercise, we will identify the melting point of an Ni cluster with 2112 atoms by increasing the temperature to 2000°C. The melting point of bulk Ni is well known, but that of an Ni cluster depends on the size of the cluster.

- Program implemented: XMD
- System: Ni cluster (2112 atoms, diameter of 3.5 nm) in a vacuum ignoring any facet formation for simplicity
- Potential: Voter and Chen's NiAl_EAM potential (Voter and Chen 1987)
- Temperature: increased by 100 K from 100 to 2000 K

3.2.1 Run file

```
#############################################################################
# Ni-cluster2112.xm

echo on
read NiAl_EAMpotential.txt

calc NX=40
calc NY=40
calc NZ=40
dtime 0.5e-15
```

```
calc Ni=1
calc MassNi=58.6934
calc A0 = 3.5238
eunit ev

box NX NY NZ
# bsave 20000 Ni-cluster-b.txt        # save box size every 20000 timesteps,
                                      Need only for bulk
# esave 20000 Ni-cluster-e.txt        # save potential energy every
                                      20000 timesteps (≡ 1 repeat)
# fill atoms into sphere of 5 unit radius centered at 20, 20, 20
fill boundary sphere 5.0 20.0 20.0 20.0
fill particle 4                       # make Ni in fcc structure
1 1/4 1/4 1/4                         # position type 1 atom (Ni)
1 1/4 3/4 3/4
1 3/4 1/4 3/4
1 3/4 3/4 1/4
fill go                               # make Ni cluster of 2112 atoms
select near 8 point 20 20 20          # select 8 atoms near to the center
fix on                                # fix the selected 8 atoms for
                                      stability of run
select type 1
mass MassNi
scale A0
# pressure clamp 1.80                 # only for bulk, bulk modulus of
                                      Ni = 1.8 Mbar (180 GPa)

calc temp=100

repeat 20
 select all
 itemp temp                           # set initial T of 100K
 clamp temp                           # set adiabatic simulation at 100 K
 cmd 20000
 calc temp=temp+100                   # increase T by 100K every repeat
 write file +Ni-cluster-T.txt temp    # write T
 write file +Ni-cluster-e.txt energy  # write energy
write xmol +Ni-cluster.xyz            # write atom coordinates accumulatively
                                      for display and animation
end
##############################################################################
```

The new command "itemp" will assign random velocities to all the particles with the appropriate Maxwell–Boltzmann distribution for the specified temperature as discussed in Section 2.4.2. Further, the velocity scale scheme as discussed in Section 2.5.1 will maintain the temperature at specified numbers. In bulk solids, thermal expansion should be taken into account by allowing the system to expand until zero external pressure is obtained. However, for a cluster in vacuum, it is not necessary.

3.2.2 Results

The output file Ni-cluster-T.txt shows temperatures on every 20,000 timesteps, and another output file Ni-cluster-e.txt shows the corresponding potential energies on every 20,000 timesteps. Figure 3.2 is the energy versus temperature curve that shows the sudden increase of potential energy at ~1500 K, which is indicative of melting. This temperature is

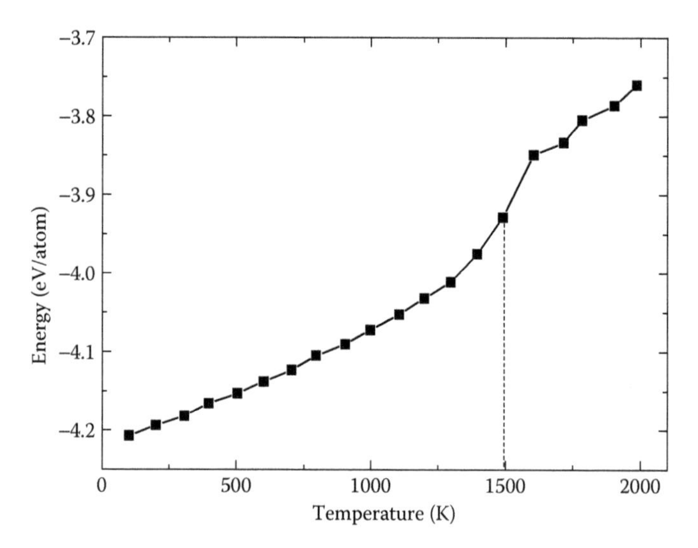

Figure 3.2 Potential energy versus temperature for an Ni cluster of 2112 atoms showing the first-order transition of melting.

much lower than the melting temperature of bulk Ni reported (1726 K). Note that the actual melting temperature of this Ni cluster will even be 20%–30% lower than ~1500 K as discussed in Section 2.6.2. Thus, it is expected that nanosized particles can lead a much lower temperature for melting due to the less-coordinated surface atoms.

3.2.2.1 Visualization with MDL ChimeSP6

There are various programs available for display and animation of the run results. We will first use a very simple online program, MDL ChimeSP6, which reads files in a standard XYZ format and is available for free download.* By double-clicking on Ni-cluster.xyz, an MDL window opens and displays/animates the system as shown as snapshots in Figure 3.3.

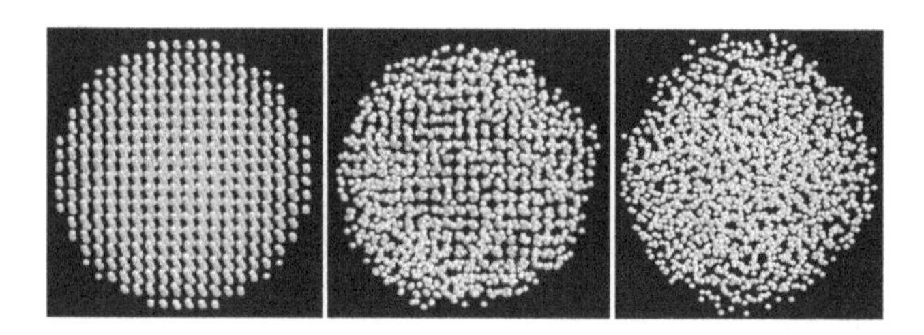

Figure 3.3 Snapshots of an Ni cluster of 2112 atoms: (from left) at 100, 1500, and 2000 K.

* MDL ChimeSP6. http://www.mdl.com/products/framework/chime/.

Note that some traces of the crystalline feature still remain in the core of the cluster at 1500 K, and these disappear completely by melting at 2000 K.

Further runs in a similar manner for Ni clusters with various sizes can reveal the melting point changes with cluster size. Normally, the melting points decrease with a decreasing cluster size because the smaller the cluster, the more the portion of surface atoms that have less bonding, and thus they easily lose the crystalline nature by heating.

3.3 *Sintering of Ni nanoparticles*

In this exercise, we will simulate the sintering phenomena with three nanoparticles of Ni:

- Program implemented: XMD.
- Potential: Voter and Chen's NiAl_EAM potential.
- Temperature: 500, 1000, and 1400 K.
- System: three nanoparticles of Ni (456 atoms, diameter of 2.1 nm), total 1368 atoms ignoring any facet formation for simplicity.
- Three atoms around the center of the second nanoparticle are fixed to give positional stability to the system.

3.3.1 *Input file*

Three nanoparticles of Ni are first prepared, and temperature is gradually increased to 500, 1000, and 1400 K to observe sintering stages and to observe temperature increase by the velocity scaling discussed in Section 2.5.1. The driving force for sintering is reduction of surface area, and the process involves atomic diffusion to the neck region via surface, grain boundary, and bulk. Thus the degree of sintering can be conveniently measured by counting the increased number of atoms at the triple-neck region.

```
###############################################################################
# Ni-3clusters-500-1000-1400K

echo on

read NiAl_EAMpotential.txt

calc NX=20
calc NY=20
calc NZ=20
dtime 0.5e-15
calc Ni=1
calc MassNi=58.6934
calc A0=3.5238
eunit ev

box NX NY NZ

ESAVE 2000 Ni-3clusters.e
```

```
fill boundary sphere 3.0 7.0 10.0 8.0      # make a first nanoparticle

fill particle 4
1 1/4 1/4 1/4
1 1/4 3/4 3/4
1 3/4 1/4 3/4
1 3/4 3/4 1/4
fill go

fill boundary sphere 3.0 10.0 10.0 13.4    # make a second nanoparticle

fill particle 4
1 1/4 1/4 1/4
1 1/4 3/4 3/4
1 3/4 1/4 3/4
1 3/4 3/4 1/4
fill go

fill boundary sphere 3.0 13.0 10 8.0       # make a third nanoparticle

fill particle 4
1 1/4 1/4 1/4
1 1/4 3/4 3/4
1 3/4 1/4 3/4
1 3/4 3/4 1/4
fill go
scale A0
write energy

# select 3 atoms at the center of the second nanoparticle and fix them to
# give stability to the system
select near 3 point 10.0*A0 10.0*A0 13.4*A0
write sel ilist                             # write the selected atoms' IDs
fix on

# select all atoms in a box at the triple-neck region and count them to
# update their number
select box 7.5*A0 5*A0 7*A0 12.5*A0 15*A0 13*A0
write sel ilist +Ni-3clusters.i            # count atoms in triple-neck region
write xmol Ni-3clusters-intial.xyz         # for snapshot
write xmol +Ni-3clusters.xyz               # for animation

repeat 5                                   # initial heat-up to 500 K
 select type 1
 mass MassNi
 itemp sel 500
 clamp sel 500
 cmd 2000
 write energy
 write ekin
 write temp
 write xmol +Ni-3clusters.xyz              # for animation
 write file +Ni-3clusters.t temp           # write temperature
select box 7.5*A0 5*A0 7*A0 12.5*A0 15*A0 13*A0
write sel ilist +Ni-3clusters-5.i          # count atoms in triple-neck region
end
write xmol Ni-3clusters-5.xyz              # for snapshot after 5 repeats

repeat 10                                  # second heat-up to 1000 K
 select type 1
 mass MassNi
 itemp sel 1000
 clamp sel 1000
 cmd 2000
 write energy
```

```
 write ekin
 write temp
 write xmol +Ni-3clusters.xyz             # for animation
 write file +Ni-3clusters.t temp          # write temperature
select box 7.5*A0 5*A0 7*A0 12.5*A0 15*A0 13*A0
write sel ilist +Ni-3clusters-15.i        # count atoms in triple-neck region
end
write xmol Ni-3clusters-15.xyz            # for snapshot after 5 + 10 repeats

repeat 15                                 # final heat-up to 1400 K
 select type 1
 mass MassNi
 itemp sel 1400
 clamp sel 1400
 cmd 2000
 write energy
 write ekin
 write temp
 write xmol +Ni-3clusters.xyz             # for animation
 write file +Ni-3clusters.t temp          # write temperature
select box 7.5*A0 5*A0 7*A0 12.5*A0 15*A0 13*A0
write sel ilist +Ni-3clusters-30.i        # count atoms in triple-neck region
end
write xmol Ni-3clusters-30.xyz            # for snapshot after 5+10+15 repeats
##############################################################################
```

Remember that all remarks with # following command syntax should be removed when executing the file for an XMD run. The remarks with # at the beginning of the line cause no problem.

3.3.2 *Results*

Figure 3.4 shows snapshots of nanoparticles of Ni at three timesteps, and a full formation of necks is observed at timestep = 10,000 (500 K). Further sintering proceeds by increasing the temperature to 1000 K, which provides sufficient thermal activation for atomic diffusion. When the temperature is increased to 1400 K, melting takes place, and the shape becomes a sphere by the action of surface tension. Note the noncrystalline nature of the sphere (Figure 3.4c).

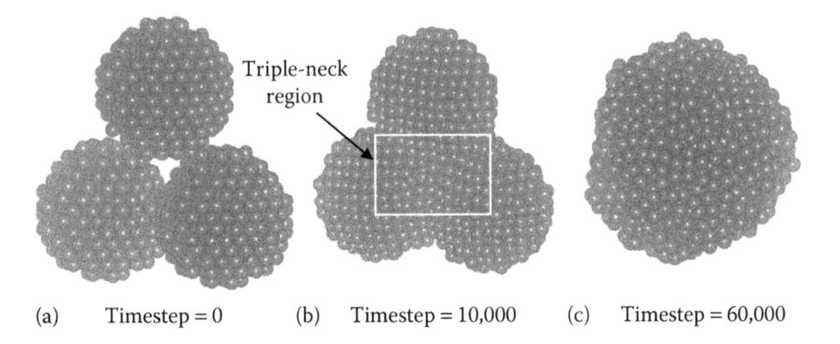

Figure 3.4 Snapshots of sintering nanoparticles of Ni at three timesteps. (a) Timestep = 0, (b) Timestep = 10,000, and (c) Timestep = 60,000.

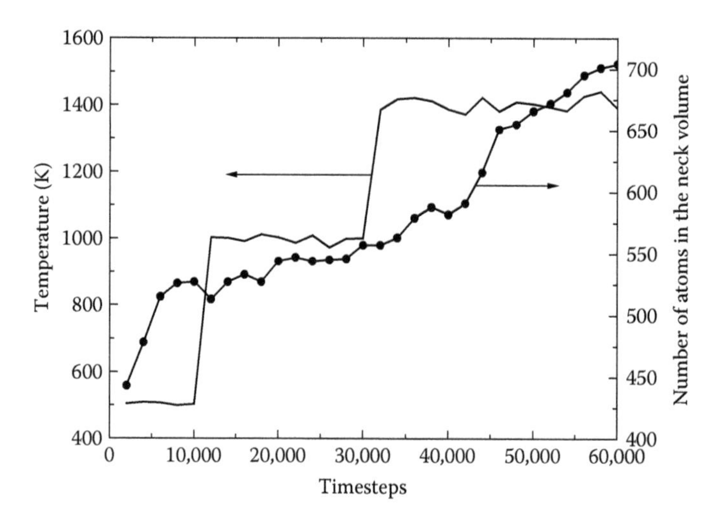

Figure 3.5 Increase in the number of atoms in the triple-neck region with timesteps. (The solid line indicates temperature.)

Figure 3.5 shows an increase in the number of atoms in the triple-neck region with timesteps accompanied by temperature increases. The number of Ni atoms in the triple-neck region is increased from 436 to 704. Because of the high surface area of nanoparticles, densification is largely finished at 1000°C. A further increase in the number of atoms in the triple-neck region corresponds to the rounding of the particle shape by surface tension, like the rounding of raindrops.

3.4 *Speed distribution of Ar gas: A computer experiment*

This exercise is provided by Nam (2010, private comm.) as a good demonstration of a computer experiment. Ar is a close-to-ideal gas, and a set of Ar atoms follows the simple ideal gas law, $PV = nRT$. Each Ar atom moves randomly without any interaction except the weak Lennard–Jones (LJ) interaction via the van der Waals force as discussed in Section 2.2.1. The behavior of an ideal gas is well described by the Maxwell–Boltzmann distribution, and the probability of atom speed, s, is given by

$$f(s) = \sqrt{\frac{2}{\pi}\left(\frac{m}{K_B T}\right)^3} s^2 \exp\left(-\frac{ms^2}{2K_B T}\right) = C_1 s^2 \exp\left(-C_2 s^2\right) \tag{3.4}$$

Note that the foregoing equation is a speed-distribution expression, and the speed is related to the velocity (see Section 2.4.2) as $s = (v_x^2 + v_y^2 + v_z^2)^{1/2}$. It provides a sound basis for the kinetic theory of gases, which explains

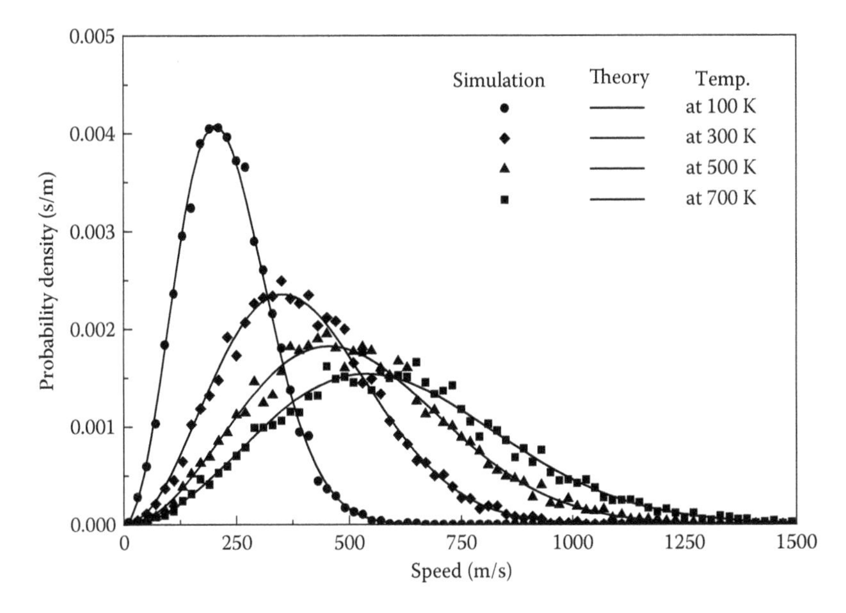

Figure 3.6 Comparison between simulation results (symbols) and the Maxwell–Boltzmann theory (lines) for Ar gas, showing probability density versus atomic speed at four different temperatures. (From Nam, H.-S., private communication, 2010. With permission.)

many fundamental gas properties, including pressure and diffusion. In Figure 3.6, the above function for Ar gas is plotted as solid lines at four different temperatures.

We will verify these theoretical results by a computer simulation using the empirical LJ potential:

- Program implemented: XMD
- Potential: LJ potential
- System: Ar gas of 8000 atoms
- Temperatures: 100–1000 K

The LJ potential for Ar is already given in Section 2.2.1, expressed in terms of interatomic distance with two parameters:

$$U_{LJ}(r) = 4\varepsilon\left[\left(\frac{\sigma}{r}\right)^{12} - \left(\frac{\sigma}{r}\right)^{6}\right] \tag{3.5}$$

where parameter ε is the lowest energy of the potential curve (≡ well depth, 1.654×10^{-21} J = 0.010323 eV), and parameter σ is the interatomic distance at which the potential is zero (3.405 Å), as shown in Figure 2.5. Note that the equilibrium distance, r_0, is $2^{1/6}\sigma = 3.822$ Å.

3.4.1 Input file

In this MD simulation, 8000 atoms of Ar gas are generated and thermally equilibrated at various temperatures (100, 300, 500, and 700 K) to observe their speed distributions:

```
##############################################################################
# Simulation of ideal gas(argon) #
# Element name & system unit #
typename 1 Ar
eunit eV

# Set-up constants for argon(Ar)
calc Ar=1                       # calculate {variable=mathematical expressions}
calc MassAr = 39.948
calc rminLJ = 1.122462048
calc epsilonAr = 0.010323       # eV
calc sigmaAr = 3.405            # Å
calc reqAr = rminLJ*sigmaAr     # 2^1/6*sigma

# Set-up for LJ potential #
potential set pair 1
calc epsilon = epsilonAr
calc sigma = sigmaAr
calc r0 = 0.1*sigmaAr
calc rcutoff = 2.0*sigmaAr

# Produce potential using formula for the Lennard-Jones potential
potential pair 1 1 100 r0 rcutoff FORMULA
4.0*epsilon*((sigmaAr/r)^12 - (sigmaAr/r)^6)

# Set box dimension (1000 Angstrom = 0.1 micrometer), box, and periodic
# boundary conditions
calc Lx = 1000
calc Ly = 1000
calc Lz = 1000
box Lx Ly Lz
surface off x y z

# generate unit cell with one Ar, make the cell dimension of 50 x 50 x 50, fill
# the whole box (8,000 Ar atoms)
fill particle 1
 Ar 10 10 10
fill cell
 Lx/20 0 0
 0 Ly/20 0
 0 0 Lz/20
fill boundary box 0.0 0.0 0.0 Lx Ly Lz
fill go

# Check initial atom configuration
write pdb ArGas0.pdb

# Set mass and timestep (approximated as 0.01*Tau, Tau=sigmaAr*sqrt(MassAr/
# epsilonAr)=2.17e-12 s)
select all
mass MassAr
dtime 20.0e-15
calc npico = 50                 # 50 dtime = 1 ps

# Set temperature and perform preliminary relaxation (NVT ensemble)
calc T0 = 100
itemp T0
```

```
clamp T0
cmd 100*npico

# repeat simulation at increasing temperatures starting from T=100 K
calc ipico=0
calc Tsim = 100
repeat 10                    # start outer loop
# preliminary relaxation of surface
 repeat 80                   # start first inner loop
  cmd npico                  # solve Newton's equations of motion
  write xmol +solidAr.xyz
  calc ipico=ipico+1
  write pdb ArGas_$(ipico)ps.pdb
end                          # end first inner loop
# MD
 repeat 20                   # start second inner loop
  cmd npico                  # solve Newton's equations of motion
  write file +E.dat energy
  write file +T.dat temp
  write file +P.dat stress
  write file +V_$(Tsim)K.dat velocity
  write xmol +solidAr.xyz
  calc ipico=ipico+1
  write pdb ArGas_$(ipico)ps.pdb
  end                        # end second inner loop

  calc Tsim=Tsim+100
  clamp Tsim
end                          # end outer loop
##############################################################################
```

The timestep is approximately $\tau \times 0.01$, where τ is defined as

$$\tau = \sigma\sqrt{\frac{m}{\varepsilon}} = 2.17 \times 10^{-12}\,\text{s} \tag{3.6}$$

The syntax for the potential command is "POTENTIAL PAIR type1 type2 ntable r_0 rcutoff FORMULA expression," which indicates potential, pair type, first atom type, second atom type, number of entries in the potential table, the starting radius in the table, the ending cutoff radius in the table, and the formula expression for the potential (Equation 3.5), respectively.

3.4.2 *Results*

Figure 3.6 compares the simulation results and the Maxwell–Boltzmann theory for Ar gas in terms of atomic speed at four different temperatures. The y-axis is in s/m so that the area under any section of the curve (which represents the probability of the speed being in that range) is dimensionless. It is apparent that the simulation results are in good agreement with the theoretical Maxwell–Boltzmann distribution. Figure 3.7 shows a snapshot of the Ar gas system where atoms move and collide randomly at a thermodynamic equilibrium (also can be animated with solidAr.xyz file).

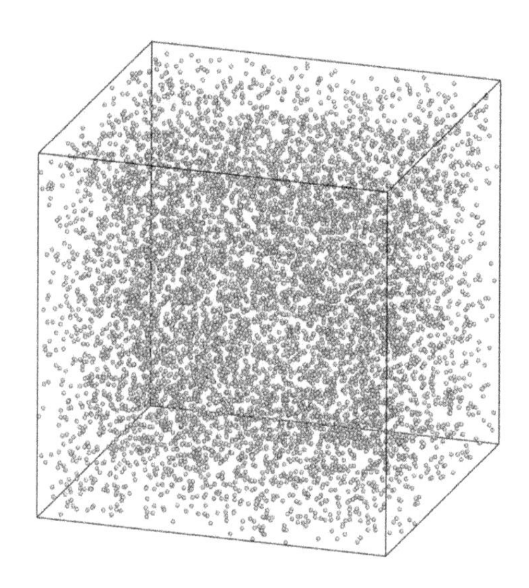

Figure 3.7 A snapshot of the Ar gas system where atoms move and collide randomly at thermodynamic equilibrium. (From Nam, H.-S., private communication, 2010. With permission.)

3.5 SiC deposition on Si(001)

This exercise is provided by Nam (2010, private comm.) and involves a thin-film deposition, which is a commonly adopted technology in many fields (e.g., PVD [physical vapor deposition] and CVD [chemical vapor deposition]). In this exercise, we will deposit C and Si atoms on the Si(001) surface and observe the formation of amorphous silicon carbide (a-SiC) film:

- Program implemented: XMD
- Potential: Tersoff potential (embedded in XMD)
- Substrate: 10 layers of Si (diamond), total 1440 atoms, fixed at 300 K
- Depositing atoms: C and Si, 300 atoms each
- Deposition: Two atoms (C and Si) per picosecond, incident energy = 1 eV/atom, incident angle = 0° (vertical)

Through this exercise, we also practice many commands available in the XMD code that can conveniently customize the system as needed.

3.5.1 Input file

```
##################################################################
# a-SiC deposition on Si(001) substrate #

potential set Tersoff        # switch on Tersoff C-Si potential

calc C=1
calc Si=2
```

```
calc MassC =12.01
calc MassSi=28.086
calc A0 = 5.432
calc DTIME=0.5e-15

typename 1 C 2 Si
eunit eV

# Simulation condition #
calc Tdepo = 300
calc Edepo = 1.0
calc angledepo = 0.0

# Simulation box & substrate dimension #
calc Lx = 6*A0
calc Ly = 6*A0
calc Lz = 40*A0
calc Lz_film = 5*A0
calc Lz_fix = 1*A0

# Set box and boundary conditions #
box Lx Ly Lz
surface off x y on z

# Si(001) substrate (diamond unit cell) creation with 1440 Si atoms #
# and 6A0 * 6A0 * 5A0 dimension #
fill particle 8
 Si 0.125*A0   0.125*A0   0.125*A0
 Si 0.125*A0   0.625*A0   0.625*A0
 Si 0.625*A0   0.125*A0   0.625*A0
 Si 0.625*A0   0.625*A0   0.125*A0
 Si 0.375*A0   0.375*A0   0.375*A0
 Si 0.375*A0   0.875*A0   0.875*A0
 Si 0.875*A0   0.375*A0   0.875*A0
 Si 0.875*A0   0.875*A0   0.375*A0

fill cell
 1.0*A0  0      0
 0     1.0*A0   0
 0     0    1.0*A0

fill boundary box 0.0 0.0 0.0 Lx Ly Lz_film
fill go

write pdb Si001_sub.pdb      # to check substrate configuration

select all
mass MassSi
set add 1

select box 0.0 0.0 0.0 Lx Ly Lz_fix
fix on
select box 0.0 0.0 Lz_fix Lx Ly Lz_film
select set 1
itemp sel Tdepo
clamp sel Tdepo
dtime 1.0e-15
```

```
# Relaxation of substrate #
cmd 1000

# estimate deposition rate: r = 2 atoms/ps = 2*(A0*A0*A0/8)/(Lx*Ly) #
# Angstrom/ps #
calc Rdepo = 2*(A0*A0*A0/8)/(Lx*Ly)

# assign velocity of depositing atoms: v = sqrt(2*energy/mass) with #
# unit conversion #
 calc vdepoC = sqrt(1*Edepo*1.602e-12/(MassC /6.022e23))
 calc vxC = vdepoC*sin(angledepo)
 calc vyC = 0
 calc vzC = -vdepoC*cos(angledepo)
 calc vdepoSi = sqrt(2*Edepo*1.602e-12/(MassSi/6.022e23))
 calc vxSi = vdepoSi*sin(angledepo)
 calc vySi = 0
 calc vzSi = -vdepoSi*cos(angledepo)

# repeat depositing C and Si atoms #
calc npico=0
repeat 300
# assign release point of depositing atoms #
 calc xdepo = rand(Lx)
 calc ydepo = rand(Ly)
 calc zdepoC = Lz_film + 5.0*A0 + Rdepo*npico
 posvel add 1
 C xdepo ydepo zdepoC vxC vyC vzC
 calc xdepo = rand(Lx)
 calc ydepo = rand(Ly)
 calc zdepoSi = Lz_film + 3.5*A0 + Rdepo*npico
 posvel add 1
 Si xdepo ydepo zdepoSi vxSi vySi vzSi

 select type C
 mass MassC
 select type Si
 mass MassSi

 cmd 3000

# remove rebound atoms #
 select box 0.0 0.0 zdepoC Lx Ly Lz
 remove

 write ekin
 write energy
 calc npico=npico+1
 write pdb SiC_depo_on_Si_$(npico).pdb
end

# final relaxation of system at 300K
select all
cmd 1000
write xmol +SiC_depo_on_Si.xyz
######################################################################
```

3.5.2 Results

The stdout on the monitor will look like the following:

```
################################################################
.....
*** NUMBER SELECTED 284
*** NUMBER SELECTED 1736
*** Current step is 901000
*** NUMBER SELECTED 0
WARNING: No atoms are selected
*** NUMBER REMAINING 2020
EKIN 3.336005e-002
EPOT -4.642033532e+000

# final relaxation of system to 300K
select all
*** NUMBER SELECTED 2020
cmd 1000
*** Current step is 902000
write xmol +SiC_depo_on_Si.xyz
.....
################################################################
```

It indicates that 284 C atoms and 296 Si atoms are deposited, and the rest of the atoms (16 C atoms and 4 Si atoms) are bounced and removed; this makes the deposited film slightly Si-rich. Figure 3.8 shows three snapshots of SiC deposition on Si(001) at the simulation timesteps = 0; 450,000; and 900,000. One clear conclusion is that SiC on the Si(001) surface forms an amorphous structure under the given condition. The saved SiC_depo_on_Si_*.pdb files (300 of them) contain atomic coordinates of

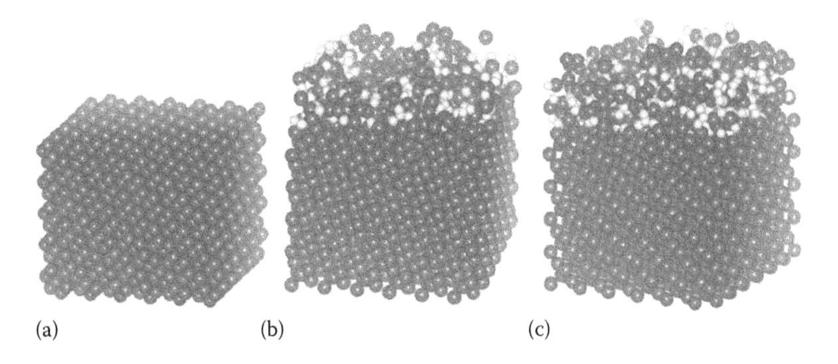

Figure 3.8 Three snapshots of SiC depositions on Si(001) at the simulation timesteps = 0 (a); 450,000 (b); and 900,000 (c). (From Nam, H.-S., private communication, 2010. With permission.)

the system at each repeat and can be visualized with a program such as AtomEye[*] when they are read in a series together.

A more intensive study of the system (Kim et al. 2004) found that the density of a-SiC films after cooling increased with increasing substrate temperature and incident energy. Extensions of this exercise to other conditions and systems will be straightforward (e.g., SiC on SiC, SiC on C).

3.6 *Yield mechanism of an Au nanowire*

This work (Park, N.-Y. et al., 2010, private comm; Park et al. 2015) is a good example to demonstrate how a simulation study becomes relevant to actual experimental observations. The aim of this MD simulation is identifying the yield mechanism of an Au nanowire under tensile deformation, focusing especially on the twin and slip formations and the surface reorientation from the {111} to the {100} plane.

- Program implemented: LAMMPS-4Jan10
- Material: Au nanowire made of 8000 atoms in an FCC structure
- Structure: wire axis oriented along the [110] direction with four {111} lateral surfaces (see Figure 3.9)

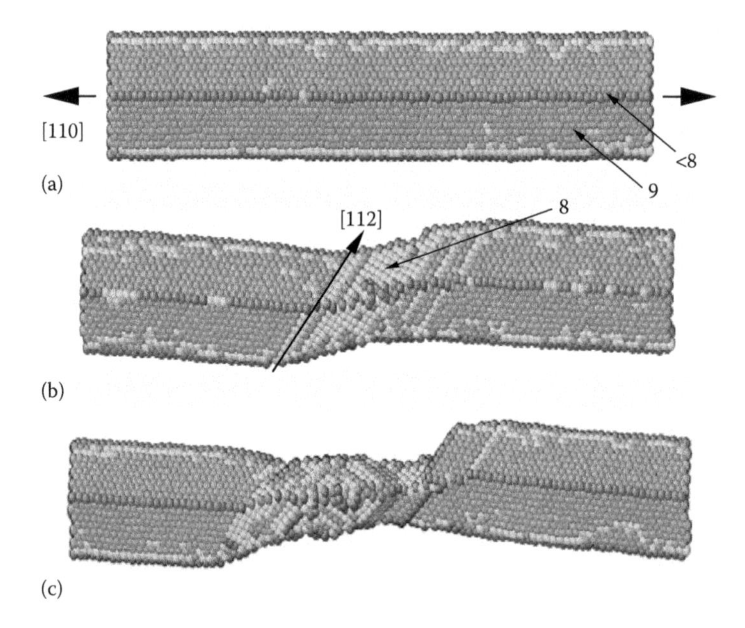

Figure 3.9 Deformation behavior of an Au nanowire during tensile loading. (a) Timestep = 0, (b) Timestep = 150,000, and (c) Timestep = 282,000. (From Park, N.-Y. et al., private communication, 2010. With permission.)

[*] AtomEye. http://mt.seas.upenn.edu/Archive/Graphics/A/.

- Size: 4 × 2.9 nm of rhombic cross section and 23 nm in length
- Potential: Foiles' EAM potential (Foiles et al. 1986)
- Temperature: 300 K
- Tensile stress: applied by increasing the *z*-axis of the simulation cell by a strain rate of 0.0001 per picosecond

3.6.1 *Input file*

The input file consists of three parts: structure generation of an Au nanowire, its thermal equilibration at 300 K, and tensile loading.

```
##############################################################################
# Deformation of Au nanowire #
# Structure generation #
units metal
atom_style atomic
boundary m m p
# define lattice with lattice parameter, origin, orientation in z, x, and y axis
lattice    fcc 4.07 origin 0 0 0 orient z 1 1 0 orient x 0 0 -1 orient y -1 1 0
region   box block 0 10 0 5 0 40 units lattice side in
create_box    1 box
create_atoms    1 box
region   1 prism 5.1 10.0 0 5 -1 1000 10 0 0
region   2 prism -1.0 4.9 0 5 -1 1000 -10 0 0
region   3 prism 14.9 20.0 0 5 -1 1000 -10 0 0
region   4 prism -20.0 -4.5 0 5 -1 1000 10 0 0
group  del1 region 1
group  del2 region 2
group  del3 region 3
group  del4 region 4
delete_atoms  group del1       # trimming corner to make a rhombic wire
delete_atoms  group del2       # trimming corner to make a rhombic wire
delete_atoms  group del3       # trimming corner to make a rhombic wire
delete_atoms  group del4       # trimming corner to make a rhombic wire
# Interatomic potential #
pair_style eam
pair_coeff                     # Au_u3.eam
neighbor 1.5 bin
neigh_modify every 1 delay 1
# Thermal equilibration at 300 K #
velocity   all   create 300 87654321 dist gaussian
velocity   all   zero linear
velocity   all   zero angular
thermo   200
thermo_style custom step atoms temp pe lx ly lz pzz press
thermo_modify lost warn norm yes flush yes
timestep 0.005                 # ps (pico-second)
dump   1 all custom 20000 pos.dump id type x y z
fix   1 all npt 300.0 300.0 10.0 aniso NULL NULL NULL NULL 0.0 0.0 10.0 drag 1.0
run   20000
# Tensile loading #
unfix  1
undump  1
reset_timestep 0
compute MyTemp all temp
compute MyPe  all pe
compute peratom all stress/atom
# taking averages of T and potential energy between step 100 ~ 200
```

```
fix ThermoAve all ave/time 1 100 200 c_MyTemp c_MyPe
thermo 200
# saving a log.lammps file with timestep, atoms, T, potential energy, zbox, #
# volume, P in z-axis #
thermo_style custom step atoms f_ThermoAve[1] f_ThermoAve[2] lz vol pzz
thermo_modify lost warn norm yes flush yes
fix   1 all nvt 300.0 300.0  10.0
fix   2 all deform 200 z erate 0.0001    # equal to strain rate of 0.0001/ps
# dump 2 all cfg 20000 pos.*.cfg id type xs ys zs # for display with Atomeye
dump 2 all xyz 50000 Au_*.xyz             # for display with MDL
# dump_modify 2 element Au                # for display with Atomeye
run   500000
#############################################################################
```

3.6.2 Results

3.6.2.1 Snapshots

Figure 3.9 shows three snapshots during the simulation, which show the deformation behavior of an Au nanowire under the given setup. In the figure, atoms are classified by their coordination numbers. The surface of the initial structure (Figure 3.9a) consists of mainly (111) planes with the coordination number of 9. At the four edges of the rhombic wire, however, atoms are much less coordinated with less than eight neighboring atoms.

With an increased length strain of about 7%, the deformed structure (Figure 3.9b) exhibits various changes, especially on the surface:

- The nucleation of Schottky partial dislocations in <112>/{111} and its propagation.
- The twinning starts and propagates through the deformed area/volume.
- The coordination number of most atoms is reduced to about 8.
- The lattice reorients from the {111} to the {100} plane, which involves an extensive slip.
- The slip takes place along the easy slip plane of (111), resulting in deformation like a sliding deck of cards. Note that only the easiest slip planes are active at this stage.

With a further increased length strain of about 14%, all deformation mechanisms become active, giving the following changes:

- The deformed structure (Figure 3.9c) becomes a mixture of all kinds of elements: twins, stacking faults, dislocations, and Shockley partial dislocations.
- The surface atoms are rearranged dominantly in (100) planes, but some amorphous arrangements of atoms are noticed at the edges and on the surface of the rhombic bar.

3.6.3 Conclusions

The spontaneous transition of the Au (100) nanofilm into the (111) nanofilm has already been observed experimentally (Kondo and Takayanagi 1997, Kondo et al. 1999). This simulation study suggests that it is also possible to have a (111) to (100) transition in an Au nanowire under tension, and a recent experimental work (Ahn, J.-P. and Seo, J.-H., 2010, private comm.) shows that the transition takes place very closely reproducing this simulation study. The calculated yield stress is in the range of 2.2–2.6 GPa, which is much higher than that of a normal bulk Au (about 0.12 GPa).

This study eloquently speaks to the fact that computational materials science is a viable tool for the study of materials. Furthermore, when it is pursued along with experimental work, a significant synergic benefit can be expected in accelerating the research and interpreting the results.

3.7 Nanodroplet of water wrapped by a graphene nanoribbon

In this exercise provided by Kim et al. (2010, private comm.), we simulate the interaction between a graphene nanoribbon (GNR) and a water nanodroplet as a representative example to demonstrate how atoms with different chemical bonds interact. Three kinds of chemical interactions are considered: the covalent bonding between the carbon atoms, the van der Waals interaction between carbon atoms and the water molecules, and the charge-associated Coulombic interactions among the water molecules. By the hybridization of interatomic potentials, this exercise demonstrates how a water nanodroplet can trigger the folding of the planar GNR, and also how the surface tension of water can be obtained from the change in potential energy during this folding process.

3.7.1 Input files

- Program implemented: LAMMPS-12Sep10
- Graphene nanoribbon: 7200 C atoms (25 × 7 nm in a honeycomb crystal lattice)
- Nanodroplet of water: a cluster of water molecules composed of 3604 of O atoms and 7208 of H atoms ($\rho = 0.98$ g/cm^3)
- Structure: zigzag (the bending direction) and armchair (perpendicular to the bending direction) structures

- Potentials: AIREBO potential for the C–C interaction (Brenner et al. 2002), TIP4P potential for H_2O (Jorgensen et al. 1983), and the LJ type of pair potential for the C and H_2O interaction (Walther et al. 2004)
- Temperature: 300 K with the temperature rescaling method

3.7.1.1 *Positions file (data.C–H_2O)*

```
######################################################################
# The header part, it typically contains a description of the file #
LAMMPS readable position file (atom style: full)
# Define number of atoms, types, bonds, and angles #
        18012 atoms
        7208 bonds
        3604 angles

        3 atom types
        1 bond types
        1 angle types
# System size in Angstrom #
     0.000 104.000 xlo xhi
     0.000 214.000 ylo yhi
     0.000 100.000 zlo zhi
Masses
        1 12.011
        2 15.999
        3 1.008
Atoms
# Position information; atom-ID molecule-ID atom-type q x y z #
1 1 1 0.0 0.000000 1.212436 10.000000
2 1 1 0.0 0.700000 0.000000 10.000000
3 1 1 0.0 2.100000 0.000000 10.000000
:
:
18010 2 2 -1.0484 48.250000 108.244845 71.250000
18011 2 3 0.5242  47.657170 107.505832 71.250000
18012 2 3 0.5242  47.657170 108.983867 71.250000

Bonds
# Position information; ID type atom1 atom2 #
1 1 6401 6402
2 1 6401 6403
3 1 6404 6405
:
:
7206 1 17207 17209
7207 1 17210 17211
7208 1 17210 17212

Angles
# Angle information; ID type atom1 atom2 atom3 #
1 1 6402 6401 6403
2 1 6405 6404 6406
3 1 6408 6407 6409
:
:
```

```
3602 1 17205 17204 17206
3603 1 17208 17207 17209
3604 1 17211 17210 17212
# End #
####################################################################
```

3.7.1.2 Input file

The input file is composed of three parts: system identification using the positions file (data.C–H_2O), setting up hybrid type of interatomic potentials, and MD calculation, including structural relaxation via conjugate gradient energy minimization.

```
####################################################################
# Nanodroplet of water on the graphene nanoribbon #
units metal
boundary p p p
atom_style full
bond_style harmonic
angle_style harmonic

# Loading position and angle, bond information #
read_data data.C-H2O

# Interatomic potential for hybrid-type #
angle_coeff 1 2.6018473 104.520000
bond_coeff 1 10.84 0.957200
pair_style hybrid airebo 2.5 0 1 lj/cut/coul/long/tip4p 2 3 1 1
0.125 9 8
pair_coeff 1 2 lj/cut/coul/long/tip4p 0.0032399 3.19
pair_coeff 2 2 lj/cut/coul/long/tip4p 0.0070575 3.16435
pair_coeff 2 3 lj/cut/coul/long/tip4p 0.0 0.0
pair_coeff 3 3 lj/cut/coul/long/tip4p 0.0 0.0
pair_coeff 1 3 lj/cut/coul/long/tip4p 0.0 0.0
pair_coeff * * airebo CH.airebo C NULL NULL
kspace_style pppm/tip4p 1.0e-4

# Wedging part #
region r_fix block INF 22.4 INF INF 9.0 11.0 units box

# Grouping #
group g_C type 1
group g_O type 2
group g_H type 3
group g_W union g_O g_H
group g_fix region r_fix
group g_MD subtract all g_fix

# Structure relaxation using energy minimization #
minimize 1e-5 1e-6 10000 100000
reset_timestep 0
compute mype all pe/atom
```

```
compute E_C g_C reduce sum c_mype
compute E_W g_W reduce sum c_mype
thermo 100
thermo_style custom step temp pe c_E_C c_E_W
# Saving a log.lammps file with timestep, T, potential energy for
Carbon #
# and water #
velocity g_MD create 300 790316 dist gaussian
timestep 0.002        # 2 femto-second (10^-15 second)

fix 1 all nve
fix 2 g_fix setforce 0 0 0
fix 3 g_MD temp/rescale 10 300.0 300.0 5.0 0.8

dump mydump all cfg 5000 dump_*.cfg id type xs ys zs c_mype
dump_modify mydump element C O H

restart 10000 restart. C-H2O_*
# Start MD calculation #
run 500000
# End #
####################################################################
```

3.7.2 Results

Figure 3.10 shows the potential energy evolution of the GNR during its interaction with the water nanodroplet. The atomic configurations at four different instances are also displayed. Because graphene is hydrophobic, the water molecules tend to avoid wetting the graphene structure and instead bind with other water molecules to form a sphere as shown in the lowest inset in Figure 3.10.

The density of the water at this stage is 0.98 g/cm^3, which can be estimated from the size of the water droplet (6.0 nm in diameter) and the number of H_2O molecules in the droplet (3604 H_2O). After the structure is relaxed, the GNR gradually bends until it wraps around the water droplet. Although the interaction between the GNR and water molecule is governed by a weak van der Waals force, these forces are sufficient to bend the GNR because the energy required for bending the graphene for this range of bending curvature is rather small (Lu et al. 2009). The driving force of the GNR wrapping is to reduce the free surface area of the droplet associated with high surface tension. As shown in Figure 3.10, the potential energy of the GNR increases with decreasing free surface area of the water droplet. Beyond 0.6 ns, the surface energy balances the bending energy of the GNR, and the potential energy of the GNR saturates. From energy balance, we obtain

$$\gamma_{H_2O} = \frac{\Delta E_{GNR}^{tot}}{S_W} \tag{3.7}$$

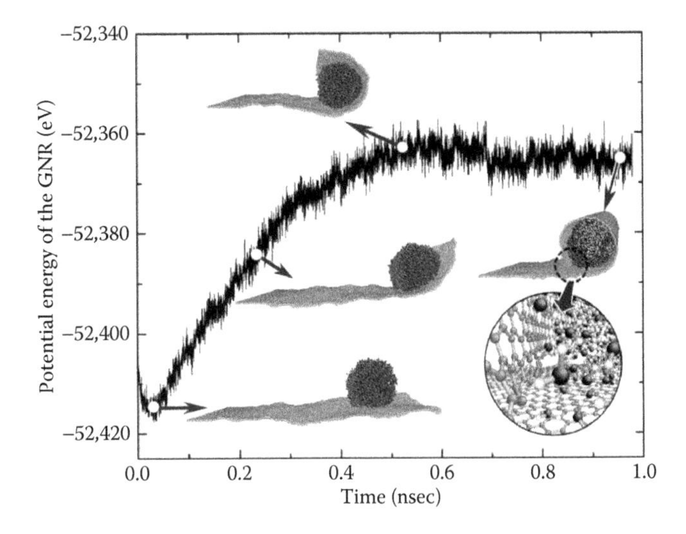

Figure 3.10 Spontaneous wrapping of a water nanodroplet by the GNR. (From Kim, S.-P. et al., private communication, 2010. With permission.)

where:

γ_{H_2O} is the surface tension of the water droplet
ΔE_{GNR}^{tot} denotes the change in the total potential energy of the GNR
S_W is the surface area covered by the GNR

From the results in Figure 3.10, ΔE_{GNR}^{tot} and S_W are calculated to be 51 eV and 92 nm^2, respectively. From Equation 3.7, the surface tension of the water is calculated as 88.64 mN/m (or 0.554 eV/nm^2), which is in fair agreement with the experimental value of 75 mN/m (Wikipedia 2010).

3.7.3 Conclusions

As demonstrated here, MD simulation is an important tool that can mimic real experiments, albeit in a much better-controlled environment. Recent advances in computational capabilities, coupled with increasing accuracy of available MD interaction potentials, have extended the use of MD to analyzing real practical situations. In particular, the use of MD in graphene-related research has uncovered insights into the theoretical assessment of graphene. Although there are still limitations of MD with regard to simulation time and length scales, MD is expected to play a pivotal role in bridging nanoscience and technology. One final remark is that the LAMMPS program is continuously upgrading and thus it may be necessary to modify some commands of the input files (Exercises 3.6 and 3.7) accordingly if one uses the newer versions.

3.8 Carbon nanotube tension

3.8.1 Introduction

This exercise is the first part of a series of runs that is for the computational development of a nanocomposite in three steps. We again follow a typical MD procedure and see how everything we learned in the previous chapter works in an actual run. We use here the 2015 version of LAMMPS on Windows for the classroom uses. Note that some minor differences are there in writing the command lines in the new version, which can be easily fixed by referring the manual. The new LAMMPS program is downloaded and installed by:

- *Download*:
  http://lammps.sandia.gov/ > Code > Download
  > Pre-built Windows executables Windows installer package >
  Latest version for 64-bit Windows this site > 64-bit Windows
  > lammps-64bits-latest.exe
- *Install*:
  lammps-64bits-latest.exe > double click
- *Confirm*:
  C: Program Files > LAMMPS 64-bit 20150724 > bin > lmp_serial.exe and lmp_mpi.exe
- *Manual download*:
  http://lammps.sandia.gov/doc/Manual.pdf
  (10 Aug. 2015 version or later)

Note that, as shown above, a series of selections on tabs between sites is marked with ">" symbols here and throughout this book.

Carbon nanotube (CNT), due to its exceptional magnetic, electrical, and mechanical properties, is a promising candidate for various technical applications ranging from nanoelectronic devices to nanocomposites. The extremely high specific strength coming from the intrinsic strength of the carbon–carbon sp^2 bond makes the CNT especially suitable for high-performance nanocomposites as a reinforcing material, and it is expected that this mechanical application of CNT could create the biggest large-scale application for the material. We here simulate the tensional behavior of a CNT.

- Program: LAMMPS prebuilt Windows executables (2015).
- System: single-walled CNT structure in an armchair configuration.
- Potentials: CH.old.airebo (adaptive intermolecular reactive empirical bond order) potential (Stuart et al. 2000).
- Goal: to evaluate the stress–strain behavior at the temperature of 300 K as a reinforcing material for composites.

3.8.2 Input file

```
###########################################################################
# in.CNT-tension.txt

### Basic setup ###
units                metal
boundary             f f p            # PBC on z boundary only
# p: periodic, f: non-periodic & fixed, s: non-periodic & wrap
# wrap: face position always encompasses the atoms
atom_style                            atomic

### Structure ###
read_data                             readdata.CNT
mass                                  1 12.01

region                                top block INF INF INF INF 0 45 units box
# units box; 0 A to 45 A
# to select the bottom 80 atoms (No. 1-80, see readdata.CNT)
region                          bottom block INF INF INF INF 116 161 units box
# units box; 116 A to 161 A
# to select the top 80 atoms (No. 1201-1280, see readdata.CNT)
# total box length in z = 161 A (see readdata.CNT)
# this creates 40 A vacuum at both ends of CNT

group                                 g_cnt type 1
group                                 g_top region top
group                                 g_bottom region bottom
group                                 g_boundary union g_top g_bottom
group                                 g_body subtract all g_boundary

### Potential ###
pair_style                            airebo 2.5 1 0
pair_coeff                            * * CH.old.airebo C C
# for pure CNT, the old CH.old.airebo with 2.0 rcmin_CC works better.

### Calculation of stress ###
compute                               body_temp g_body temp
compute_modify                        body_temp dynamic yes
compute                               body_pe all pe/atom
compute                               body_st all stress/atom body_temp pair bond
compute                               strAll all reduce sum c_body_st[3]
compute                               strC g_cnt reduce sum c_body_st[3]
# c_body_st[3]; compute body stress of zz in one column of array
variable             szzC equal c_strC/(8.784*count(g_cnt))*(10^-4)
variable             szz equal v_szzC

thermo               100
thermo_style         custom step temp c_body_temp pe etotal pzz c_strAll vol
thermo_modify        lost ignore norm yes

### NVE relaxation at 300 K ###
timestep             0.001

velocity             g_body create 300 4928459 dist gaussian rot yes units box
velocity             g_body zero linear
# zero; make g_body's momenta zero by velocity adjust

fix                                   1 all nve
fix                                   2 g_body temp/berendsen 300.0 300.0 0.1
fix                                   3 g_boundary setforce 0.0 0.0 NULL
# top and bottom's x, y fixed
```

```
dump                                relaxAll all custom 2000 relaxAll_*.dat
                                    id zs c_body_pe c_body_st[3]
fix                                 4 all ave/time 1 100 100 c_strAll file
stress_relax_All.dat
# In stress_relax_All.dat, z stress converges close to 0.
# 1 atm = 1.01325 bar = 1.01325 x 10^5 Pa = 1.01325 x 10^5 N/m^2
dump                 snapshotrelax all xyz 5000 CNT_relax_*.xyz

run                                 10000
write_restart                       restart.10000

undump                              relaxAll
undump                              snapshotrelax
unfix                               3
unfix                               4

### Applying tension ###
reset_timestep       0

fix                  3 g_boundary setforce 0.0 0.0 0.0
velocity             g_bottom set 0.0 0.0 0.2 units box
velocity             g_top set 0.0 0.0 -0.2 units box
# eng. strain rate = 0.005/ps, system length = 78.975 A
# Tensional strain on both ends (top and bottom) of CNT = 0.4 A/ps
# strain = timestep x strain rate
# lammps's atomic stress must be divided by atomic volume.
# For crystal, atomic volume = atomic mass/density
# For solid deformation, need ionic radius or van der Waals radius.
# initial system V (40.5 x 40.5 x 80 A^3).

dump                 tensileAll all custom 5000 CNT_tensile_*.dat id zs
                     c_body_pe c_body_st[3]
dump                 snapshot all xyz 500 CNT_tensile_*.xyz

fix                  22 all ave/time 1 100 100 c_strAll file stress_tensile_Body.dat
# total atomic stress of all atoms to have units of stress (pressure)
fix                  33 all ave/time 1 100 100 v_szz file stress_tensile_GPa.dat

run                  40000
write_restart        restart.40000
##############################################################################
```

Note that the unit of timestep is fs (femtosecond), which makes the number always less than 1, and thus the Tayler series-based algorithms become valid for the finite difference integration.

3.8.3 *readdata.CNT*

The *xyz* coordinates for a CNT can be generated from either of the following two sites:

- http://turin.nss.udel.edu/research/tubegenonline.html.
- http://nanofun.kaist.ac. kr/cnt_gen/cnt_generator.html.

It is edited to fit into the LAMMPS's reading data format, resulting in the readdata.CNT of 10 × 10 chirality and 32 cells along the CNT axis with 1280 atoms (dia.: 13.751 Å, length: 78.566 Å):

```
################################################################
LAMMPS data file

1280 atoms

1 atom types

0.000000      80.500000    xlo xhi
0.000000      80.500000    ylo yhi
0.000000      161.000000   zlo zhi

Masses

1 12.011000

Atoms

1 1     47.125494     40.250000     41.217088
2 1     46.975248     41.679496     41.217088
3 1     45.812391     44.291314     41.217088
…..
1278 1     47.087829     39.531315     119.782912
1279 1     46.788983     42.374644     119.782912
1280 1     46.204352     43.687747     119.782912
################################################################
```

Note that the system is 80.5 × 80.5 × 161.0 Å, and the CNT occupies x = from 33.374506 to 47.125494 Å, *y* = from 33.374506 to 47.125494 Å, and *z* = from 41.217088 to 119.782912 Å (see readdata.CNT).

3.8.4 CH.old.airebo

The original CH.airebo potential (Stuart et al. 2000) is selected for this CNT run, because it works better for a pure CNT. It is renamed as CH.old.airebo and saved in the LAMMPS/Potentials directory. Note that its rcmin_CC parameter is 2.0 instead of 1.7 in the LAMMPS-provided CH.airebo potential.

```
################################################################
# AIREBO Brenner/Stuart potential
# Cite as S. J. Stuart, A. B. Tutein, J. A. Harrison,
# "A reactive potential for hydrocarbons with intermolecular
interactions",
# J. Chem. Phys. 112 (2000) 6472-6486.
2.0     rcmin_CC
1.3     rcmin_CH
1.1     rcmin_HH
…..
################################################################
```

3.8.5 Results

Many output files are written in the directory, and the CNT_tensile_*.xyz files have the structural evolution of a CNT under the applied tension. Figure 3.11 shows CNT structures visualized on VESTA at timesteps of 0; 20,000; and 40,000 from the corresponding CNT_tensile_*.xyz files, respectively. Just like a spring, the CNT is stretched and fractured at a timestep of about 32,000.

The stress_tensile_GPa.dat file has tensile stress data developed every 100 steps:

```
############################################################
# Time-averaged data for fix 33
# TimeStep v_szz
100 1.1089
200 2.23678
300 1.99073
…..
31900 105.469
32000 105.662
32100 105.621
…..
############################################################
```

Figure 3.12 shows the plot in terms of timestep, indicating the maximum tensile stress of slightly over 105 GPa. This value is in good agreement with the experimental tensile strength of about 100 GPa (Eatemadi et al. 2014). Note that the up and down bounce of a CNT after fracture is due to the release of stored elastic energy. We conclude that this highly elastic

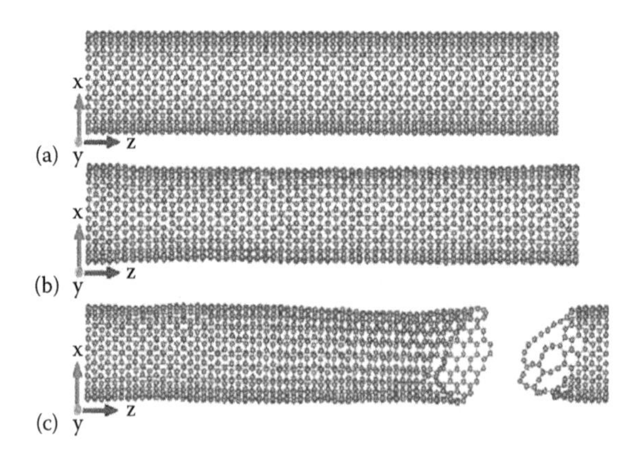

Figure 3.11 Fracture of a CNT under tension at timesteps of (a) 0; (b) 20,000; and (c) 40,000.

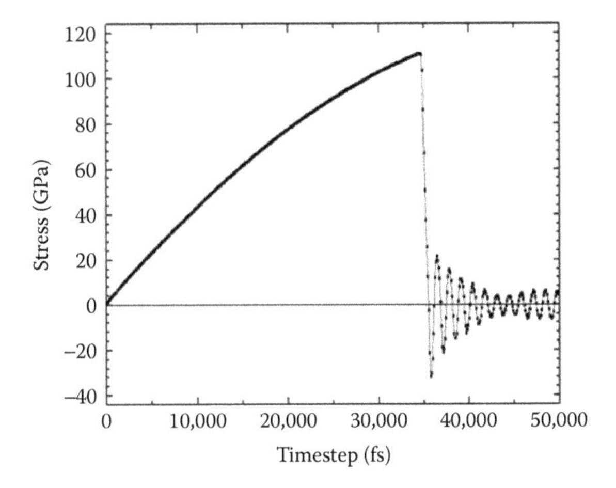

Figure 3.12 Stress–strain curve of a CNT under tension.

behavior of the CNT with high fracture strength is ideal for a reinforcing material for composites.

3.9 *Si-tension*

This exercise is the second part of a series of runs that is for the computational development of a nanocomposite in three steps. We deal with the matrix system, bulk silicon. Silicon is chosen because it is a typical covalent solid, and is very brittle (normally toughness K_{IC} = ~1 MPa. $m^{1/2}$ [Petersen 1982]). We expect that, by incorporating with the CNT as studied in the previous section, silicon's brittle nature could be improved.

3.9.1 *Introduction*

- System: Si bar with 1661 atoms under tension.
- Potential: SiC_Erhart-Albe.tersoff.
- Tensional strain is applied along the *y*-direction of the Si bar (corresponds to the [100] direction of the Si crystal).
- Goal: to evaluate the stress–strain behavior at the temperature of 300 K as a matrix material for composites.

3.9.2 *Input file*

```
##############################################################################
# Si-smallCrack.txt

### Basic setup ###
units               metal
boundary            s s p
```

```
# p: periodic, f: non-periodic & fixed, s: non-periodic & wrap
# wrap: face position always encompasses the atoms
atom_style         atomic

### Structure ###
lattice            diamond 5.431
region             box block 0 20 0 10 -0.5 0.5
create_box         5 box
# create five types of Si atoms in the box
create_atoms       1 box
# create type 1 atom in the box (other types will be set later)

mass               1 28
mass               2 28
mass               3 28
mass               4 28
mass               5 28

region             1 block INF INF INF 1.25 INF INF
group              lower region 1
region             2 block INF INF 8.75 INF INF INF
group              upper region 2
group              boundary union lower upper
group              mobile subtract all boundary
region             leftupper block INF 2 5 INF INF INF
group              leftupper region leftupper
region             leftlower block INF 2 INF 5 INF INF
group              leftlower region leftlower
set                group leftupper type 2
set                group leftlower type 3
set                group lower type 4
set                group upper type 5
# to see each group on VESTA, set 1 grey, 2 red, 3 blue, 4 green, 5 yellow

### Potential & Neighbor list ###
pair_style         tersoff
pair_coeff         * * SiC_Erhart-Albe.tersoff Si Si Si Si Si
# all atoms by sw potential, 5 types of Si
neighbor           1.0 bin
# neighbor list skin 1.0 thick with bin algo (scales linearly with N/P)
neigh_modify       every 1 delay 5 check yes
# check any atom moved more than half the skin, list, delay 5 steps, list,.....

### Relaxation at 300 K ###
compute            new mobile temp
velocity           mobile create 300 887723 temp new

fix                1 all nve
fix                2 mobile temp/berendsen 300 300 0.1
# T for mobile region only, fixed upper & lower boundary
fix                3 lower setforce 0.0 0.0 0.0
fix                4 upper setforce 0.0 0.0 NULL
compute            1 mobile temp
compute_modify     1 dynamic yes

compute            body_pe all pe/atom
compute            body_st all stress/atom NULL pair bond
# compute the symmetric per-atom stress tensor for each atom
# ke or NULL, pair, and bond contributions

timestep           0.001
thermo             200
```

```
thermo_style        custom step temp c_1 pe pzz vol
thermo_modify       lost ignore norm yes
# if an atom is lost, ignore it and normalize data with the new number of
atoms
dump                snapshotrelax all xyz 5000 Si_smallCrack_relax_*.xyz

run                 10000
write_restart       restart.10000

undump              snapshotrelax
unfix               3
unfix               4
### Applying tension ###
reset_timestep      0

fix                 3 boundary setforce 0.0 0.0 0.0
velocity            upper set 0.0 0.1 0.0
# x-dimension fixed, y- dimension extended
# 0.1 lattice unit = 0.5431 A/ps, 0.1 A/ps if units box
velocity            mobile ramp vy 0.0 0.1 y 1.25 8.75 sum yes

fix                 1 all nve
fix                 2 mobile temp/berendsen 300 300 0.1
# T for mobile region only

fix                 3 boundary setforce 0.0 0.0 0.0

compute             strSi all reduce sum c_body_st[2]
# compute for the 2nd tensor yy out of 6

variable            syySi equal c_strSi/(20.09*count(mobile))*(10^-4)
# atomic V = 20.09 (A^3) = 12.1 (cm3/mol), (10^-4) to convert GPa
variable            syy equal v_syySi

### Run ###
timestep            0.001
thermo              200
thermo_modify       temp new
thermo_style        custom step temp c_1 pe etotal pyy vol

neigh_modify        exclude type 2 3
# pair interactions shut off between type 2 and 3 atoms to create a crack

fix                 33 all ave/time 1 100 100 v_syy file Si_smallCrack_stress_
                    GPa.dat
# save timestep vs. v_syy (GPa) data
dump                1 all atom 5000 Si.smallCrack.dat
dump                2 all xyz 5000 Si.smallCrack_*.xyz
# to visualize the snapshots on VESTA
dump                3 all atom 10 Si.smallCrack.lammpstrj
# to visualize the whole simulation on VMD

run                 30000
############################################################################
```

Note that the presence of surface cracks is a typical feature on a brittle material such as Si, and the pairwise interactions were shut off between left-upper and left-lower blocks to create a crack.

3.9.3 Results

Figure 3.13 shows the snapshots at timesteps of 0, 15,000, and 25,000. The figure indicates a typical brittle fracture of crack propagation without any noticeable plastic deformation except right at the fractured surface of crack. The Si_smallCrack_stress_GPa.dat file has tensile stress data developed every 100 steps:

```
############################################################
# Time-averaged data for fix 33
# TimeStep v_syy
100 0.148749
200 0.384504
300 0.58315
.....
18500 30.7028
18600 30.713
18700 30.5116
.....
29800 0.377279
29900 0.390561
30000 0.443635
############################################################
```

Figure 3.14 shows the plot in terms of timestep, indicating the maximum tensile stress of slightly over 30 GPa when the crack initiation layer is small (10.86 Å thick). If we increase the initial crack sizes by 50% and 100% by adjusting the x-direction thickness of left-upper (type 2) and left-lower (type 3) regions, we have stress–strain fracture strengths in the range of 20–30 GPa. It is known that the strength of brittle solids decreases in proportion to the square root of the crack size. Considering this fact, these values are in reasonable agreement with that and also the DFT calculated

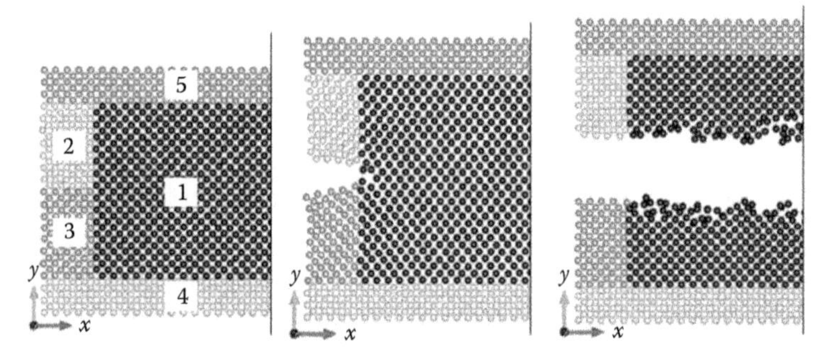

Figure 3.13 Si bar with a small crack-initiating block under tension: at timesteps of (a) 0, (b) 15,000, and (c) 25,000. Note that only half of the bar is shown.

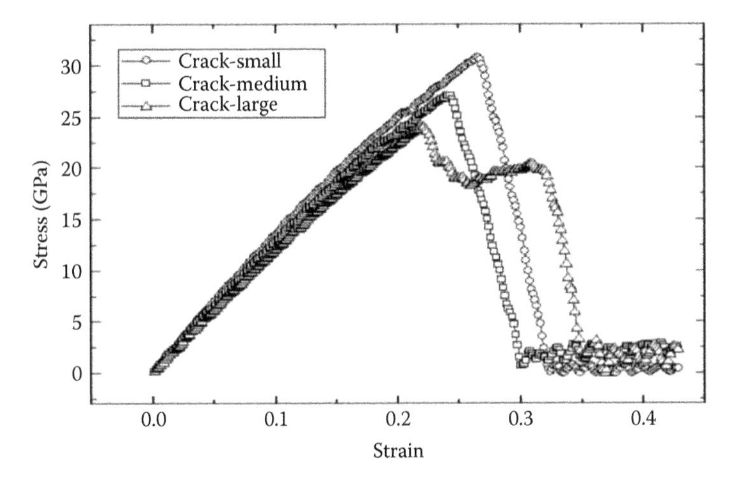

Figure 3.14 Stress–strain curves of Si bars with various initial crack sizes.

fracture strength of about 30 GPa along the [100] direction (Lazar 2006). When the crack initiation layer is too thick (large crack case), the assumption of brittle fracture is no longer valid, and some plastic behavior is observed after reaching the fracture strength.

3.10 Si–CNT composite under tension

This exercise is the final part of a series of runs that is for the computational development of a nanocomposite in three steps (Kim et al. 2012). It is known that there are two key issues for improving the mechanical properties of CNT-reinforced nanocomposites: the interfacial bonding between CNTs and matrix and uniform dispersion of CNTs into matrix. We focus on the first issue in this exercise.

Not like normal fiber-reinforced nanocomposites, it is very difficult to study CNT-reinforced nanocomposites experimentally as clearly indicated by rather scattered data reported by various studies. Difficulties in uniform mixing and in formation of proper interfacial bonding between the matrix and the CNT are considered as the main causes for the inconsistency. An MD approach, however, can provide an alternative to the problem and can generate fundamental information such as the stress–strain behavior that is vital to tailor the nanocomposites for a specific use.

3.10.1 Introduction

We investigate the effects of interfacial bonding on the mechanical properties in the Si–carbon nanotube (CNT) nanocomposite. We initiate a crack and propagate it to fracture via elastic deformation, slip, plastic

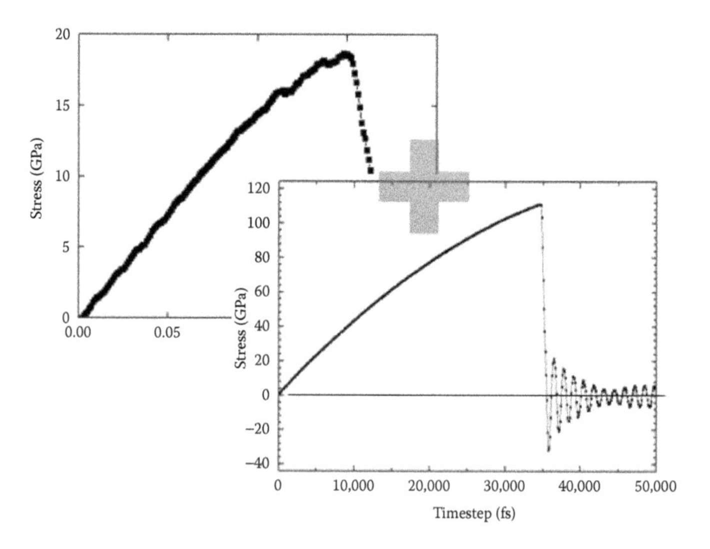

Figure 3.15 Design concept for the Si–CNT composite system.

deformation, crack deflection, and so on. The role of the CNT and Si–CNT interface will be evaluated to optimize structural design for the Si–CNT composite.

- System: Si–CNT composite based on the two previous runs.
- Figure 3.15 shows the design concept for the Si–CNT composite.
- Goal: calculation of the stress–strain behavior at the temperature of 300 K and identification of the fracture mechanism in terms of the bonding between Si and the CNT.

We use a simulation model (total 17,081 atoms) shown in Figure 3.16 that consists of a Si matrix (54.31 × 54.31 × 108.62 Å with 15,961 atoms) reinforced with a single-wall CNT (dia.: 13.751 Å, length: 68.589 Å with 28 hexagons of 1120 atoms, armchair type with chirality of (10, 10)). Here, Si was chosen as a typical brittle matrix to be improved in toughness by the CNT.

To implement different interactions and loading conditions, the Si matrix is divided into five regions: (a) fixed lower region, (b) extending upper region, (c) left-lower and (d) left-upper crack-initiating regions, and (e) main Si around region. Regions (c)–(e) are defined as the mobile region because it is the part responding to tension with a dynamic movement. Periodic boundary conditions are applied in both y- and z-directions that correspond to a plane strain condition under tension. A crack is initiated between regions (c) and (d), by preventing interactions between those regions.

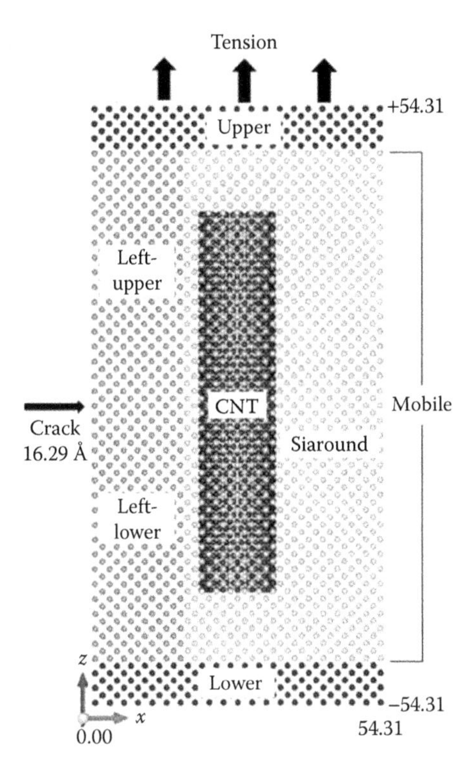

Figure 3.16 Starting structure of the Si–CNT composite system.

3.10.2 Potentials

To simulate the system appropriately, we need three potentials to describe two different materials and their interface as shown in Figure 3.17. We use a hybrid potential that includes the SiC_Erhart-Albe.tersoff potential (Erhart and Albe 2005) for Si–Si in the Si matrix, the CH.airebo (adaptive intermolecular reactive empirical bond order) potential (Stuart et al. 2000) for C–C in the CNT, and the LJ 6–12 potential (Jones 1924) for the Si–C interface, which is embedded in the LAMMPS program. The LJ potential will perform the deciding role in this exercise.

3.10.3 Input files

We have two input files: an in.*** file and a separate atomic position file. The initial equilibrium was established by relaxing the system for 10,000 steps with the use of a Berendsen thermostat to allow the system to fluctuate close to 300 K. A timestep of 1 fs was used in all runs.

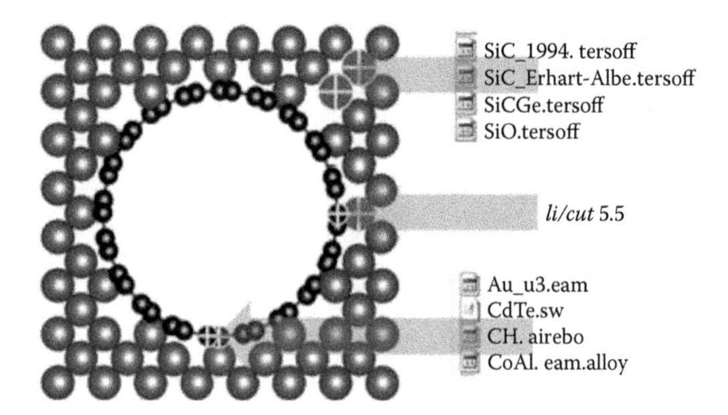

Figure 3.17 Three potentials adopted for the study to describe Si, CNT, and their interface. Note that only atoms near the interface are shown.

Tensile loading of the Si–CNT system is implemented by keeping one end (bottom) fixed, while slowly displacing the other end (0.25 Å/ps) in the *z*-direction, as shown in Figure 3.16. This is many orders higher than in real experimental conditions, obviously due to the fundamental limitation on the time scale of MD simulations. However, we confirmed that the strain rate was satisfied to represent the quasistatic loading, that is, the total energy difference in a timestep is less than 10^{-5}%. The displacement was so controlled that a gradient toward zero is accomplished throughout the main matrix to the fixed end.

```
#############################################################################
# in.Si-CNT-tension.txt

### Basic setup ###
units                    metal
boundary                 f p p
atom_style               atomic

### Structure ###
read_data                readdata.SiCNT

region                   box block 0 54.31 0 54.31 -54.31 54.31 units box
region                   1 block INF INF INF INF INF -47.52125 units box
group                    lower region 1
region                   2 block INF INF INF INF 47.52125 INF units box
group                    upper region 2
group                    boundary union lower upper
group                    mobile subtract all boundary
region                   leftupper block INF 16.293 INF INF 0 INF units box
region                   leftlower block INF 16.293 INF INF INF 0 units box
group                    leftupper region leftupper
group                    leftlower region leftlower

region                   void cylinder z 27.155 27.155 7.875 -36 36 units box
group                    CNT region void
# to delete_atoms in group void
group                    Sicrack union leftupper leftlower
```

```
group                       Siaround subtract mobile Sicrack

set                         group Siaround type 1
set                         group leftupper type 2
set                         group leftlower type 3
set                         group lower type 4
set                         group upper type 5
set                         group CNT type 6
# to identify each group on VESTA, set 1 grey, 2 red, 3 blue, 4 green, 5 yellow

### Potentials ###
mass        1 28
mass        2 28
mass        3 28
mass        4 28
mass        5 28
mass        6 12.01

pair_style                  hybrid airebo 2.5 1 0 tersoff lj/cut 5.5
pair_coeff                  1 6 lj/cut 0.50 1.2
# Si-C(CNT) by lj, original value; 0.038 2.96
pair_coeff                  * * tersoff SiC_Erhart-Albe.tersoff Si Si Si Si
                            Si NULL
# Si-Si by tersoff
pair_coeff                  * * airebo CH.airebo NULL NULL NULL NULL NULL C
# C(CNT)-C(CNT) by airebo
pair_coeff                  2 * 5 6 none

neighbor                    1.0 bin
neigh_modify                every 1 delay 5 check yes

### Relaxation at 300 K ###
compute                     new mobile temp
velocity                    mobile create 300 887723 temp new

fix                         1 all nve
fix                         2 mobile temp/berendsen 300 300 0.1
# T for mobile region only, fixed upper & lower boundary

fix         3 lower setforce 0.0 0.0 0.0
fix         4 upper setforce 0.0 0.0 NULL

compute                     1 mobile temp
compute_modify              1 dynamic yes

compute                     body_pe all pe/atom
compute                     body_st all stress/atom NULL pair bond

timestep                    0.001
thermo                      200
thermo_style custom step temp c_1 pe pzz vol
thermo_modify               lost ignore norm yes

run                         10000
write_restart               restart.10000

unfix                       3
unfix                       4

### Applying tension ###
reset_timestep              0

fix                         3 boundary setforce 0.0 0.0 0.0
```

```
# x-dimension fixed, y-direction extended

velocity              upper set 0.0 0.0 0.25 units box
velocity    mobile ramp vz 0.0 0.25 z -47.52125 47.52125 sum yes units box

compute               strCNT CNT reduce sum c_body_st[3]
compute               strSi mobile reduce sum c_body_st[3]
compute               strSiCNT all reduce sum c_body_st[3]

variable    szzSi equal c_strSi/(20.09*count(mobile))*(10^-4)
variable    szzCNT equal c_strCNT/(8.784*count(CNT))*(10^-4)
variable              szz equal "v_szzSi + v_szzCNT"
# variable szzAl equal c_strAl/(16.60*count(g_al))*(10^-4)
# atomic volume = 20.09 (A^3) = 12.1 (cm3/mol)
# (10^-4) to convert GPa
# compute for the 2nd tensor yy out of 6

neigh_modify exclude type 2 3
# to initiate crack, pairwise interactions are shut off

dump                  1 all atom 5000 SiCNT.crack.dat
dump        2 all xyz 5000 SiCNT_*.xyz

fix         33 all ave/time 1 1000 1000 v_szz file stress_tensile_GPa.dat

run                   160000
write_restart         restart.160000
##############################################################################
```

The following is a separate atomic position file: "readdata.SiCNT." This file will be read by the above file to generate the structure as shown in Figure 3.16.

```
##############################################################
LAMMPS data file

17081 atoms

6 atom types

-10.000000   64.310000    xlo xhi
0.000000     54.310000    ylo yhi
-64.310000   100.000000   zlo zhi

Masses

      1 28
      2 28
      3 28
      4 28
      5 28
      6 12.01

Atoms
```

```
1     4    0          0          -54.31
2     4    0          2.7155     -51.5945
3     4    2.7155     0          -51.5945
4     4    2.7155     2.7155     -54.31
5     4    1.35775    1.35775    -52.9522
6     4    1.35775    4.07325    -50.2368
.....
############################################################
```

3.10.4 Run

This run requires more than 5 hours if it is run on a single-processor computer. We can do that way or run it faster with an MPI way with a multiprocessor computer using the lmp_mpi.exe program. Note that by changing parameters of pair_coeff for the LJ potential (e.g., 0.50 1.2 in 1 6 LJ/cut 0.50 1.2), various interfacial bonding strengths can be created. In this exercise, however, we choose parameter ε ($\equiv$ well depth, cohesive energy) = 0.5 at a fixed parameter σ (bond distance) = 1.2.

3.10.5 Results

Figure 3.18 shows snapshots of the Si–CNT composite system at its optimum interfacial bonding (ε = 0.5 eV) for strain of 10.5%, 31.5%, and 42%. In this case, all beneficial events take place: load transfer, crack deflection from (100) crack to easier (111) slip, and pull-out of SNT. The {111} planes/[112] direction in Si is known to be the weakest planes/direction by 30% compared to the {100} planes/[100] direction. Note that the CNT is just pulled-out maintaining it integrity even after the system fracture.

Figure 3.19 shows all stress–strain curves of the Si–CNT nanocomposites with various interfacial bonding strengths. It shows that fracture strength and toughness (defined here as the area under the curve) increase with increasing interfacial strength up to ε = 0.5 in the LJ potential. Compared to Figure 3.14, a marked increase in both fracture strength and toughness is observed. The calculated Young's modulus and maximum strength were about 480 and 58 GPa, respectively. The average distance of Si–C bonding at the interface was about 1.98 Å, which corresponds to that of the crystalline SiC.

A fully bonded case is simulated by allowing a full interaction between Si atoms and the CNT at the interface described by the Erhart/Albe-Tersoff potential (Erhart and Albe 2005) for Si–C bonds. It shows considerable defects formed on the CNT after the relaxation due to strong interfacial interaction although the general feature of the CNT structure is maintained.

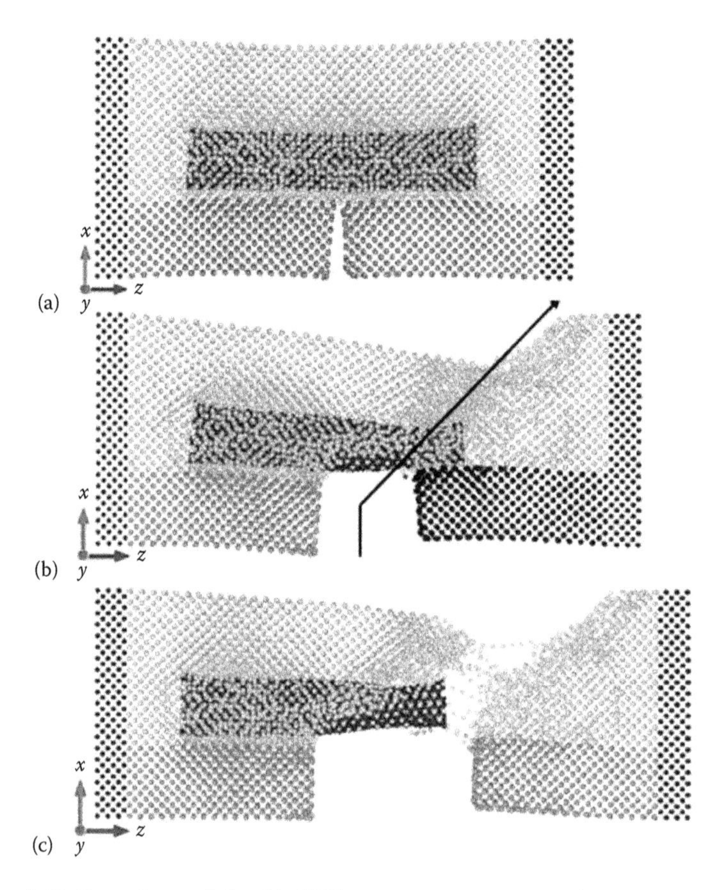

Figure 3.18 Snapshots of the Si–CNT composite system with ε = 0.5 eV under tension for various strains: (a) 10.5%, (b) 31.5%, and (c) 42%. Beneficial events taking place in the Si–CNT composite system with an optimum interfacial bonding (bonding strength = 0.5): load transfer, crack deflection, plastic deformation via (111) slip, and pull-out of SNT.

3.10.6 Conclusions

An MD approach with a hybrid potential properly describes the effects of interfacial bonding in the Si–CNT nanocomposites. We observe that the mechanical properties such as Young's modulus, maximum strength, and toughness increase steadily with increasing bonding strength at the interface of the Si matrix and CNT.

It is worth noting that CNT pull-out and load transfer on the strong CNT are identified as the main mechanism for the improved mechanical properties. At maximum bonding, however, only load transfer is operative and the fracture returned to brittle mode. At optimum bonding, crack front is deflected around the CNT and the fracture proceeds in plastic

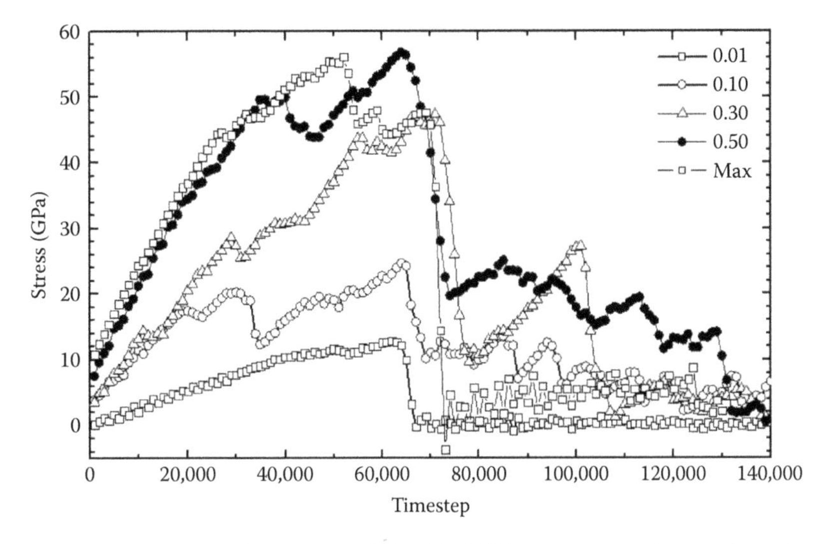

Figure 3.19 Stress–strain curves of the Si–CNT nanocomposites with various interfacial bonding strengths.

mode through the Si matrix due to the strong support by the CNT, and resulted in a further enhancement of toughness.

We suggest that a strong interface as long as the CNT maintains its structural integrity is desirable to realize the optimum result. This exercise clearly demonstrates that MD is an established tool in materials science.

3.11 ZrO_2-Y_2O_3-MSD

This exercise is kindly provided by Kim and Kim (2015, private comm.) for calculation of MSD (mean-squared displacement) and related properties from atomic trajectories during an MD run.

3.11.1 Introduction

- System: Y_2O_3-stabilized zirconia (8 YSZ) with 26 Y, 230 Zr, and 499 O atoms (total 755 atoms).
- Two Y_2O_3 create an oxygen vacancy to maintain electroneutrality in the system, and these oxygen vacancies provide easier route for oxygen diffusion.
- Program: LAMMPS/compute msd command.
- Temperature: 1000–2000 K.
- Goal: calculation of the MSD of O atoms, the total squared displacement, that is, (dx*dx + dy*dy + dz*dz), summed and averaged over all O atoms, diffusion coefficients, barrier energy for oxygen vacancy migration.

3.11.2 Input files

The following file is the in. ZrO_2-Y_2O_3-MSD.txt file prepared for the run at 2000 K, as an example:

```
##############################################################################
### Basic setup ###
variable                        t equal 2000
units                           metal
dimension                       3
boundary                        p       p       p
atom_style                      charge
newton                          on

### Structure ###
read_data                       ZrO2.data
neighbor                        3 bin
comm_modify                     vel yes

### Potentials ###
# Zr-O and O-O: A. Dwivedi, A. N. Cormack, Philos. Mag. A 1990, 61, 1.
# Y-O: G. V. Lewis, C. R. A. Catlow, J. Phys. C 1985, 1149.
kspace_style                    ewald/disp 1e-4
pair_style                      buck/coul/long 10
pair_coeff             * * 0.00 10 0.00
pair_coeff                      1 2 985.9499 0.3760 0.000000
# A. Dwivedi, A. N. Cormack, Philos. Mag. A 1990, 61, 1.
pair_coeff                      2 2 22762.26 0.1490 27.8931
# A. Dwivedi, A. N. Cormack, Philos. Mag. A 1990, 61, 1.
pair_coeff                      2 3 1345.212 0.3491 0.00000
# G. V. Lewis, C. R. A. Catlow, J. Phys. C 1985, 1149.

### Group ###
group                  Zr           type 1
group                  O            type 2
group                  Y            type 3

### Minimization ###
thermo                              100
thermo_style           custom step time vol atoms temp ke pe lx press
fix                    1 all box/relax iso 0.0 vmax 0.001
min_style              cg
minimize               1e-25 1e-25 100000 100000
unfix 1

### NPT1 ###
reset_timestep         0
dump                   1 all xyz 1000 YSZ_xyz_NPT1_$t
dump                   2 all atom 1000 YSZ_atom_NPT1_$t
thermo                 1000
thermo_style           custom step dt time atoms temp ke pe vol press lx ly lz
velocity               all create $t 78432 dist gaussian
fix                    NPT all npt temp $t $t 10 iso 1.013 1.013 100 drag 0.2
timestep               0.001
run                    250000

undump                 1
undump                 2
```

```
### Mean square displacement ###
reset_timestep      0
compute             msd O msd com yes
fix                 vec all vector 1 c_msd[4]

### NPT2 ###

dump                1 all xyz 1000 YSZ_xyz_NPT2_$t
dump                2 all atom 1000 YSZ_atom_NPT2_$t
thermo              1000
thermo_style        custom step temp ke pe vol press lx c_msd[4]

timestep 0.001
run                 500000

unfix                             vec
unfix               NPT
undump              1
undump              2
uncompute                         msd
#############################################################################
```

The following file is the ZrO_2.data file prepared for the structure of the ZrO_2–$6Y_2O_3$ (6 atomic %) system, as shown in Figure 3.20.

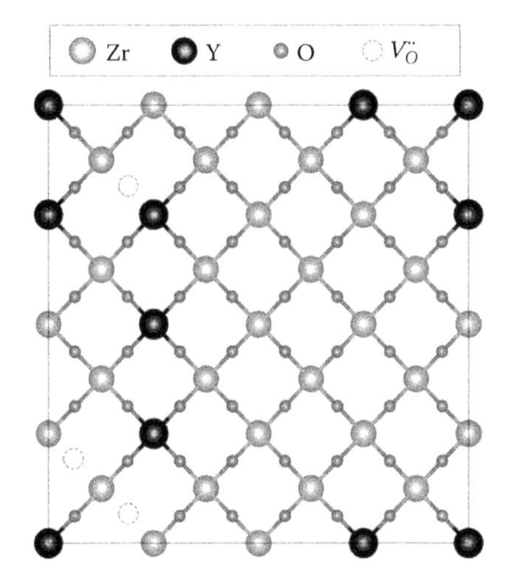

Figure 3.20 Sliced structure of the ZrO_2–$8Y_2O_3$ system showing a metal (Zr and Y) layer and an oxygen layer along the [001] direction (large gray ball: Zr, large dark ball: Y, small gray ball: O, large dotted white ball: oxygen vacancy). Six Y atoms and three oxygen vacancies are randomly distributed and shown in the structure.

```
################################################################
YSZ

755 atoms

3 atom types

0.0 20.5200004578 xlo xhi
0.0 20.5200004578 ylo yhi
0.0 20.5200004578 zlo zhi

Masses

1 91.22400
2 15.99940
3 88.90585

Atoms

1 2 -2 1.2825000290 1.2825000290 1.2825000290
2 2 -2 3.8475000860 3.8475000860 1.2825000290
3 2 -2 3.8475000860 1.2825000290 3.8475000860
…..
753 3 3 15.3900003430 10.2600002290 15.3900003430
754 3 3 5.1300001140 15.3900003430 15.3900003430
755 3 3 10.2600002290 17.9550004010 17.9550004010
################################################################
```

3.11.3 Run

The run could be as usual as any normal LAMMPS calculation. One problem is that it takes long time in a serial way, because the whole number of timesteps is 750,000. We better do this job in an MPI way with at least 16 cores.

3.11.4 Results

In the log.lammps file, we can collect the fourth MSD data, which are the total squared displacement (dx*dx + dy*dy + dz*dz), summed and averaged over all O atoms. Figure 3.21 is the MSD plot in terms of time in the range of the last 500,000 timesteps for collection of MSD data at various temperatures from 1000 to 2000 K. A very steady increase of MSDs is recorded, and diffusion coefficients are calculated from the linear-fitted slopes of each curve. Plotting diffusion coefficients versus $1/T$, we can calculate the barrier energy of O migration as 0.56 eV by the Arrhenius equation for diffusion.

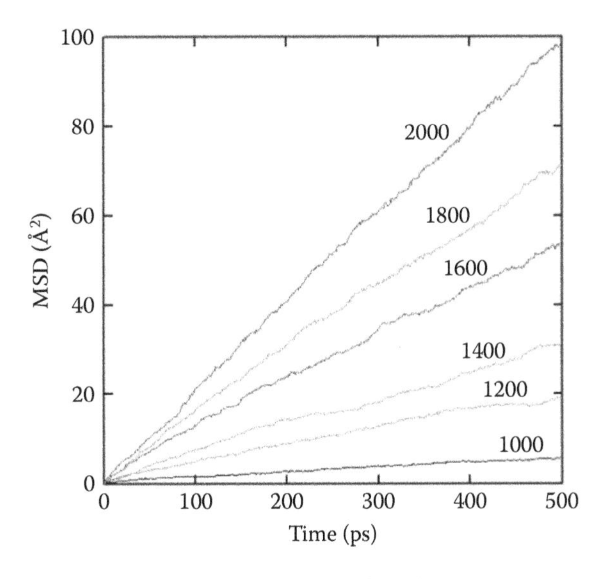

Figure 3.21 MSD plot with time in the range of the last 500,000 timesteps at various temperatures from 2,000 to 1,000 K.

Homework

3.1 Referring to Section 3.1, obtain a potential curve for Si (cubic-diamond structure with an equilibrium lattice parameter of around 5.43 Å) by an XMD run, calculate its bulk modulus, and compare with the available experimental value.

3.2 Referring to Figure 3.2, the melting points will increase and the potential energy curves will go down if we increase the size of the Ni cluster in a series of MD runs. Explain why it is.

3.3 In an MD run, we always notice fluctuations on instantaneous quantities such as temperature, kinetic and potential energies, and so on with iteration. It is known that these fluctuations will be reduced with the increasing system size as fluctuation $\propto 1/\sqrt{N}$, where N is the number of atoms. Explain why it is.

3.4 Choosing the optimum timestep size is an important issue in an MD run. Referring to Section 3.1, write down an input file with the timestep size as the variable for an XMD run so that we can identify the optimum timestep size for a particular system.

3.5 This homework is provided by B.-H. Kim and K.-R. Lee at KIST, Korea (Kim and Lee, 2011, private comm.) In processing of semiconductors, Si oxidation is an important step and we will try a very preliminary modeling. Construct a slab model for the process, simulate it under the following conditions, plot the number of oxygen atoms along the *z*-axis, and discuss the significant findings.

Program: LAMMPS-5Dec10 or newer

Potential: Reactive-force-field potential (ReaxFF) ffield.reax from the LAMMPS-27Aug09 package

Define the potential in the input file as:

```
#############################################################
# define the interatomic potential, ReaxFF
pair_style      reax
pair_coeff      * * ffield.reax 3 6            # 3; O, 6; Si
#############################################################
```

Substrate: a slab of 4 × 4 × 4 Si(001), 512 atoms with the bottom layer fixed

Vacuum above the slab (5.43 Å × 4 lattice thick) and PBC only along the *x*- and *y*-axes

Simulation sequence: relaxation of Si slab for 5 ps at 700 K, generation of 200 free oxygen atoms in random in vacuum, and oxidation for 30 ps at 700 K

If any oxygen atom leaves through the top boundary, ignore them and normalize corresponding quantities with the new atom number

References

Brenner, D. W., O. A. Shenderova, J. A. Harrison, S. J. Stuart, B. Ni, and S. B. Sinnott. 2002. A second-generation reactive empirical bond order (REBO) potential energy expression for hydrocarbons. *J. Phys.: Condens. Matter* 14:783–802.

Eatemadi, A. et al. 2014. Carbon nanotubes: Properties, synthesis, purification, and medical applications. *Nano. Res. Lett.* 9:393–406.

Erhart, P. and K. Albe. 2005. Analytical potential for atomistic simulations of silicon, carbon, and silicon carbide. *Phys. Rev. B* 71:035211.

Jones, J. E. 1924. *Proc. R. Soc. London, Ser. A* 106:463.

Foiles, S. M., M. I. Baskes, and M. S. Daw. 1986. Embedded-atom-method functions for the fcc metals Cu, Ag, Au, Ni, Pd, Pt, and their alloys. *Phys. Rev. B* 33:7983–7991.

Gaudoin, R., W. M. C. Foulkes, and G. Rajagopal. 2002. Ab initio calculations of the cohesive energy and the bulk modulus of aluminum. *J. Phys.: Condens. Matter* 14:8787–8793.

Jorgensen, W. L., J. Chandrasekhar, and J. D. Madura. 1983. Comparison of simple potential functions for simulating liquid water. *J. Chem. Phys.* 79:926–935.

Kim, B.-H. et al. 2012. Effects of interfacial bonding in the Si-CNT nanocomposite: A molecular dynamics approach. *J. Appl. Phys. Mater.* 112:044312.

Kim, J.-Y., B.-W. Lee, H.-S. Nam, and D. Kwon. 2004. Molecular dynamics analysis of structure and intrinsic stress in amorphous silicon carbide film with deposition process parameters. *Mat. Sci. Forum.* 449–452:97–100.

Kondo, Y. and K. Takayanagi. 1997. Gold nanobridge stabilized by surface structure. *Phys. Rev. Lett.* 79:3455–3458.
Kondo, Y., Q. Ru, and K. Takayanagi. 1999. Thickness induced structural phase transition of gold nanofilm. *Phys. Rev. Lett.* 82:751–754.
Lazar, P. 2006. Ab initio modelling of mechanical and elastic properties of solids. PhD Dissertation, Department of Physics, University of Vienna, Vienna, Austria.
Lu, Q., M. Arroyo, and R. Huang. 2009. Elastic bending modulus of monolayer graphene. *J. Phys. D: Appl. Phys.* 42:102002–102007.
Murnaghan, F. D. 1944. The compressibility of media under extreme pressures. *Proc. Natl. Acad. Sci. USA* 30:244–247.
Park, N. Y., H. S. Nam, P. R. Cha, and S. C. Lee. 2015. Size-dependent transition of the deformation behavior of Au nanowires. *Nano Research* 8(3):941–947.
Petersen, K. E. 1982. Silicon as a mechanical material. *Proc. IEEE* 70:420–457.
Stuart, S. J., A. B. Tutein, and J. A. Harrison. 2000. A reactive potential for hydrocarbons with intermolecular interactions. *J. Chem. Phys.* 112:6472–6486.
Voter, A. F. and S. P. Chen. 1987. Accurate Interatomic potentials for Ni, Al, and Ni3Al. *Mat. Res. Soc. Symp. Proc.* 82:175–180.
Walther, J. H., R. L. Jaffe, E. M. Kotsalis, T. Werder, T. Halicioglu, and P. Koumoutsakos. 2004. Hybrophobic hydration of C_{60} and carbon nanotubes in water. *Carbon.* 42:1185–1194.
Wikipedia. 2010. Surface tension of water. http://en.wikipedia.org/wiki/File:Temperature_dependence_surface_tension_of_water.svg.

chapter four

First-principles methods

> All matter is merely energy condensed to a slow vibration.
> We are all one consciousness experiencing itself subjectively.
> There's no such thing as death, life is only a dream, and we are the imagination of ourselves.
>
> **Bill Hicks (1961–1994)**

Some say that all matter has evolved from "the Big Bang," whereas others insist that "the divine programmer" let the event proceed that particular way or created all matter at once. What really happened, we may never know. We do, however, know that there exist four basic forces acting on matter that probably originated from that particular event: electromagnetic force, gravitational force, and strong and weak nuclear forces acting inside atomic nuclei. Fortunately, we need to be concerned with only the first one since studies on materials rarely encounter the other three forces.

The electromagnetic force is the very governing force of nuclei and electrons. With it are built atoms, molecules, and materials. Whatever happens in materials comes ultimately from this basic interaction. In previous Chapter 2 and 3, we have already encountered this force in the form of empirical potential acting on atoms. In this chapter, we will go down further to the level of positively charged nuclei and negatively charged electrons from which the electromagnetic forces originate.

If an accurate description of these interactions is possible, then all experimentally relevant properties and phenomena for materials should surface accordingly from our calculations. The governing law for these small particles is known as *quantum mechanics*. The calculation methods based on quantum mechanics, therefore, do not use any fitting parameters from the experimental data as used in MD and are based solely on the basic laws of physics. Thus, they are named as first-principles (or *ab initio*) methods. In principle, the only information the methods require is the atomic numbers of the constituent atoms. Then, the electromagnetic properties and bond-forming/-breaking that are unobtainable in MD can be calculated with these first-principles methods.

In dealing with quantum mechanics, a rather drastic change in our classical habits of thinking is required. The subject is especially difficult for materials people since we have accustomed ourselves to relying on instruments to amplify the subatomic phenomena into some form of signal that we can observe macroscopically. The quantum world is a magical world in the sense that it can make an object disappear at the first floor and reappear at the third floor at the same time, or make it penetrate a high wall and place it outside the wall. Note that quantum acts are real things, whereas magic is all well-designed and manipulated illusions. Furthermore, it is fascinating that the quantum world is seamlessly interwoven with the classical world.

In this chapter, quantum mechanics will be introduced as briefly as possible in a materials-science fashion, and the early first-principles methods of computation will also be briefly reviewed, avoiding too many mathematical procedures but following conceptual developments to the ends to which they lead. The reader may be tempted to dismiss all of this as needless words better suited for physicists. Remember, however, that knowing the no-need-to-know is often needed to know the must-know. I believe spending some time on this chapter will be rewarded when we move to the next chapter and deal with the density functional theory (DFT), the core of computational materials science.

For the sake of simplicity, several omissions are made throughout this chapter:

- The word *operator* and the operator notation (the hat, $\wedge$) are omitted for all operators except $\hat{H}$.
- The coordinate notation $\mathbf{r}$ is defined to include both three-dimensional position (x, y, z) and spin $(\uparrow, \downarrow)$ coordinates, although the spin part is largely ignored and is not explicitly written down unless indicated otherwise.

In addition, nuclei and electrons are considered as point-like particles when we treat them as a particle in terms of masses, charges, positions, and so on.

4.1 Quantum mechanics: The beginning

> Anyone who is not shocked by quantum theory has not understood it.
>
> **Niels Bohr (1885–1962)**

> I think I can safely say that nobody understands quantum mechanics.
>
> **Richard Feynman (1918–1988)**

Quantum mechanics was the most talked about and yet least understood theory at the beginning. It was not until the early twentieth century that scientists, who believed that Newton's laws were all they needed to describe matters, realized that they were all wrong. Classical mechanics, the "almighty" law of the universe for nearly three centuries, was found to be inadequate for explaining phenomena in the subatomic scale. For example, red-heated hydrogen gas was supposed to emit energies in continuous fashion like the jumping atom in Figure 2.3. Instead, by emitting radiations at specific frequencies, it produces a distinctive energy spectrum of its own like a fingerprint.

Furthermore, having second thoughts on the atomic structure in general made scientists more puzzled and even embarrassed. They had pictured the atom consisting of electrons circulating around a central nucleus and were amazed at this planetary phenomenon in such a small-scale system. However, they soon realized that charged electrons in rotating motion should radiate energy according to the well-known electromagnetic theory. It implied that electrons eventually should lose all their energy and collapse onto the nucleus. The consequence, of course, jeopardizes even the very existence of matter. These and other inconceivable cases forced scientists to think differently and search for a new theory. Unlike classical mechanics and the theory of relativity that were established single-handedly by two individuals, quantum mechanics was completed by fitting many pieces of knowledge together over many decades by many scientists. The following are four highlights in the development of the new theory, quantum mechanics.

4.1.1 Niels Bohr and the quantum nature of electrons

The ancient Greeks imagined that all matter was composed of small indivisible particles called atoms. After the discovery of electrons, however, the picture of an atom had to be modified whenever a new finding was revealed. And it was about to be renewed once more. Niels Bohr (1913) proposed that the electron in a hydrogen atom is not circulating on any arbitrary orbit but on specific ones (later named *orbitals*) for which the electron does not radiate energy (see Figure 4.1).

In these orbitals, he showed that only specific energies are allowed as

$$E = \frac{-13.6\,\text{eV}}{n^2} \tag{4.1}$$

where n is the numbering of orbitals. For example, an electron in the lowest energy level of hydrogen ($n = 1$), therefore, has –13.6 eV; the next energy level ($n = 2$) is –3.4 eV, and so forth. The energies are, of course, negative because these are bound states of an electron with nucleus with reference to the zero-energy state when they are infinitely apart. He further

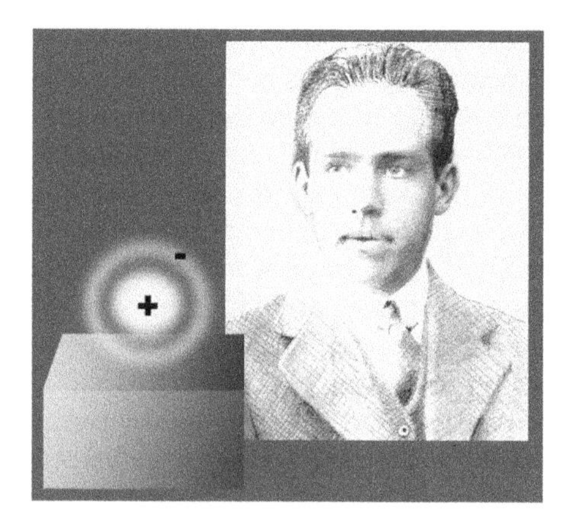

Figure 4.1 Young Bohr and his hydrogen atom.

proposed that electrons can jump up or down between orbitals by absorbing or emitting not any energy but a quanta of energies, ΔE, which is the difference in energy between two orbitals, n_1 and n_2, involved:

$$\Delta E = -13.6\left(\frac{1}{n_1^2} - \frac{1}{n_2^2}\right) \tag{4.2}$$

This is analogous to the operation of an elevator that can go only to floor 1, floor 2, and so forth, but not to floor 1.9, 2.0, 2.1, and so on, similar to a normal escalator as shown in Figure 4.2. Of course, the elevator takes some time to lift people upward or downward, but the electrons move in no time. This new picture, although not completely correct, successfully

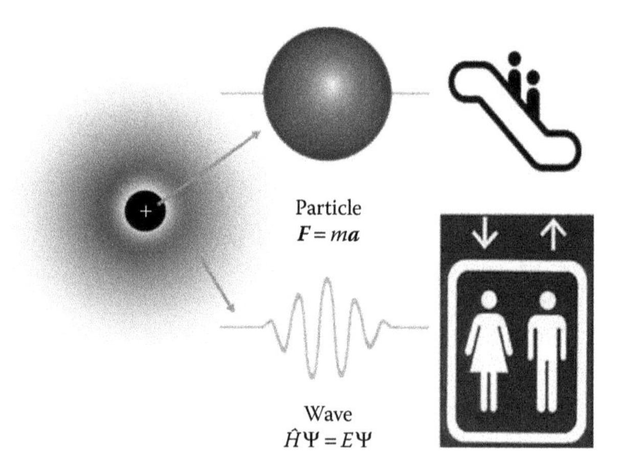

Figure 4.2 Energy changes in classical and quantum worlds.

explained the hydrogen spectrum and prevented the electron from falling down on the nucleus by assigning a specific ground-state energy (later named *1s-orbital energy* of −13.6 eV). If the electron is in the next orbit of $n = 2$, a hydrogen atom will have one fourth of the ground-state energy; this state is called the *excited* state. And, with an increasing n, potential energy goes up to zero, and the electron is completely separated from the nucleus, becoming a free electron.

In fact, the quantum nature of radiation energy, formulated as early as 1900 by Max Planck, had been unnoticed by others for many years. He reported that blackbody radiation could be explained nicely if the radiation energies were not continuous as the classical mechanics would predict, but they could only lose or gain energies in quanta, that is, $E = nh\nu$, where n is an integer number, ν is frequency of the blackbody radiation, and h is Planck's constant. It is clear that both Planck and Bohr were talking about the same thing from different perspectives. And the long-hidden side of nature was about to be unveiled.

4.1.2 De Broglie and the dual nature of electrons

The notion of radiation as a wave and electron as a particle was once a popular belief for all. For example, electrons displayed enough evidence to be a particle: a definite mass and charge, changing velocity, and a seemingly straight traveling path. Louis de Broglie (1924) imagined differently, and redefined a particle and wave as a single entity "matter wave." He declared with a stunning statement that the momentum of a particle is inversely proportional to the wavelength (see Figure 4.3) and the frequency is directly proportional to the particle's kinetic energy:

$$p = \frac{h}{\lambda} \tag{4.3}$$

Figure 4.3 Young de Broglie and his particle–wave duality.

$$E = h\nu \tag{4.4}$$

where:

p is momentum
h is the Planck constant
λ is wavelength
ν is frequency

Now it is clear why we do not notice the wave nature of our everyday objects such as a bowling ball or a billiard ball. When an object's momentum is very large relative to the Planck's constant, then its wavelength is so small that the wave nature can hardly be noticed, and the quantum world becomes the same as the classical world. For example, a 4 kg bowling ball traveling at 1.65 m/s will have an extremely small wavelength:

$$\lambda = \frac{h}{p} = 10^{-34} m \tag{4.5}$$

No matter how big or how small it is, the seemingly different particles and waves are of the same kind after all in the alien world of quantum mechanics.

4.1.3 *Schrödinger and the wave equation*

Soon, scientists realized that small particles such as electrons could only be described as a wave to express their characteristics. And expressing the wave nature of electrons in terms of energy in mathematical form was the prime concern of scientists at that time. In other words, people wanted to know how God runs the universe other than the classical way.

As shown in Figure 4.4, Erwin Schrödinger (1926), allegedly inspired by Swiss winter scenery and other things, came up with an equation that belongs to the level of (or surpassing) Newton and Einstein. The wave equation of Schrödinger is as simple as the following:

$$\hat{H}\Psi(\mathbf{r}_i, \mathbf{r}_I, t) = E\Psi(\mathbf{r}_i, \mathbf{r}_I, t) \tag{4.6}$$

Here, $\hat{H}$, Ψ, and E are the Hamiltonian operator, wave function, and system energy, respectively. The variables, $\mathbf{r}_i$ and $\mathbf{r}_I$ are the coordinates of the electron and nucleus, respectively.

Let us leave what is what in this equation to the next section and just note that this innocent-looking equation of two letters and two mathematical symbols can be used to treat every wave and particle. With the introduction of the wave function, the long-disputed hydrogen atom can now be represented as shown in Figure 4.5.

Figure 4.4 Young Schrödinger and his quantum cat and wave equation.

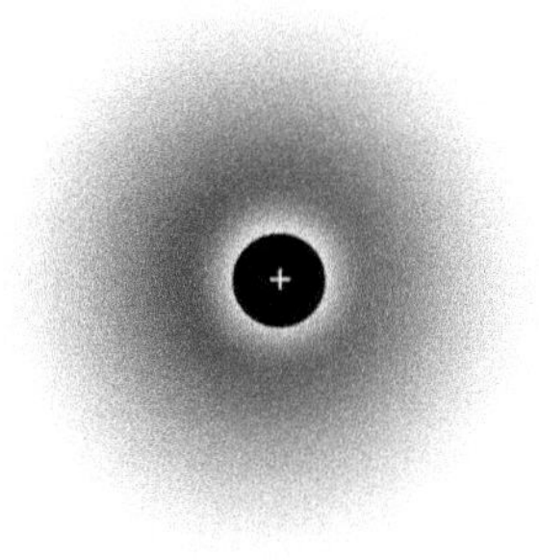

Figure 4.5 The Schrödinger's version of a hydrogen atom showing a nucleus at the center and electron wave around it.

4.1.4 *Heisenberg and the uncertain nature of electrons*

> Do you know how fast you were going, sir?
> No, but I know exactly where I am!
>
> **A highway police officer and a physicist**

Once we accept the wave nature of electrons, we have to realize that the notion of positions and momenta are rather different from those of particles. When a wave spans into space, we can tell how fast it propagates, but its position is unknown since it is everywhere. If we make the wave very short to locate its position, then the wave has no momentum to talk about. This is an oversimplified picture of the uncertainty principle (see Figure 4.6) meditated on and formulated by Heisenberg (1927):

$$\Delta x \Delta p > \frac{\hbar}{2} \quad \text{or} \quad \Delta E \Delta t > \frac{\hbar}{2} \tag{4.7}$$

Figure 4.6 Young Heisenberg and his uncertainty principle.

where Δx, Δp, ΔE, and Δt are uncertainties in position, momentum, energy, and time, respectively. In other words, quantum mechanical quantities are interrelated and delocalized in phase space, and thus simultaneous measurements of them always accompany a certain amount of uncertainty in the order of Planck's constant. This blurring nature of the quantum world is drastically different from the determinism and certainty manifested by classical mechanics. The conclusion is as simple as this: The only certainty in the quantum world is the uncertainty.

This implies that, when we deal with electrons and thus the first-principles methods, everything has to be expressed only as a set of probabilities. For example, electron positions are represented as electron densities that are probabilities of finding electrons in a finite volume element (see Section 5.1.1). Fortunately, during the actual calculations of first-principles methods, the uncertainty principle never intervenes since the time variable is generally out of the picture, and the genetic errors of the methods are far greater than the uncertainty involved. Note that the Planck constant is a very small number, close to 0.

4.1.5 Remarks

The rules of the quantum world are bizarre and cannot be derived, but they have been justified up to the level of laws by their logical consistency and by agreements with experiments over the years. They have never been disproved nor has evidence been found to the contrary. In fact, they are one of the main driving forces today. Remember all those IT-related devices in our hands, pockets, and sacks and on desks. They are all based on semiconductor technologies and, in turn, based on a quantum phenomenon, the band gap formation. The puzzle once confronted by physicists at the turn of the twentieth century is now common knowledge for everyone.

4.2 *Schrödinger wave equation*

> I don't like it (quantum mechanics), and I'm sorry
> I ever had anything to do with it.
> I wished I had never met that "cat."
>
> **Erwin Schrödinger (1887–1961)**

In this section, we will first make the problem on hand as simple as possible to make the solution of Schrödinger wave equation much easier and the conceptual background much more clear. And we will look at the equation in more detail and see what the three terms ($\hat{H}, \Psi$, and E) in Equation 4.6 really mean.

4.2.1 *Simplifying the problem*

The system we are supposed to deal with according to Schrödinger wave equation is schematically shown in Figure 4.7. Note that all entities including nuclei, electrons, and even time variable are active participants in this model via the wave functions. This system is in fact an extremely complex one that can be solved only for hydrogen-like systems. Unless we do some drastic measures on the model, the equation has no use in practice. The following is a list of approximations adopted to deal with the first-principles approach.

4.2.1.1 *Forget about gravity, relativity, and time*

Here, the first two simplifications—forgetting about gravity and relativity—are obvious, considering that an electron's mass is so small and its speed is much slower than that of light. For heavy atoms, however, there are significant relativistic effects, but these can be incorporated

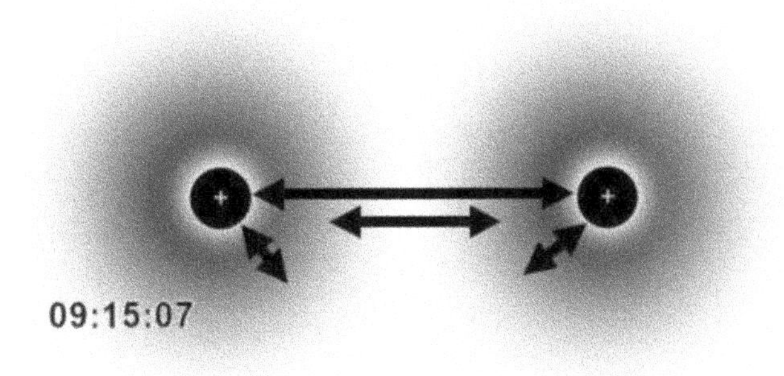

Figure 4.7 A very schematic of the Schrödinger system for an n-electron system with all quantum interactions indicated by black arrows.

in the construction of the pseudopotentials (see Chapter 6). During the formulation of pseudopotentials, the core electrons and the nucleus are treated together, and thus the relativistic effects, which affect mostly the core, can be neglected. The third one, forgetting about time, is possible if we restrict our interest to only the ground-state energy of electrons; then, the potential energy of the system is constant in time. This can remove the t in Equation 4.6 and we have $\hat{H}\Psi(\mathbf{r}_i,\mathbf{r}_I)=E\Psi(\mathbf{r}_i,\mathbf{r}_I)$.

4.2.1.2 *Forget about nuclei and spin*

Nuclei are far more massive than electrons (1836 times for hydrogen and about 10,000 times for semiconductors), and thus, whenever a nucleus moves, the electrons respond instantaneously to nuclear motion and always occupy the ground state of that nuclear configuration. It is analogous to the monkey king and Buddha tale. No matter how fast the monkey king zips around, he always remains in the palm (the ground state) of the Buddha as illustrated in Figure 4.8.

This means that the positions of nuclei are considered to be "frozen" like the Buddha and become not variables but only parameters from the electron's view. This decoupling of nuclear and electronic dynamics is called the Born–Oppenheimer approximation (Born and Oppenheimer 1927), expressing the total energy of an atom as the sum of nuclear and electronic energies:

$$E_{\text{atom}} = E_{\text{nucleus}} + E_{\text{electron}} \tag{4.8}$$

Figure 4.8 The monkey king and Buddha analogy to the Bohn–Oppenheimer approximation.

Then the wave function depends only on electronic positions $\boldsymbol{r}_i$, and Equation 4.6 is simplified to $\hat{H}\Psi(\boldsymbol{r}_i) = E\Psi(\boldsymbol{r}_i)$. We can skip the spin consideration for the sake of simplicity because spin variable can be switched on anytime if it is necessary.

4.2.1.3 *Forget about the excited states*

We will limit our interest only within the ground states of electrons at 0 K, which provides a most practical solution for the problems encountered in materials science. Since the ground-state energy is independent of time, we can use a much simpler time-independent wave equation. Of course, all first-principles methods can be extended to the excited states, which is conceptually straightforward but involves more implications (see Section 5.6.1).

4.2.1.4 *Use of atomic units*

Unless otherwise written, we will use convenient atomic units as shown in Appendix 6. These units make many quantum quantities to 1 and thus make equations simpler.

4.2.2 *Time-independent electronic wave equation*

After the approximations described in the previous Subsection 4.2.1, the system is much simpler as illustrated in Figure 4.9. Note that all nuclei are fixed, and the time variable is no longer active.

The resulting equation is called the *time-independent electronic wave equation* (to be called just the *wave equation* from now on), newly defined by

$$\hat{H}(\boldsymbol{r})\Psi(\boldsymbol{r}) = E\Psi(\boldsymbol{r}) \tag{4.9}$$

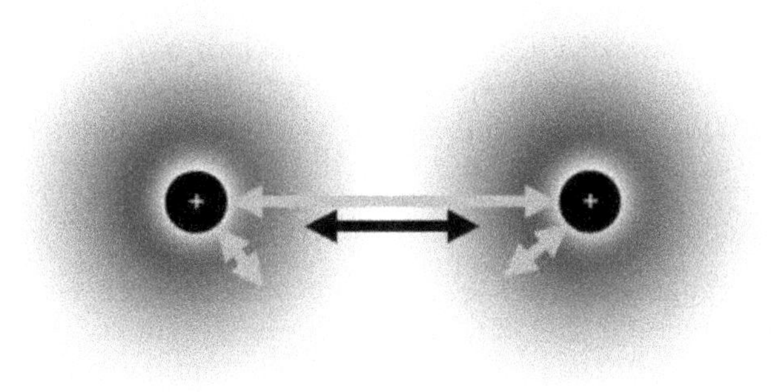

Figure 4.9 Schematic of the Schrödinger model for an n-electron system after the BO approximation, indicating the electron–electron interaction (black arrow) as the remaining quantum mechanical entity.

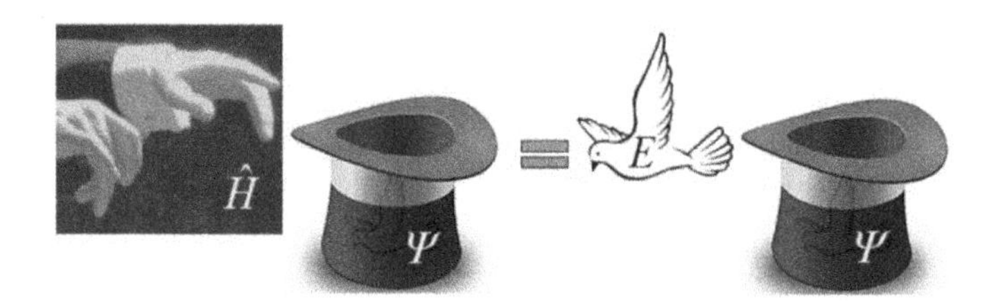

Figure 4.10 The analogy between the quantum-mechanical operation and an act of a magic.

where $\hat{H}(\boldsymbol{r})$, $\Psi(\boldsymbol{r})$, and E are the Hamiltonian energy operator, wave function, and the total energy of the system, respectively, and $\boldsymbol{r}$ is the electronic coordinate. This is the starting point for computational materials science, and the ground-state properties of any system can be determined by solving this equation, if we have a method to solve it.

Let us look at how the equation works, assuming that the $\hat{H}$ and Ψ are known. In principle, it is surprisingly simple: the energy operator $\hat{H}$ operates on the wave function Ψ (left side of the equation), and it becomes the right side of the equation, $E\Psi(\boldsymbol{r})$. Note that $\Psi(\boldsymbol{r})$ remains as $\Psi(\boldsymbol{r})$, and energy, E, comes out as a constant, which is called the *eigenvalue*, which is a physically observable quantity as energy and is what we want to have.

The whole process is like a magician (≡ operator) taking a dove (≡ E) out of a supposedly empty hat (≡ $\Psi(\boldsymbol{r})$), as shown in Figure 4.10. We know that the dove was somewhere there where we could not see it, and the magician simply moves her into the hat in a tricky way and shows her in front of us. Note that there is no change to the hat by the magic act. Similarly, all necessary information about E is contained in Ψ, and it shows up only after the operation of $\hat{H}$. The equation with this unique property is specifically called the *eigenvalue problem*. If we (untrained nonmagicians) do the same trick, the dove may never show up. Stated quantum-mechanically, if either $\hat{H}$ (≡ magician) or $\Psi(\boldsymbol{r})$ (≡ the supposedly empty hat) was not properly prepared, they would not fit into the eigenvalue equation (Equation 4.9), and its solution would not be possible.

4.2.3 *Energy operator: Hamiltonian* Ĥ

The Hamiltonian operator, $\hat{H}$, is a sum of all energy terms involved: kinetic and potential energies. The kinetic energies are of the nuclei (E_I^{kin}) and electrons (E_i^{kin}). The potential energies are coming from the Coulomb interactions by nucleus–electron (U_{Ii}), electron–electron (U_{ij}), and nucleus–nucleus (U_{IJ}):

$$\hat{H} = E_I^{\text{kin}} + E_i^{\text{kin}} + U_{Ii} + U_{ij} + U_{IJ} \tag{4.10}$$

In the previous Subsection 4.2.1, we reasoned that terms connected with nuclei can be skipped and therefore

$$\hat{H} = E_i^{\text{kin}} + U_{Ii} + U_{ij} \tag{4.11}$$

The first term, kinetic energies of n electrons, is given in an atomic unit by

$$E_i^{\text{kin}} = -\frac{\hbar^2}{2m}\sum_i^n \nabla_i^2 = -\frac{1}{2}\sum_i^n \nabla_i^2 \tag{4.12}$$

where ∇^2 is the Laplacian operator ("delsquared") where

$$\nabla^2 = \frac{\partial^2}{\partial x^2} + \frac{\partial^2}{\partial y^2} + \frac{\partial^2}{\partial z^2} \tag{4.13}$$

Here, the kinetic energy in Equation 4.12 has a negative sign but becomes positive as it should after this Laplacian operation on wave function. The second term, attractive potential energies of n electrons due to the nuclei, is given by

$$U_{Ii} = -\sum_I^N \sum_i^n \frac{Z_I}{|\boldsymbol{r}_{Ii}|} \tag{4.14}$$

where N and n are the number of nuclei and electrons in the system, and Z_I are the charges of the nuclei. The double sum indicates that the interaction is from all electrons to all nuclei. Electrons therefore move within this static external potential from an array of fixed nuclei. Note that, from the electron's view, forces from nuclei are considered "external." Thus the positions of nuclei are involved, but only as parameters and any derivative with respect to nuclear coordinates will conveniently disappear. Here, the position vector, $\boldsymbol{r}$, can be expressed in Cartesian coordinates as $|\boldsymbol{r}| = (x^2 + y^2 + z^2)^{1/2}$.

The third term, repulsive potential energies of n electrons between each other, is given with the usual correction factor of ½ for double-counting correction:

$$U_{ij} = \frac{1}{2}\sum_{i \neq j}^n \frac{1}{|\boldsymbol{r}_{ij}|} \tag{4.15}$$

Summing all together, the Hamiltonian operator, $\hat{H}$, is now given by

$$\hat{H} = -\frac{1}{2}\sum_i^n \nabla_i^2 - \sum_I^N \sum_i^n \frac{Z_I}{|\boldsymbol{r}_{Ii}|} + \frac{1}{2}\sum_{i \neq j}^n \frac{1}{|\boldsymbol{r}_{ij}|} \tag{4.16}$$

For example, the hydrogen molecule, H_2, has the wave equation of

$$\left[-\frac{1}{2}\sum_{i=1}^{2}\nabla_i^2-\sum_{i=1}^{2}\frac{Z_I}{|r_i|}+\frac{1}{|r_{12}|}\right]\Psi(r_1,r_2)=E\Psi(r_1,r_2) \tag{4.17}$$

Even for this simple system of two nuclei and two electrons, solving the above equation is quite demanding due to the third electrostatic term. We already discussed in Chapter 2 that, whenever more than two particles (atoms or electrons) are involved, the solution is almost impossible, and we are facing the same problem here. In addition, $\Psi(r_1,r_2)$ couples the two coordinates together and makes the above equation a very complicated partial differential equation in six coordinates.

In this chapter and the next, we will see that the history of first-principles calculation is full of approximations for $\hat{H}$ and $\Psi(r)$. One considers that electrons are independent and another considers the electron wave in the form of a determinant. One finally replaces the whole system with an n one-electron system. We will follow the main developments in this chapter and reach the final scheme, the DFT, in Chapter 5.

4.2.4 *Waves and wave function*

As MD concerns positions and momenta of atoms, first-principles methods concern wavelengths, wave vectors, and wave functions of electrons. Not all wavelike functions can be qualified as a wave function in the quantum world, but only special ones. For example, let us assume that we have only the kinetic energy term in Equation 4.17. If an operator such as ∇^2 acts on a general wave function, the resulting function is usually a completely different form. However, there are cases such that the result is a multiple of the original wave function. These special wave functions are called *eigenvectors* (or eigenstates or Ψ), and the multiples are called *eigenvalues* (or energy E). Further observation in relation with Equation 4.17, by quantum mechanics, reveals that

- Ψ Should be continuous, square integrable (remember ∇^2), and single valued for all coordinates to fit into Equation 4.17.
- The real physical meaning of Ψ can be realized only when it is squared as $|\Psi|^2$.
- Each Ψ must be orthogonal with other wave functions in a given system. This ensures that each wave function represents distinct physical states (1s, 2s, 2p, …) of its own corresponding to its unique eigenvalue. Mathematically speaking, this means that the two

wave functions have no net overlap, and their inner product is zero:

$$\int \psi_i \psi_j d\boldsymbol{r} = 0 \text{ (if } i \neq j\text{): orthogonality} \tag{4.18}$$

Since an electron must be somewhere in space, the probability of finding this electron must add up to unity (normalized):

$$\int \psi_i^* \psi_j d\boldsymbol{r} = 1 \text{ (if } i = j\text{): normality} \tag{4.19}$$

Ψ Must be antisymmetric (it is simply a law of nature) with respect to the exchange of any pair of electron coordinates (including the spin variables) because electrons are Fermions. This requirement is equivalent to the usually stated Pauli's principle: no two electrons of the same spin can occupy the same orbital.

In Chapter 2, we learned that position and momentum are all we need to describe a classical atom at a particular time. In the quantum world, all information required to define a quantum state is contained in $\psi_i(\boldsymbol{r})$. In other words, a wave function can represent everything about an electron and thus about the system. Thus, by solving the wave equation for a system, a set of wave functions, $\{\psi_i(\boldsymbol{r})\}$, comes out as solutions (one solution per each electron i).

4.2.4.1 *Plane wave*

One basic wave function is the plane wave (PW), which propagates perpendicularly to wave fronts with a constant frequency as shown in Figure 4.11 in one dimension.

Fortunately, this wave function is the simplest one and the most frequently used one, especially for solids. This wave has the general

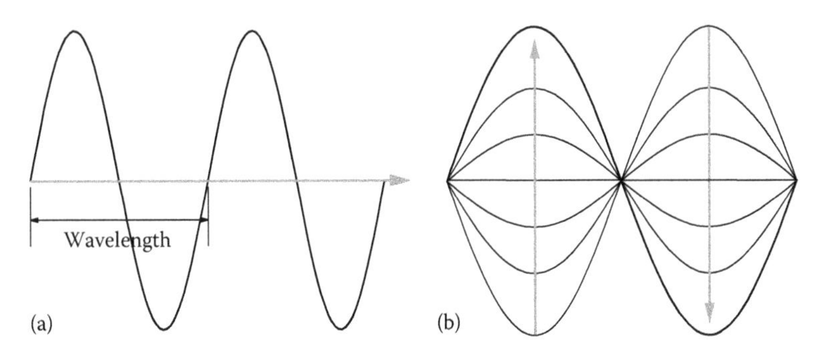

Figure 4.11 A plane wave (a) and a standing wave (b).

mathematical form with cosine and sine terms that can be combined together as an exponential expression:

$$\psi_{PW}(x) = \cos(kx) + i\sin(kx) = \exp(ikx) \tag{4.20}$$

where $i = \sqrt{-1}$ and k = wave vector. Note that the function sin(kx) for some real number k has an arbitrary wavelength $\lambda = 2\pi/k$. Functions such as cos(kx), sin(kx), and exp(ikx) are all eigenfunctions that calculate out $-k^2$ as the eigenvalue after double differentiation:

$$\frac{d^2}{dx^2}\sin(kx) = -k^2 \sin(kx) \tag{4.21}$$

$$\frac{d^2}{dx^2}\cos(kx) = -k^2 \cos(kx) \tag{4.22}$$

$$\frac{d^2}{dx^2}\exp(ikx) = -k^2 \exp(ikx) \tag{4.23}$$

Free electrons travel this way, and valence electrons in metals do similarly. More important, as discussed in Chapter 6, we can expand any wave function with many of these PWs, like building a dinosaur with LEGO bricks.

4.2.4.2 *Standing wave*

Another basic wave form is the standing wave, which does not proceed but vibrates up and down in place as shown in Figure 4.11. Vibrating string (in one dimension), vibrating drum skin (in two dimensions), and diffracting electron waves in solids (in three dimensions) are good examples of standing waves. In an one-dimensional box of length L that has very high walls at both ends, the solution of standing waves can take the general form as

$$\psi(x) = A \sin\left(\frac{\pi n}{L}x\right) \tag{4.24}$$

where A is amplitude and n is an integer. The standing wave is not so much our concern except on two occasions: when we solve a model system of one electron in a well (see Section 4.2.6) and when electron waves meet solid lattices and form band gaps (see Section 6.4.3).

4.2.4.3 *Superposition principle of waves*

Waves have a special property called the *superposition principle*. Unlike particles, one PW plus another PW depends on how their phases match each other as shown in Figure 4.12. If two waves are completely out of

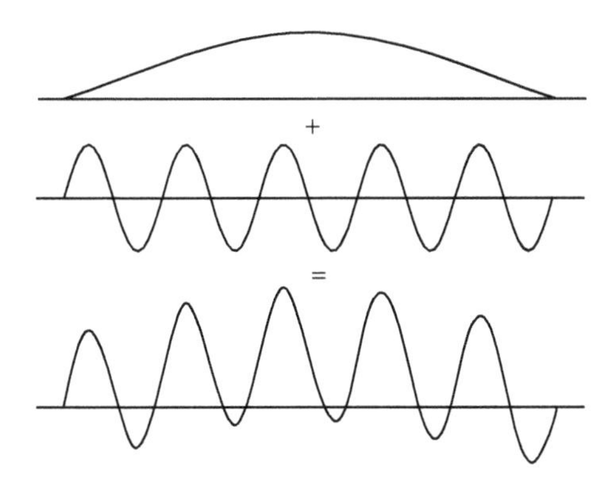

Figure 4.12 A simple example of superposition principle of waves.

phase, for example, that will result in no wave. Thus, interference takes place when two PWs meet each other and diffraction takes place between waves and solid lattice.

4.2.4.4 *Indistinguishability of electrons*

Macroscopic objects such as tennis balls can be mixed, taken out, and identified one by one if we mark them or write down tiny differences ahead of time. Even microscopic objects such as atoms can be identified one by one if we trace them as we did in MD. A mixed wave of electrons, however, forgets all its previous identities and becomes a new entity since electrons are exactly the same as each other with no marks, no tags, or any difference. This means that we have no way of noticing when two electrons have switched their positions. This is called the *indistinguishability* of an electron, and thus, a wave function describing two electrons must have a symmetric probability distribution:

$$\left|\Psi(x_1, x_2)\right|^2 = \left|\Psi(x_2, x_1)\right|^2 \tag{4.25}$$

This is simply how they are created and exist in nature.

4.2.5 *Energy* E

E is the eigenvalue of $\hat{H}$ associated with the wave function, Ψ, and it can tell us a lot about the system. If we calculate various ground-state energies over possible positions of nuclei, it will generate a potential energy surface in which each point represents a structural configuration. Its global minimum can tell us the most stable structure including lattice parameters, bond lengths, bond angles, cohesive energies, and defect formation energies. In addition, the landscape between one point and another provides us the barrier energy between two configurations. Furthermore, the ground-state configuration is the starting point for a variety of other calculations such as band structure, DOS (density of states), and thermodynamic quantities.

4.2.6 Solutions of Schrödinger wave equation: An electron in a well

The Schrödinger wave equation can be exactly solved for only a few simple cases. One case is an electron in a well, which can be solved analytically. However, one should not be misled by the simplicity of the subject since all relevant and important features of an atom or quantum devices (e.g., quantum well FETs in cell phones, quantum well lasers in DVD players, and gate oxide layers in flash memory) are closely related to this system.

4.2.6.1 An electron in a one-dimensional infinite well

As shown in Figure 4.13, an electron is trapped inside the well of width L where the potential $U(x) = 0$ inside and $U(x) = \infty$ outside, and the bottom of the well is set as the potential zero for convenience. Thus, there is, no electron–electron interaction, and the electron's energy is purely kinetic.

Intuitively, we can guess that the lowest energy state (the ground-state solution) has to be a wave function, which

Is zero at the two walls ($\psi(0) = \psi(L) = 0$) and nonzero in between.
Has no node (points where the wave function crosses through zero) to make the energy minimum.

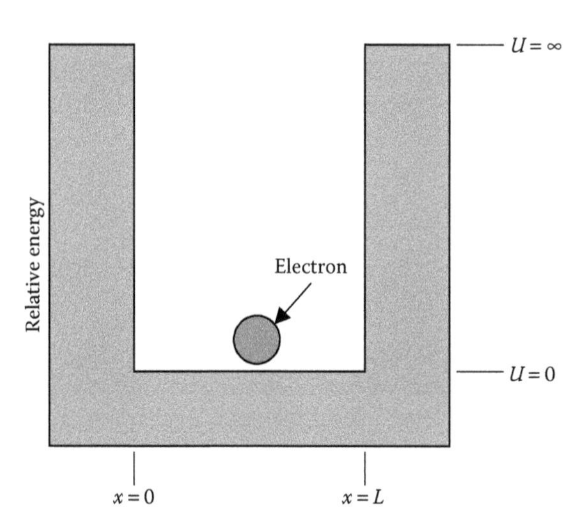

Figure 4.13 An electron trapped inside the well of zero potential and an infinite wall.

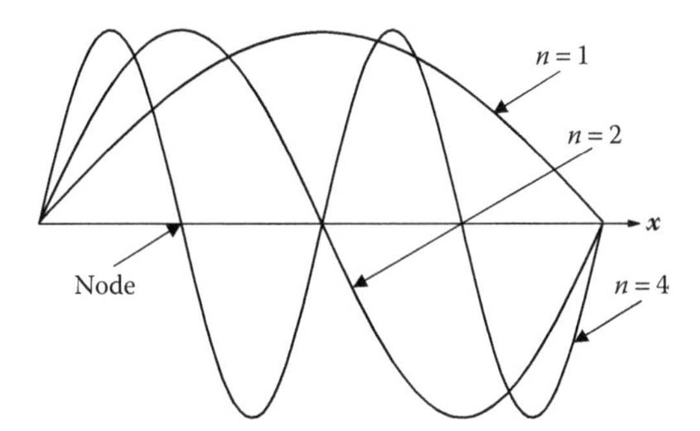

Figure 4.14 Three wave functions in the model system of an electron in an infinite well. Note that the three sinusoidal waves are orthogonal to each other.

This corresponds to half a period of a sine curve with the least curvature. The next wave function with a higher energy will have one node and is still zero at the two walls as shown in Figure 4.14. And it goes on similarly, that is, higher energies are produced by wave functions with more nodes. By examining the curves, one can conclude that the wave functions for the system cannot be any wave but must be waves that satisfy the following sine equation with the argument of $n\pi$:

$$\Psi(x) = A \sin\left(\frac{n\pi}{L} x\right) \tag{4.26}$$

where:

- A is a constant (amplitude)
- n is an integer number that is indicative of the quantum nature of the system

These solutions correspond to standing waves with a different number of nodes within the well, which is equivalent to shorter wavelengths and a higher kinetic energy of the electron. Note that the second curve (n = 2) has one node to be orthogonal with the first curve, whereas the third curve (n = 4) has three nodes to be orthogonal with all three underlying curves. (The curve of n = 3 is not shown.)

Mathematically, the above procedure proceeds as

$$\hat{H}\Psi(x) = -\frac{\hbar^2}{2m}\frac{d^2\Psi(x)}{dx^2} = E\Psi(x) \tag{4.27}$$

If we put all constant together as k^2, the equation becomes a standard second-order differential equation with the corresponding general solution

$$\frac{d^2\Psi}{dx^2} = -k^2\Psi(x) \tag{4.28}$$

$$\Psi(x) = A\sin kx + B\cos kx \tag{4.29}$$

Normalizing Equation 4.29 to 1 ($A = \sqrt{2/L}$) and substituting into Equation 4.27, we obtain energies in which the integer n clearly indicates the quantum energies at each energy levels:

$$E_n = \frac{\hbar^2}{2m}\left(\frac{\pi n}{L}\right)^2 \tag{4.30}$$

In three dimensions, the argument is precisely analogous, and the solution will be a combination of three solutions at each dimension. The wave function is the product of one-dimension solutions and the energy is the sum of one-dimension energies:

$$\Psi(\boldsymbol{r}) = \Psi(x)\Psi(y)\Psi(z) \tag{4.31}$$

$$E = E_x + E_y + E_z \tag{4.32}$$

The wave function can be rewritten as

$$\Psi = \left(\frac{1}{L}\right)^{3/2} \exp(i\boldsymbol{k}\cdot\boldsymbol{r}) \tag{4.33}$$

where $\boldsymbol{r}$ is the position vector and $\boldsymbol{k}$ is now identified as wave vector with the three components:

$$\boldsymbol{k} = \boldsymbol{k}_x + \boldsymbol{k}_y + \boldsymbol{k}_z \tag{4.34}$$

Then the energy can be rewritten in two forms as

$$E_{\text{kin}} = \frac{1}{2}mv^2 = \frac{p^2}{2m} = \frac{\hbar^2 k^2}{2m} \tag{4.35}$$

$$E_{\text{kin}} = \frac{\hbar^2}{2m}\left(\frac{\pi}{L}\right)^2 \left(n_x^2 + n_y^2 + n_z^2\right) \tag{4.36}$$

where n_x, n_y, and n_z are the quantum numbers (1, 2, …). Note that the above equation has all squared terms, and thus an electron cannot have a negative or zero energy but will have some zero-point energy even at the ground state. The spacing between the energy levels is extremely small since Planck's constant is so small. However, the important fact is that energies are discrete in the quantum world.

4.2.6.2 *An electron in a one-dimensional well with a finite potential*

If the walls have a finite potential of U_0, the wave equation becomes

$$\hat{H} = -\frac{\hbar^2}{2m}\frac{d^2}{dx^2} + U_0 \quad (4.37)$$

$$-\frac{\hbar^2}{2m}\frac{d^2\Psi(x)}{dx^2} + U_0\Psi(x) = E\Psi(x) \quad (4.38)$$

If we assume $U_0 > E$, the electron is largely bound to the well but displays another purely quantum mechanical phenomenon, quantum tunneling. Since Ψ cannot change slope discontinuously at the walls, there is a nonvanishing probability of finding the electron outside the well, which is totally forbidden in the classical world. It is known that the solution curves inside the well go smoothly into a linear combination of exponentially increasing and decreasing terms into the walls at both sides.

4.2.6.3 *Hydrogen atom*

The hydrogen atom consists of a nucleus (charge $+e$) and an electron (charge $-e$). The system is similar to the case of an electron in a well but has the Coulomb potential between the nucleus and the electron:

$$U_{Ii} = -\frac{1}{\boldsymbol{r}} \quad (4.39)$$

where $\boldsymbol{r}$ is the position vector of the electron with respect to the nucleus. This is the simplest atomic system, and the solution of the wave equation is possible analytically. Due to the potential, however, solving the equation is more complicated, and we write down only the results here. Readers may refer to any physics textbook for the derivation. The wave function at the ground state, $\Psi(\boldsymbol{r})$, is

$$\Psi(\boldsymbol{r}) = C\exp(-a\boldsymbol{r}) \quad (4.40)$$

where C and a are constants. The energy for the nth state, E_n, is

$$E_n = \frac{-13.6}{n^2} \quad (4.41)$$

where n is the principal quantum number. The energy at $n = 1$, –13.6 eV, is called the ground-state energy of hydrogen, that is, the lowest state of energy.

4.2.6.4 Degenerate states

Note that, by a combination of n in Equation 4.36, it is possible to have multiple wave functions corresponding to the same total energy. This is called *energy degeneracy* that happens in actual systems, too. This arises from the fact that the linear combinations of wave functions of the same energy return wave functions that still have that same energy. Therefore, the nth energy level has $2n^2$ states of the same energy because of different orbitals (n, m, and l) and two spin states.

4.3 Early first-principles calculations

> The observer, when he seems to himself to be observing a stone, is really, if physics is to be believed, observing the effects of the stone upon himself.
>
> **Bertrand Russell (1872–1970)**

In this section, we will review the early first-principles calculations, the Hartree method and the Hartree–Fock (HF) method, that are based on the Schrödinger wave equation.

4.3.1 n-electron problem

Despite the innocent look of the Schrödinger wave equation, solving it and calculating a property is a very difficult task. In fact, the underlying physics of the first-principles calculations is not very complex. It works nicely for simple systems such as an electron in a well, hydrogen, or helium (although not exact). Remember, however, that we are talking about materials of up to several hundred atoms that easily contain several thousand electrons. Then the calculations for these n-electron systems are completely out of the question. This is the so-called many-body problem that we already discussed for the N-atom systems in MD. Again, what makes it so difficult is the size of the system. Dealing with n electrons that interact with all other electrons at the same time is just too complex to solve even numerically.

Just imagine 10 ladies who love the same man happening to be in the same place at the same time. Their emotions and feelings toward each other, definitely repulsive just like electrons, will be so complicated and obscure that not even 10 Shakespeares could possibly describe them. It does not matter whether the system consists of n atoms, or n electrons, or n ladies. Any many-body system where each one interacts with others at the same time cannot be treated as it is. The conclusion is obvious: One has to resort to various levels of approximation without sacrificing the parameter-free nature of the first-principles methods.

4.3.2 Hartree method: One-electron model

To bring the problem down to the tractable level, there is no alternative but to assume that each body is independent and interacts with others in an averaged way. This means that, for an n-electron system, each electron does not recognize others as single entities but as a mean field like a baseball player feels the booing stadium in an away game. Hence, an n-electron system becomes a set of noninteracting one-electrons where each electron moves in the average density of the rest as shown in Figure 4.15.

With this simplified model, Hartree (1928) treated one electron at a time and introduced a procedure he called the *self-consistent field method* to solve the wave equation:

$$\left(-\frac{1}{2}\nabla^2 + U_{\text{ext}}(\boldsymbol{r}) + U_H(\boldsymbol{r})\right)\Psi(\boldsymbol{r}) = E\Psi(\boldsymbol{r}) \tag{4.42}$$

where:

U_{ext} is the attractive interaction between electrons and nuclei

U_H is the Hartree potential coming from the classical Coulomb repulsive interaction between each electron and the mean field

Since electrons are independent, the total energy is the sum of n numbers of one-electron energies:

$$E = E_1 + E_2 + \cdots + E_n \tag{4.43}$$

Then, the n-electron wave function can be simply approximated as the product of n numbers of one-electron wave functions (see Equation 4.31):

$$\Psi = \psi_1 \times \psi_2 \times \cdots \times \psi_n \tag{4.44}$$

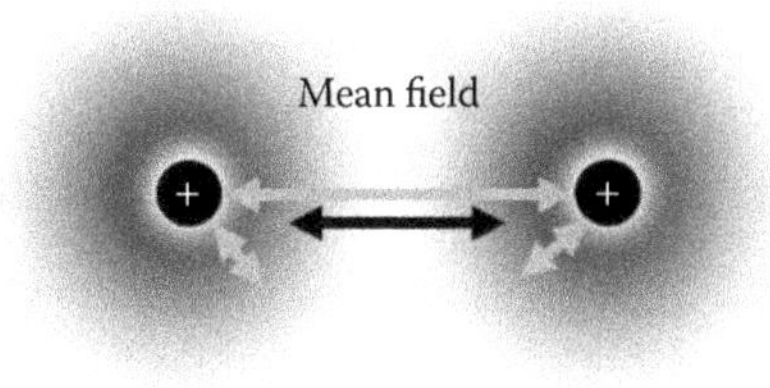

Figure 4.15 Schematic of the Hartree model with the mean field approximation for electrons.

This Hartree model fits exactly with a hydrogen atom of just one independent electron, and the exact ground-state energy analytically calculates out as –13.6 eV, that is, exactly the same as the experimental value. It seemed that the wave equation passed the first test very well, and this method became a starting point for all other first-principles methods. In other systems, however, the Hartree method produced only crude estimates of energy due to these oversimplifications:

- It does not follow two basic principles of quantum mechanics: the antisymmetry principle and thus Pauli's exclusion principle.
- It does not count the exchange and correlation energies coming from the n-electron nature of the actual systems.

The Hartree method, therefore, was soon refined into the HF method.

4.3.3 Hartree–Fock method

Based on the one-electron and the mean-field approaches by Hartree, Fock (1930) enhanced the method to higher perfection. This time, the key of the move was on the wave function:

Expressing it better so that the missing pieces in the Hartree method could be properly described
Improving it with the variational process

4.3.3.1 Expression for $\Psi(r)$

In the HF method, the true n-electron wave function is approximated as a linear combination of noninteracting one-electron wave functions in the form of a Slater determinant. For example, the wave function of an He atom with two electrons is

$$\Psi(r_1, r_2) \approx \frac{1}{\sqrt{2}} \begin{vmatrix} \psi_1(r_1) & \psi_2(r_1) \\ \psi_1(r_2) & \psi_2(r_2) \end{vmatrix} = \frac{1}{\sqrt{2}} \left[\psi_1(r_1)\psi_2(r_2) - \psi_2(r_1)\psi_1(r_2) \right] \quad (4.45)$$

Here, $1/\sqrt{2}$ is the normalization factor. This does not look very impressive, but actually it is because it exactly expresses the actual electron that changes its sign whenever the coordinates of any two electrons are exchanged:

$$\Psi(r_1, r_2) = -\Psi(r_2, r_1) \quad (4.46)$$

This is called the antisymmetry principle, and the determinant follows it by changing the sign when two rows or columns are exchanged. The one-electron wave function in the form of a Slater determinant always guarantees the total wave function to be antisymmetric and thus correctly accounts for one missed piece in the earlier Hartree method.

Before we go to the next point, let us remember that the two wave functions, $\psi_1(r_1)$ and $\psi_2(r_2)$ in Equation 4.45, have additional spin variables (one up-spin and one down-spin). Then, the more exact expression for the Slater determinant is a combination of $\psi_1(r_1,\uparrow)$, $\psi_1(r_1,\downarrow)$, $\psi_2(r_2,\uparrow)$, and $\psi_2(r_2,\downarrow)$.

$$\Psi(r_1,r_{2,\uparrow,\downarrow}) = \frac{1}{\sqrt{2}} \begin{vmatrix} \psi_1(r_{1,\uparrow}) & \psi_2(r_{1,\downarrow}) \\ \psi_1(r_{2,\uparrow}) & \psi_2(r_{2,\downarrow}) \end{vmatrix} \tag{4.47}$$

Now, notice that any determinant with two identical rows or columns is equal to zero. This implies that, if two electrons occupy the same spin wave functions (≡ spin orbital), such wave function just does not exist, and thus the Pauli's principle is satisfied. The general expression of the Slater determinant for an n-electron system, if we neglect the spin variable, is

$$\Psi(r_1,r_2,\ldots,r_n) = \frac{1}{\sqrt{n!}} \begin{vmatrix} \psi_1(r_1) & \psi_2(r_1) & \cdots & \psi_n(r_1) \\ \psi_1(r_2) & \psi_2(r_2) & \cdots & \psi_n(r_2) \\ \vdots & \vdots & \vdots & \vdots \\ \psi_1(r_n) & \psi_2(r_n) & \cdots & \psi_n(r_n) \end{vmatrix} \tag{4.48}$$

where $1/\sqrt{n!}$ is the normalization factor for an n-electron system. Note that, for closed shells of even numbers of electrons, one Slater determinant is sufficient to describe the wave function fully. For open shells of odd numbers of electrons, a linear combination of more than two Slater determinants is needed.

4.3.3.2 *Orthonormality of wave functions*

Let us restate the orthogonal and normality conditions that we have already defined in Section 4.2.4. Both conditions are universal ones for any wave function applied to a quantum system. Thus, as illustrated in Figure 4.16, all elements in the Slater determinant, ψ_i, must follow the first condition as

$$\int \psi_i \psi_j \, d\mathbf{r} = 0\text{: orthogonality } (i \neq j) \tag{4.49}$$

Further, all elements in the Slater determinant, ψ_i, must also follow the second condition as

$$\int \psi_i \psi_j \, d\mathbf{r} = 1\text{: normality (if } i = j) \tag{4.50}$$

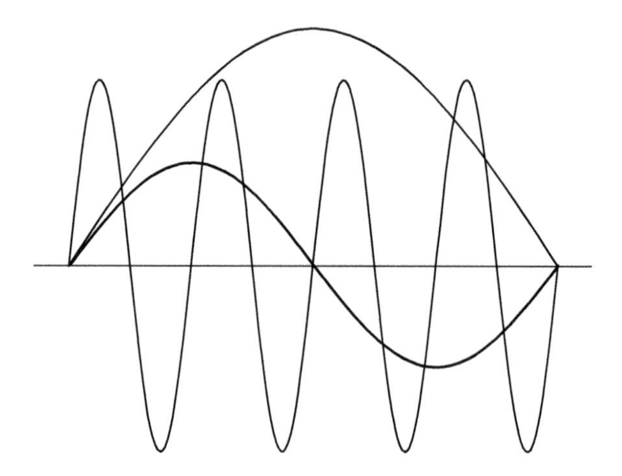

Figure 4.16 Three orthogonal wave functions in sinusoidal shape.

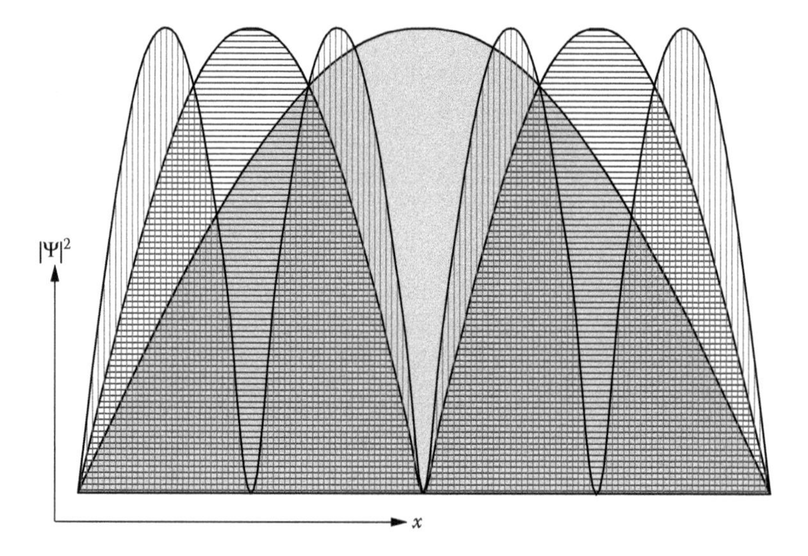

Figure 4.17 Three orthogonal wave functions in sinusoidal shape after normalization.

Therefore, if we take the probability of finding an electron ($\psi_i \times \psi_i$) and integrate it over the whole space, the outcome should be 1 as schematically shown in Figure 4.17. The above two conditions are equivalent to the often-stated overlap integral between wave functions becoming 1.

4.3.3.3 *Expression for* E

Now, let us rewrite the wave equation with the wave function consisting of a Slater determinant form as

$$\left(-\frac{1}{2}\nabla^2 + U_{\text{ext}}(\mathbf{r}) + U_{ij}(\mathbf{r})\right)\Psi(\mathbf{r}) = E\Psi(\mathbf{r}) \tag{4.51}$$

Here, $U_{ij}(\mathbf{r})$ represents the true electron–electron interaction, and we only rewrite it here to remind us that it contains both classical and quantum terms:

$$U_{ij} = \frac{1}{2}\sum_{i \neq j}^{n} \frac{1}{|\mathbf{r}_{ij}|} \tag{4.52}$$

From these equations, we want to calculate E, which is something we can observe and measure in a lab. For example, the expectation value of E corresponding to the operator $\hat{H}$ is obtained by the quantum manipulation:

First, let us start with the simpler electronic wave equation:

$$\hat{H}\Psi = E\Psi \tag{4.53}$$

Next, both sides of the wave equation are multiplied by Ψ^* from the left to give physical meaning (probability) to the wave function:

$$\Psi^* \hat{H} \Psi = \Psi^* E \Psi \tag{4.54}$$

Here, Ψ^* is the complex conjugate to take care of complex numbers since these wave functions may produce complex numbers and the result has to be a real number.

It is also integrated over the whole space giving the energy equation

$$\int \Psi^* H\Psi d\mathbf{r} = \int \Psi^* E\Psi d\mathbf{r} \tag{4.55}$$

$$\therefore E = \frac{\int \Psi^* H\Psi d\mathbf{r}}{\int \Psi^* \Psi d\mathbf{r}} \tag{4.56}$$

If the wave functions are normalized, the denominator becomes 1 since the probability in the whole space is unity and the equation is reduced to

$$E = \int \Psi^* \hat{H}\Psi d\mathbf{r} \tag{4.57}$$

The same procedures work for any expectation value of the observable corresponding to the quantum mechanical operator $\hat{O}$:

$$O = \int \Psi^* O \Psi dr \tag{4.58}$$

4.3.3.4 *Calculation for* E

Now, we insert $\Psi(\boldsymbol{r})$ in the form of the Slater determinant and $\hat{H}$ of three energy terms into the energy equation to calculate the total energy:

$$E = \sum_{i,j} \int \Psi_i^*(\boldsymbol{r}) \left(-\frac{1}{2} \nabla_i^2 + U_{\text{ext}}(\boldsymbol{r}) + U_{ij}(\boldsymbol{r}) \right) \Psi_i(\boldsymbol{r}) d\boldsymbol{r} \tag{4.59}$$

The first two terms in the parentheses depend on only the one-electron coordinate, and the energy calculation requires a simple single integration.

However, the last term depends on two electronic coordinates, and its energy calculation is in the form of a complicated double integral, resulting in two energy terms: the Coulomb energy for the interactions between electron i with j, which is the Hartree energy as we discussed in the previous Subsection 4.3.2, and the exchange energy coming from the antisymmetric nature of wave function in the Slater determinant form. Then, the final expression for the total energy becomes a sum of all these contributions:

$$E = E_{\text{kin}} + E_{\text{ext}} + E_H + E_x \tag{4.60}$$

Although we skipped the detailed derivation of the last two terms (it involves many terrible-looking equations), it would not cause any harm to what we are up to. Note that, compared to the Hartree equation, the preceding equation considers a new term, E_x, and the resulting energy will be that much lower (that much closer to the true ground-state energy) since E_x has negative energy. This in fact implies that like-spin electrons avoid each other, keeping some distances (called exchange hole, see Figure 4.18) between them and thus reduce the corresponding repulsive energy.

In practice, the Hartree energy is approximated by the mean-field approach adopted by Hartree:

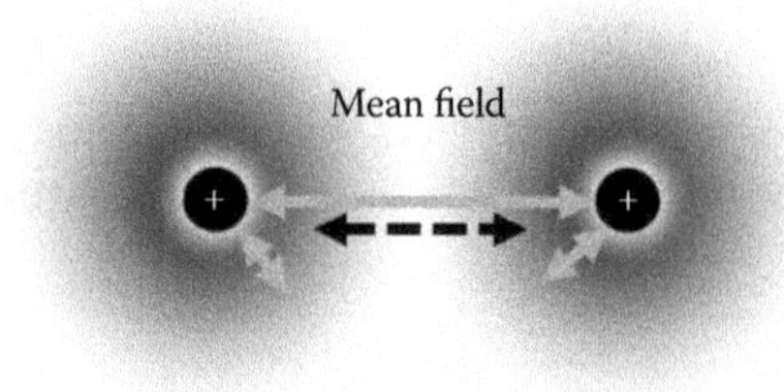

Figure 4.18 Schematic of the Hartree–Fock model with the exchange energy counted by the Slater determinant expression for the wave function.

$$E_H = \frac{1}{2}\iint \frac{\rho(\boldsymbol{r})\rho(\boldsymbol{r}')}{|\boldsymbol{r}-\boldsymbol{r}'|} d\boldsymbol{r}d\boldsymbol{r}' \tag{4.61}$$

In this scheme, an electron at $\boldsymbol{r}$ feels other electrons at $\boldsymbol{r}'$ as a smooth distribution of negative charges (field). The new exchange energy is subject to further discussion in Chapter 5, and here we only emphasize its origin and role:

- It appears automatically as a result of the antisymmetrization of electron wave functions in the form of the Slater determinant and can be calculated exactly. Thus, total wave function in the HF method changes its sign when two electron indices are "exchanged," from which the terminology "exchange energy" is originated.
- It corrects the overestimated Hartree energy and makes the HF model closer to the actual system as shown in Figure 4.18.

4.3.3.5 *Variational approach to the search for the ground-state energy*

Since electrons are interacting, expressing the wave function as a single Slater determinant is only an approximation, which is subject to a systematic improvement to the exact function. The process is specifically based on the variational principle.

In quantum mechanics, there is only one ground-state energy in a given system under given conditions if there is no degenerated state, and any energy lower than this value simply does not exist. This implies that, if we minimize the system energy with respect to wave function as far as we can, the energy at the end will be the ground-state energy. This is called the *variational principle* and is written as

$$0 = \delta\left(\int \Psi \hat{H} \Psi d\boldsymbol{r}\right)\Big|_{\text{at ground state}} \tag{4.62}$$

Through this variational process, the initial approximated wave functions in the Slater determinant will approach the true ones that calculate the ground-state energy. This approach is a very fundamental principle and thus is adaptable to all first-principles methods.

4.3.3.6 *Self-consistent procedure*

Due to the nonlinearities introduced by the HF method, it adopts the iterative self-consistent procedure for the solution of a wave equation, which is

- Choose a proper set of n wave functions, $\{\psi_i\}$ such as wave functions of hydrogen for an atomic calculation and a linear combination of atomic orbitals for solids.

- Calculate electron density, $\rho(\boldsymbol{r}')$.
- Calculate the three energy terms of $\hat{H}$ for n electrons.
- Insert these values into the wave equations and solve to have a set of $\{E_i\}$ and a new set of $\{\psi_i\}$.
- Repeat the above process until self-consistency is reached and input $\{\psi_i\}$ and output $\{\psi_i\}$ become the same in a predetermined range of error.

Once we cannot go down any further, we know that we have hit the ground state. The obtained $\{E_i\}$ will be the ground-state energies of each electron, and its sum will be the total energy of the system. For a simple system like He with one proton and two electrons, an analytical solution is possible, and an HF calculation results in the ground-state energy of –77.5 eV, which is very close to the experimental value of –78.98 eV.

4.3.3.7 Remarks

The same-spin electrons are kept away from each other by the antisymmetry requirement, and the HF treatment fully counts this quantum effect by the exchange energy. The electrons of different spins also have a tendency to stay away from each other by the same charges they have, and this is called the correlation. This electron correlation is still missing in the HF method and will be accounted for in the next chapter where we deal with the DFT. Note also that the scaling of the HF method is roughly $O(N^4)$ and thus a doubling of system size will increase the computer-time about 16 times.

Homework

4.1 For a hydrogen atom, the wave function at the ground state is in the form of $\Psi = A\exp(-x/\alpha)$. Find the normalization factor, A, and the radial probability density for finding the electron, $\rho(x)$.

4.2 Explain the uncertainty principle in terms of position and momentum with the case of a free electron.

4.3 In the model of an electron in an infinite well, the solutions are given as $\Psi(x) = A\sin(\pi nx/L)$. Draw two waves as an example and show that all solutions to the system are orthogonal to each other graphically.

4.4 We know that solving the Schrödinger equation for a hydrogen atom is described in any undergraduate physics textbook. However, solving it for a helium atom requires some computational efforts since we face an n-electron problem and r_{12} term between two electrons makes spherical symmetry not valid any more.

Variational principle, however, provides a sound base for finding the answers for a helium atom and for all other systems. To see how the principle works, we start with the hydrogen atom with the ground-state energy of –13.6 eV and assume that the presence of an additional electron in He will screen and change the nuclear charge Z (= 2) to Z′ (<2). After some mathematical manipulation, the He energy is given as

$$E = -13.6\left(-2Z'^2 + 4ZZ' - \frac{5}{4}Z'\right)$$

Using the variational principle with respect to Z′, find the best Z′.
Using the best Z′, find the corresponding energy.
Compare with the experimental value of –78.98 eV.

4.5 Helium is an inert element, and it surprisingly exists in liquid phase even down to absolute zero temperature while other matters generally condense into a solid. It is impossible to explain it from a classical view but it is possible only from a quantum mechanical understanding. Referring to Section 4.2.6, explain how it is possible.

References

Bohr, N. 1913. On the Constitution of atoms and molecules, Part I. *Phil. Mag.* 26:1–24, Part II. Systems containing only a single nucleus. *Phil. Mag.* 26:476–502, Part III. Systems containing several nuclei. *Phil. Mag.* 26:857–875.

Born, M. and J. R. Oppenheimer. 1927. Zur Quantentheorie der Molekeln. *Ann. Physik.* 84:457.

de Broglie, L. 1924. Recherches sur la théorie des quanta (Researches on the Quantum Theory). Thesis (Paris), Université en cours d'aectation.

Fock, V. 1930. Näherungsmethode zur Lösung des quantenmechanischen Mehrkörperproblems. *Z. Phys.* 61:126–148; 62:795.

Hartree, D. R. 1928. The wave mechanics of an atom with a non-Coulomb central field. Part I. Theory and methods. *Proc. Camb. Phil. Soc.* 24:89–110.

Heisenberg, W. 1927. Über den anschulichen Inhalt der quantentheoretischen Kinematik und Mechanik. *Z. Phys.* 43:172–198.

Planck, M. 1900. Entropy and temperature of radiant heat. *Ann. Physik* 1(4):719–737.

Schrödinger, E. 1926. Quantisierung als Eigenwertproblem; von Erwin Schrödinger. *Ann. Physik* 79:361–377.

Further reading

Finnis, M. 2003. *Interatomic Forces in Condensed Matter.* Oxford: University Press.

Kantorovich, L. 2004. *Quantum Theory of the Solid State.* Boston, MA: Kluwer Academic.

Martin, R. M. 2004. *Electronic Structure: Basic Theory and Methods.* Cambridge: University Press.

chapter five

Density functional theory

> Together with my new postdoctoral fellow, Lu J. Sham, we derived from the HK variational principle what are now known as the Kohn–Sham equations.
>
> **Walter Kohn (1923–2016)**

The Schrödinger wave equation, $\hat{H}\Psi = E\Psi$, is definitely a work of genius based on the very origin of physics. Unfortunately, the equation is useful in practice only for the simplest systems due to the underlying difficulties of the many-body quantum effect. People often say that first-principles methods require only a single set of input data, the atom number, and nothing else for the calculation. This is true in principle, but the reality is not as simple as the statement sounds. The reason is that the wave equation is a partial differential equation that depends on the $3n$ coordinates of n electrons. Even if we knew how to do such a calculation, a computer that can handle such a massive task is not available, and this will be true no matter how fast computers may be in the future.

As we have seen in the previous chapter, by introducing the Slater determinant for wave function and the mean-field approximation, the Hartree–Fock (HF) method was able to ease the calculation maintaining the parameter-free nature. However, its practical application is still limited to small systems with atoms of several tens, which are far from the regime of materials. The breakthrough finally happened when Hohenberg and Kohn (1964) presented two theorems concerning electron density and energy functionals, and Kohn (see Figure 5.1) and Sham (1965) came up with an extraordinary scheme called the *density functional theory* (DFT).

In this chapter, we will follow how Kohn and Sham developed the DFT based on a fictitious system and eventually made computational materials science possible. We will first review electron density and its roles with a renewed perspective. Then we will go over the basic ideas and concepts underlying the theory and how it works in terms of electron density. Next, the approximations used for the so-called exchange-correlation (XC) energies will be discussed in some detail. Finally, we will discuss practical methods to solve the Kohn–Sham (KS) equations.

Materials science students may find this chapter rather demanding, but note that all materials in this chapter are the main ingredients for

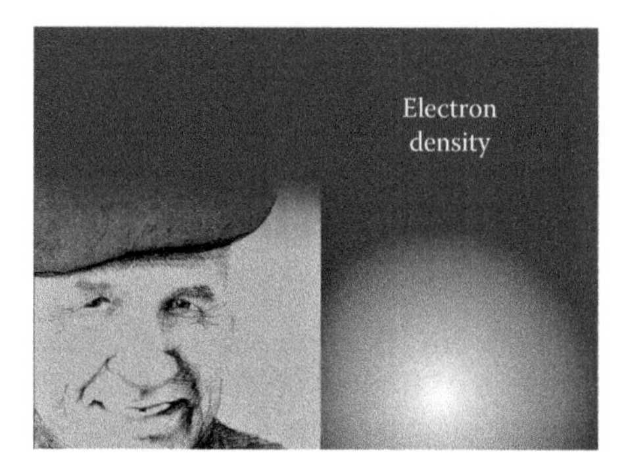

Figure 5.1 Walter Kohn, who developed two theorems and the KS equations in the framework of the density functional theory (DFT).

the DFT and are required for better DFT runs. Again, for the sake of simplicity, several omissions are made throughout this chapter, which are as follows:

- The word operator and the operator notation (the hat, ∧) are omitted for all operators except $\hat{H}$.
- The coordinate notation $\boldsymbol{r}$ is defined to include both three-dimensional position (x, y, z) and spin (↑, ↓) coordinates, although the spin part is largely ignored and is not explicitly written down unless indicated otherwise.
- Factors resulting from normalization and summations over the orbitals and $\boldsymbol{k}$-points are often omitted from the derived formulas.

For an in-depth understanding of the subject, see the "Further Reading" section.

5.1 Introduction

The Kohn and Sham approach used the once-tried electron density as the main variable to solve the n-electron problem, as schematically illustrated in Figure 5.2. One can immediately see the big difference between the previous first-principles methods in Chapter 4 (e.g., Figure 4.18) and this new approach: the problem has changed from a $3n$-dimensional equation (e.g., a description of 10 electrons requires 30 dimensions) to n separate three-dimensional ones with the use of electron density. Note that there is no individual electron but only a three-dimensional density of electrons, $\rho(x, y, z)$. Note also that the electron density is an observable quantity and does not depend on the number of electrons after it is constructed in a system.

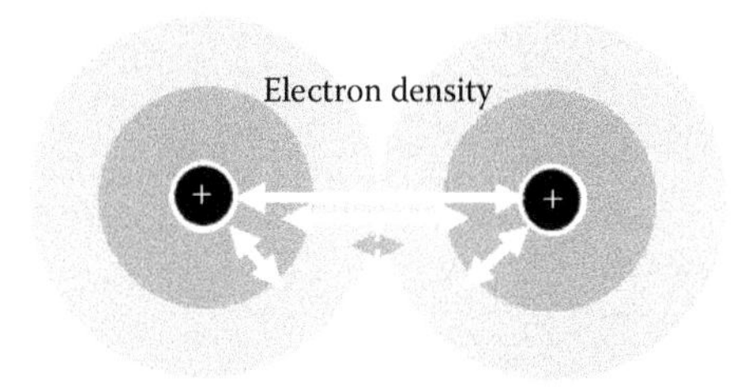

Figure 5.2 Schematic illustration of the Kohn-Sham's density functional theory (DFT) model based on electron density with the XC energy approximated.

The important point is that a quantum mechanical system is now an electron-density-dependent problem. This will reduce the computational effort drastically to the level of practical use for molecules, solids, and materials. In addition, the skipped correlation energy in the HF method is accounted by approximation, and thus, all necessary energy terms are present in the DFT.

The DFT description eventually has progressed to treat systems containing up to thousands of electrons and made computer experiments possible for materials. These days, the methods have been so successful in both accuracy and efficiency that scientists are using them routinely in a wide range of disciplines. For the moment, this much of an outline may be sufficient, and important topics, including the newly introduced XC energy, will be explained in due course.

5.1.1 Electron density

In Chapter 4, it was stated that all information required to define a quantum state is contained in the wave function. In this chapter, a similar statement will be applied to the electron density, $\rho(\mathbf{r})$: the electron density decides everything in an n-electron quantum system. Let us first define electron density, the central player in the DFT, and move on to the two theorems on which the DFT is based.

If we take squared wave functions at a point, it will build an electron density at that particular point. Specifically, the electron density, $\rho(\mathbf{r})$, in an n-electron system is the number of electrons per unit volume at a given point $\mathbf{r}$. For example, Figure 5.3 shows the electron density of an Si atom. It is evident that the core-electron densities are localized very near to the nucleus, and the valence-electron (3s and 3p electrons) densities are rather diffused around about 2 a.u. Figure 5.4 shows another way of presenting electron densities, now as an isosurface of electron densities, in Si bulk (diamond structure). Here we can see that the electron densities of each Si atom accumulate at the bond centers and build up the covalent bonds.

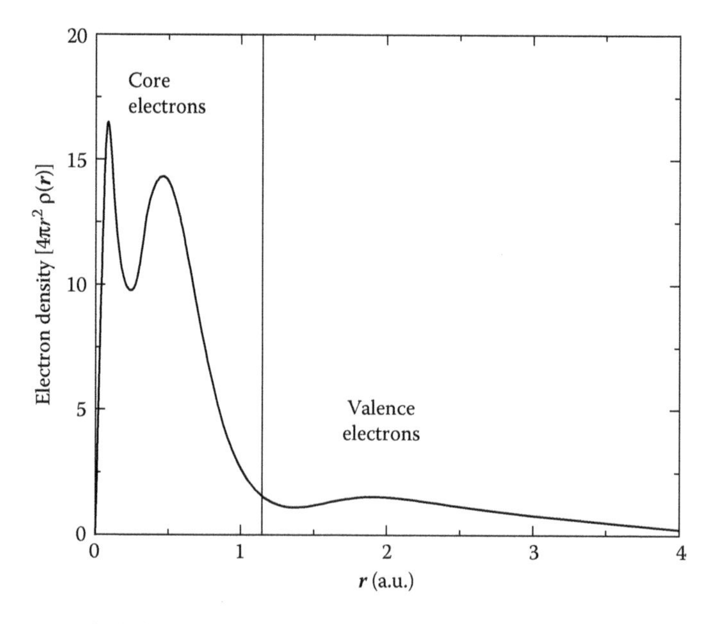

Figure 5.3 Radial electron densities of Si atom showing electron densities in the core and valence regions.

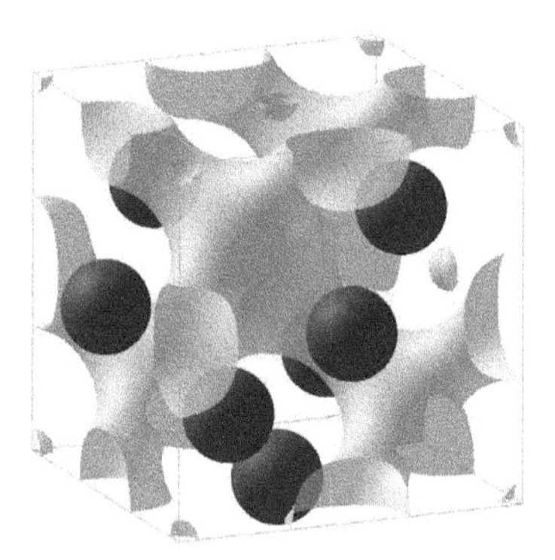

Figure 5.4 Isosurface of electron densities in an Si bulk.

These two figures are simple illustrations just to show the observable nature of electron densities. On the contrary, the wave functions are not easy to present in such a simple way. Throughout this chapter, we will see how important the electron density is in quantum systems, especially in a bulk system formed by a collection of atoms. Therefore, if we can adopt this electron density as the sole variable for the calculation of electronic systems, it will be much more realistic and convenient than the wave-function-based treatments.

5.1.1.1 *Electron density in DFT*

In the DFT scheme, we first assume that the electrons do not interact with one another. In this noninteracting reference system with decoupled coordinates, the electron density is written as a simple sum over a set of squares of (or occupied) noninteracting orbitals ϕ_i:

$$\rho(\boldsymbol{r}) = \sum_i |\phi_i(\boldsymbol{r})|^2 = 2\sum_i^{\text{occ}} |\phi_i(\boldsymbol{r})|^2 \tag{5.1}$$

Note that the usual wave functions, ψ_i are replaced by orbitals ϕ_i, implying that ϕ_i are now the so-called KS orbitals in a noninteracting reference system. In the above equation, the amplitudes (positive or negative) of each orbital are converted to a positive density of electrons. If we add up all the electron densities over the whole space, it will naturally return the total number of electrons, n:

$$\int \rho(\boldsymbol{r}) d\boldsymbol{r} = n \tag{5.2}$$

Referring to Figure 5.3, this equation indicates that the area under the curve equals to n (= 14) since the curve is plotted as a radial density of $4\pi r^2 \rho(\boldsymbol{r})$, weighted with $4\pi r^2$. In addition, if we add up all the overlapping electron densities of atoms, it will build up closely to the electron densities of solids, as partially illustrated in Figure 5.4. This suggests that, if we know an atomic electron density, we can roughly generate the electron density for a solid made of that atom. As we shall see in the following section, the electron density in a system not only represents wave function, orbital, and the total number of electrons, but is also directly related to potentials, energies, and thus all properties.

5.1.2 *Hohenberg–Kohn theorems*

Scientists knew that electron density could play a decisive role in electronic calculations in some way. However, it was subject to formal verification until Hohenberg and Kohn (1964) finally proved it with two theorems, and thus, provided a sound foundation for the designation of electron density as the key player in the DFT. The theorems thus completed the links between electron density, external energy, Hamiltonian, and wave function.

5.1.2.1 *Electron density as central player*

The first theorem states that there is a unique external potential, U_{ext}, determined solely by the ground-state electron density. Let us look at how closely the electron density and the U_{ext} are related to each other from the well-known facts:

The $U_{ext}(\equiv U_{Ii})$ is, by definition, the measure of the interaction between electrons and nuclei.
The number of electrons that are interacting with nuclei is defined by the integral of the electron densities as shown in Equation 5.2.
In an electron-density profile around an atom, the cusp shows the location of the nucleus (see Figure 5.3), and the height of the cusp indicates the relative magnitude of the nuclear charge.

Therefore, it is evident that there will be a direct relationship between electron density and the external potential. Remember that the term *external* in the U_{ext} refers to the fact that, from the electron's viewpoint, the Coulomb attraction by nuclei is external and is thus system dependent. The system-independent internal potential (the electronic kinetic energy plus the electron–electron potential) is independent of this external potential and therefore has a universal character, meaning that, once it is known, it can be applied to any other system.

Let us summarize the flow of logic of this first theorem:

- In a given system at the ground state, the electron density alone can define the external potential and vice versa.
- Since the internal energy is system independent and does not depend on the external potential, a density-dependent internal energy should be there as a universal functional $F[\rho(\boldsymbol{r})]$ although its explicit formula is unknown. Note that the mathematical form of $F[\rho(\boldsymbol{r})]$ should be the same for all systems, whereas the external potential varies from one system to another depending on the kind of nuclei.
- Then, different Hamiltonians differ only by their external potential, and, if there were two different external potentials that yield the same ground-state electron density, it leads to an apparent contradiction.
- Thus, electron density defines external potential, Hamiltonian, wave function, and all ground-state properties of the system in turn.

The conclusion is as simple as this: different external potentials will always generate different electron densities and, if we confine our interests only within the ground-state properties of a system, the sole knowledge of the electronic density at a given external potential is sufficient to deduce the total energy or any other properties. This is the base of all conventional DFT calculations.

5.1.2.2 *Search for the ground-state energy*

The following analogy will be helpful in understanding the second theorem. Suppose we play a guessing game in which we try to deduce the size of a bird under the condition that we can only ask about the relative size of the bird. It may go as follows:

Is it smaller than an eagle? Yes.
Is it smaller than a dove? Yes.
Is it smaller than a sparrow? Yes.
Is it smaller than a humming bird? No.

At this point, we can assume that the answer is a bird with size comparable to that of a humming bird. If there is an additional condition (constraint) such as *the color of the bird should be yellow* the size of the bird goes down, but the color of the bird should remain yellow throughout the guessing process.

The second theorem, like the guessing game, identified a way to find the minimum energy of a system and proved that the ground state of a system could be searched by using the variational principle. At a given U_{ext}, if we minimize the system energy as much as we can with varying electron density, then we will reach the very bottom of the energy well, yet not below it. This is called the *variational principle* in the framework of DFT, and the electron density that minimizes the system energy is the ground-state electron density, ρ_0:

$$E[\rho(\boldsymbol{r})] = F[\rho(\boldsymbol{r})] + E_{ext}[\rho(\boldsymbol{r})] \geq E_{gs} \tag{5.3}$$

The theorem offers us a very flexible and powerful means of finding the ground-state energy and other properties. Remember that we did a similar approach for the search of HF energy with respect to wave function in Chapter 4.

Figure 5.5 shows schematically the search for the ground-state energy, E_{gs}, by the variational principle. It provides both the cane (electron density)

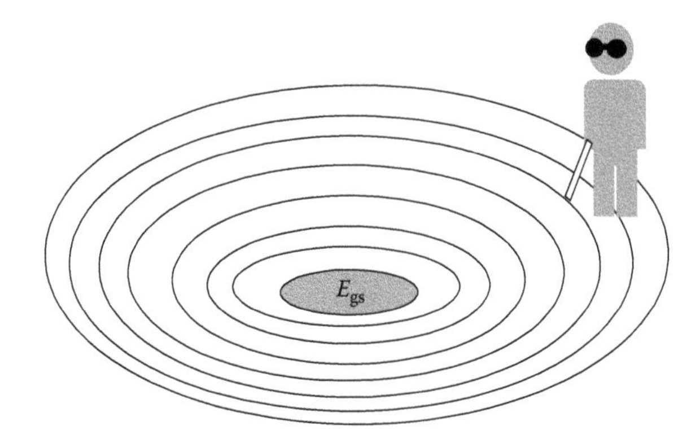

Figure 5.5 Schematic illustration of searching the ground-state energy by the variational principle.

and the downward-only instruction (variational principle) for the search. However, we cannot see where the ground-state energy is; we only know that it exists there somewhere. One additional point is that the search can start from any point (energy calculated by any educated guess for the electron density) in any possible way. In practice, for a solid, we normally start with the energy calculated at the electron density generated by overlapping atomic densities.

5.2 Kohn–Sham approach

Earlier attempts to adopt electron density without the use of any wave functions in the first-principles calculations were not very successful. The main reason is the poorly written electronic kinetic energy in terms of electron density. Given the two theorems of Hohenberg and Kohn (1964), Kohn and Sham (1965) constructed a fictitious system of one-electrons. In this section, their "divide-and-conquer" approach will be outlined, hopefully without involving too much mathematics. We will also see how their one-electron representations correspond to the actual n-electron system.

5.2.1 One-electron representations

Remember we are trying to solve the time-independent Schrödinger wave equation of a n-electron system using a full Hamiltonian:

$$\hat{H} = -\frac{1}{2}\sum_{i=1}^{n}\nabla_i^2 - \sum_{I=1}^{N}\sum_{i=1}^{n}\frac{Z_I}{|\boldsymbol{r}_i - \boldsymbol{r}_I|} + \frac{1}{2}\sum_{i \neq j}^{n}\frac{1}{|\boldsymbol{r}_i - \boldsymbol{r}_j|} \tag{5.4}$$

where:

$\boldsymbol{r}_i$ and $\boldsymbol{r}_j$ are coordinates of electrons
$\boldsymbol{r}_I$ and Z_I are coordinates and charges of the nuclei

In the above equation, the first term represents kinetic energy, the second term represents external potential, and the last term represents the Hartree potential with the correction factor of 1/2 for double-counting. The summation of the last term covers all cases where $i \neq j$ to exclude any self-interaction. A major difficulty lies in the last term, the coupled interactions between all n electrons. This last term contains various interactions that are difficult to formulate in calculable equations. Kohn and Sham (1965) apparently had no choice but to take a detour around this problem.

Let us write down the starting point, the interacting n-electron energy of the HF approach with the four familiar terms of kinetic, external, Hartree, and exchange energies:

$$E = E_{\text{kin}} + E_{\text{ext}} + E_H + E_x \tag{5.5}$$

Kohn and Sham first assumed that each electron was noninteracting since no solution was in sight otherwise and further assumed that the system was at the ground state. Then, they decomposed the energy of n-electron into that of n one-electrons. What they eventually came up with was mapping of the n-electron system (interacting) on the one-electron system (noninteracting) under the given external energy. All the interacting effects are identified as follows:

$$E_{\text{kin}} = E_{\text{kin}}^{\text{non}} + E_{\text{kin}}^{\text{int}} \tag{5.6}$$

$$E_H + E_x \rightarrow E_H + E_x + E_c^{\text{int}} \tag{5.7}$$

where $E_{\text{kin}}^{\text{non}}$ and $E_{\text{kin}}^{\text{int}}$ represent noninteracting and interacting (correlating) kinetic energies, respectively. Note that a new correlation energy is counted as E_c^{int}, which is neglected in the HF method.

5.2.2 *One-electron system replacing* n*-electron system*

Now, let us regroup all the interacting terms together as a single term called the *exchange-correlation energy*, E_{xc}:

$$E_{xc} = E_x + E_c^{\text{int}} + E_{\text{kin}}^{\text{int}} = E_x + E_c \tag{5.8}$$

Note that $E_{\text{kin}}^{\text{int}}$ and E_c^{int} sum up to be E_c, the correlation energy, since both are energies due to correlation. If we define each term again for clarity, E_x is the exchange energy we have discussed in Chapter 4, and E_c is the correlation energy representing the correlating part of the kinetic and electron–electron interaction terms. Then, the final expression of the total energy in the framework of the DFT consists of four energy terms:

$$E = E_{\text{kin}}^{\text{non}} + E_{\text{ext}} + E_H + E_{xc} = F[\rho(r)] + E_{\text{ext}} \tag{5.9}$$

The first three terms are relatively easy ones to calculate, while the last term is unknown and thus subject to approximation. Finally, the repulsive interaction energy between the nuclei will be added as a constant within the Born–Oppenheimer approximation.

In summary, Figure 5.6 shows schematically how Kohn and Sham decomposed the energies of n electrons and regrouped them in the framework of independent electrons. Note that the classical E_H (positive) finally becomes close to the true and quantum electron–electron interaction energy by accounting the quantum E_{xc} (negative) in the KS system. For the unknown E_{xc}, we may just approximate it and stay away from the problems of the n-electron. That is, in fact, what the DFT is all about. The corresponding Hamiltonian is

$$\hat{H}_{KS} = E_{\text{kin}}^{\text{non}} + U_{\text{ext}} + U_H + U_{xc} = -\frac{1}{2}\nabla^2 + U_{\text{eff}} \tag{5.10}$$

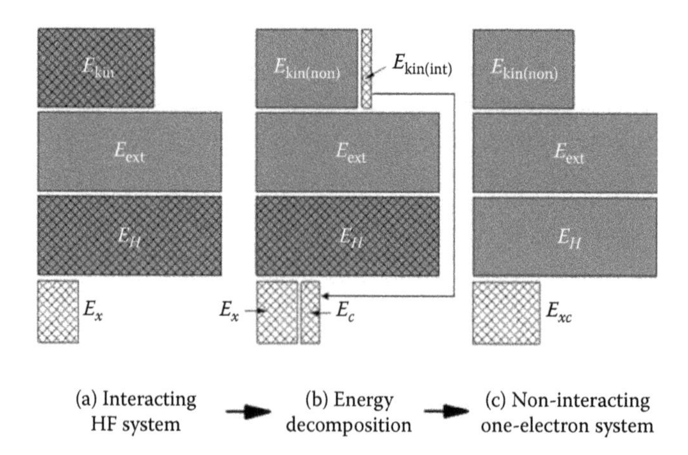

Figure 5.6 Schematic illustration of the DFT approach showing (a) the HF system, (b) the energy decomposition, and (c) the fictitious KS system (The interacting quantum parts are meshed).

Here, U_{eff} is the effective potential including three potential terms, and it manipulates the ground-state electron density of the noninteracting system to be identical with that of the true interacting system.

It is expected that this reformulation and dealing with independent electrons will provide a much easier and effective way of calculation. It has been proven over the years that the scheme in fact mimics the true ground-state density and is thus able to describe the interacting system quite accurately.

To further illustrate the KS ansatz, an analogy is shown in Figure 5.7. Here, a complex n-people system is replaced by an artificial n-robot system. The auxiliary system is now easily and fully calculable owing to its simplicity, and if the XC functional is accurately known, the electron density is the same for both systems. In other words, the noninteracting electron density and the effective potential, $U_{eff}(\mathbf{r})$, in the KS scheme are consistent with each other and are designed to return the true (or interacting) density and energy. The remaining task now is to figure out how to express energies in terms of electron density and how to deal with the unknown E_{xc} term.

5.3 Kohn–Sham equations

In this section, we will write out the energy functionals and take its derivatives with respect to electron density (or orbital in the case of kinetic energy) to derive the final KS equations. First, let us look at each energy

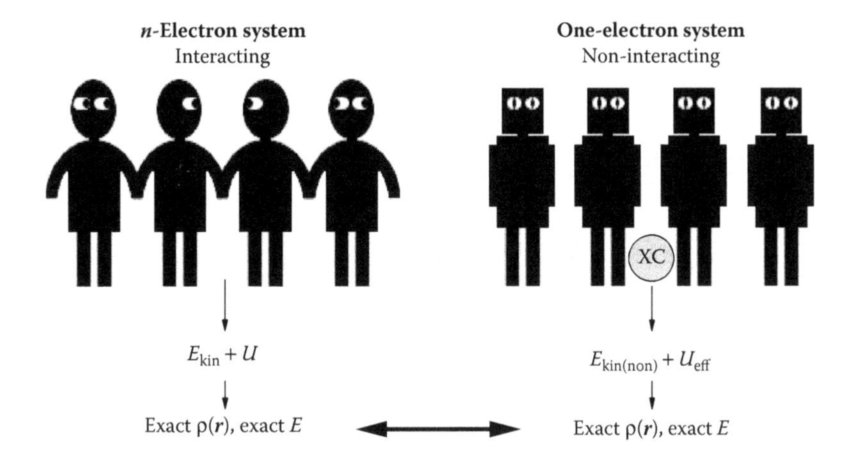

Figure 5.7 Schematic illustration of the Kohn–Sham ansatz.

term more closely and express each in terms of electron density so that we can apply the variational principle with respect to electron density.

5.3.1 Energy terms

5.3.1.1 Kinetic energy

In an interacting system, the kinetic energy term is formally expressed as

$$E_{kin} = -\frac{1}{2}\int \psi_i^*(\boldsymbol{r})\nabla^2\psi_i(\boldsymbol{r})d\boldsymbol{r} \tag{5.11}$$

where the asterisk indicates a complex conjugate. In a noninteracting system of the KS scheme, however, kinetic energy can be expressed in summation form with the KS orbitals ϕ_i:

$$E_{kin}^{non} = -\frac{1}{2}\sum_{i=1}^{n}\phi_i^*(\boldsymbol{r})\nabla^2\phi_i(\boldsymbol{r}) \tag{5.12}$$

Due to the second-order derivative operation on the orbitals, $\nabla^2\phi_i(\boldsymbol{r})$, writing the orbital and its conjugate as an electron density (a norm square, $|\phi|^2$) is not allowed. Thus, the KS scheme uses Equation 5.12 for calculation of the kinetic energy. Although orbitals are involved here, it is directly related to electron density via Equation 5.1, and thus the name of the DFT is still valid.

In a highly simplified system, kinetic energy can be directly related with electron density in a simple analytical form. For example, if we assume a system of noninteracting and locally homogeneous electrons, the kinetic energy can be given in terms of electron density alone:

$$E_{\text{kin}}^{\text{hom}} = \int \varepsilon_{\text{kin}}^{\text{hom}}[\rho(\boldsymbol{r})]d\boldsymbol{r} = C\int \rho^{5/3}(\boldsymbol{r})d\boldsymbol{r} \tag{5.13}$$

where C is a constant. Throughout this book, C is used as a constant without any specification. The above equation was the one used in the earlier version of DFT for the calculation of kinetic energy. Of course, the approach eventually failed in practice due to the oversimplification employed for kinetic energy of real systems. The KS approach, on the other hand, treats kinetic energy in terms of the orbital under the one-electron scheme, and thus can evaluate kinetic energy precisely. This noninteracting kinetic energy accounts for most of the total kinetic energy, and the neglected interacting kinetic energy will be included in the XC term, E_{xc}.

5.3.1.2 *External energy*

As we discussed previously, the external potential, $U_{\text{ext}}(\boldsymbol{r})$, comes from the interaction between an electron and other nuclei. Its energy, the expectation value of $U_{\text{ext}}(\boldsymbol{r})$, is determined by quantum manipulation, multiplying $U_{\text{ext}}(\boldsymbol{r})$ from left with the orbitals and integrating it over all space:

$$E_{\text{ext}}[\rho(\boldsymbol{r})] = \int \phi^*(\boldsymbol{r})U_{\text{ext}}(\boldsymbol{r})\phi(\boldsymbol{r})d\boldsymbol{r} = \int U_{\text{ext}}(\boldsymbol{r})\rho(\boldsymbol{r})d\boldsymbol{r} \tag{5.14}$$

Here, the external energy, $E_{\text{ext}}[\rho(\boldsymbol{r})]$, is a functional, meaning that it is a function of $\rho(\boldsymbol{r})$, which in turn is a function of $\boldsymbol{r}$ (electronic coordinates). Therefore, a functional takes a function as input and outputs a number. Square brackets such as in $[\rho(\boldsymbol{r})]$ are used to distinguish a functional from a regular function. Therefore, external energy is formally expressed in terms of electron density as the only variable.

5.3.1.3 *Hartree energy*

In a noninteracting charge distribution, the Hartree potential comes from the interaction between an electron at $\boldsymbol{r}$ and the mean electron density at $\boldsymbol{r}'$ in a mean-field approximation:

$$U_H(\boldsymbol{r}) = \int \frac{\rho(\boldsymbol{r}')}{|\boldsymbol{r}-\boldsymbol{r}'|}d\boldsymbol{r}' \tag{5.15}$$

This equation requires an integral on $\boldsymbol{r}'$ for the evaluation of the Hartree potential at $\boldsymbol{r}$. With reference to this potential, the Hartree energy can be expressed as the expectation value:

$$E_H[\rho(\boldsymbol{r})] = \int U_H(\boldsymbol{r})\rho(\boldsymbol{r})d\boldsymbol{r} = \frac{1}{2}\iint \frac{\rho(\boldsymbol{r})\rho(\boldsymbol{r}')}{|\boldsymbol{r}-\boldsymbol{r}'|}d\boldsymbol{r}d\boldsymbol{r}' \tag{5.16}$$

One convenient way is to represent the potential in terms of electron density via the Poisson equation:

$$\nabla^2 U_H(\mathbf{r}) = -4\pi\rho(\mathbf{r}) \tag{5.17}$$

The mean-field description in Equation 5.16 requires double integrals between an electron and the mean-field generated by all other electrons. The energy so calculated is purely classical and Coulombic. Since any electron is included in *all other electrons*, the energy is double-counted and results in an unphysical self-interaction. Unlike humans, an electron does not interact with itself (it neither suffers from depression nor commits suicide), and thus this self-interaction will later be corrected with the exchange energy term. Again, the noninteracting Hartree energy is expressed as a functional of electron density.

5.3.1.4 *Exchange-correlation energy*

The final term is XC energy, which consists of all quantum effects, and is approximated in terms of electron density as we will see in the next section:

$$E_{xc} = E_x + E_c \tag{5.18}$$

Here E_x is the exchange energy between electrons with the same spin, and E_c is the correlation energy between electrons with a different spin. E_x is associated with the Pauli exclusion principle, which is equivalent to the wave function's antisymmetric nature with respect to the exchange of any two electrons' coordinates. The resulting exchange energy is the sum of the four-center integrals as a function of the single-particle orbitals:

$$E_x = -\frac{1}{2}\sum_{ij}^{n}\iint \frac{\phi_i(\mathbf{r})^*\phi_j^*(\mathbf{r}')\phi_i(\mathbf{r}')\phi_j(\mathbf{r})}{|\mathbf{r}-\mathbf{r}'|}\,d\mathbf{r}d\mathbf{r}' \tag{5.19}$$

This will lead to less overlapping of electron densities around the reference electron, and thus the electron–electron repulsion energy will be reduced. The net effect is attractive as indicated by the negative sign in the above equation. In principle, E_x can be calculated exactly with the above equation, but if we do that, we go back to the HF method losing the original spirit of the DFT because the HF calculation scales badly with the number of electrons ($>O(n^3)$). In practice, therefore, E_x is always approximated for computational convenience.

On the other hand, two electrons with different spins can occupy the same orbital. However, they also repel each other because of the same negative charges they have. This is called *electronic correlation*, and it also results in less overlapping of electron densities, generating a small attractive energy. This is the unknown n-electron effect that was missed in the HF method and is subject to approximation.

For materials at typical electron densities, the correlation energy is much smaller than the exchange energy. However, at very low electron densities, the correlation energy becomes very important since the exchange interaction becomes less active in these sparse electron densities. The subject of XC energy will be explained later in more details (see Section 5.4).

5.3.1.5 *Magnitudes of each energy term*

Let us see how big each energy term really is in order to comprehend the general energy spectrum of the electron world. Calculation (Huang and Umrigar 1997) from exact electron densities showed that the He atom has a total energy of –2.904 Hartree, made up of the following:

- Noninteracting kinetic energy $\left(E_{\text{kin}}^{\text{non}}\right)$ of 2.867 Hartree
- External energy (E_{ext}) of –6.753 Hartree
- Hartree energy (E_H) of 2.049 Hartree
- XC energy (E_{xc}) of –1.067 Hartree, including two components of exchange energy (E_x) of –1.025 Hartree and correlation energy (E_c) of –0.042 Hartree

The energy proportion is plotted in Figure 5.8. It is evident that atomic stability is decided by various competing energies: positive energies for repulsive contribution (will drive the atom unstable) and negative energies for attractive contribution (will drive the atom stable). We can easily tell that the major contribution for the stability of an He atom comes from the attractive interaction of E_{ext} between nuclei and electrons. Note that the exchange energy cancels one-half of the Hartree energy, which is wrongly included in the Hartree energy due to the self-interaction in this two-electron system.

One might speculate as to why we are so concerned about such small XC energies, especially the electronic correlation energies that are only

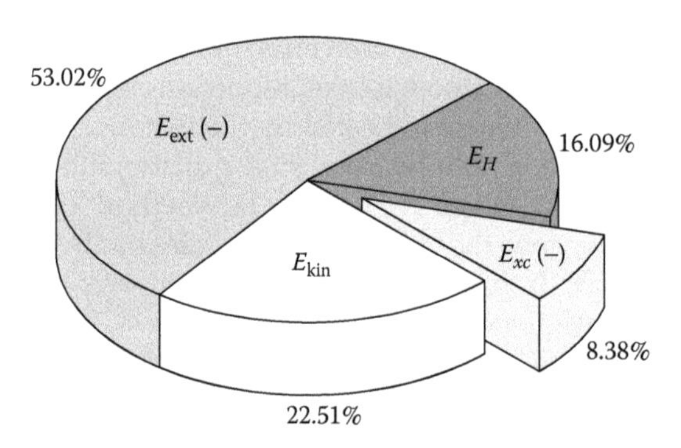

Figure 5.8 Magnitudes of energy terms of an He atom.

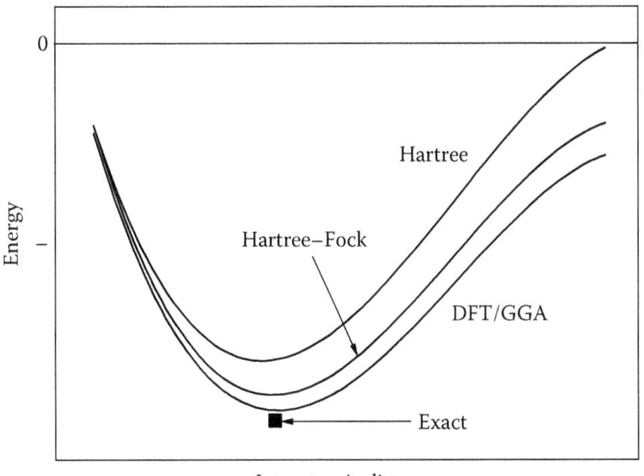

Figure 5.9 Schematic illustrations of typical potential curves as a function of interatomic separation calculated by the Hartree, HF, and DFT methods.

about 0.3% in this example of a He atom. However, the XC energy is actively involved in the changes happening in the atomic, molecular, or solid systems. It crucially determines, for example, whether BCC or FCC is stable, whether an adatom prefers a top or a hollow site on a metal surface, and whether a magnetic material becomes ferromagnetic or antiferromagnetic (with spin density considered). Therefore, having an accurate XC energy becomes a key factor for the successful operation of the DFT. A gallery of XC functionals is reported, and the accuracy of DFT improves whenever a better description of XC energy becomes available. The XC functionals will be discussed in more detail in the next section.

As a summary, Figure 5.9 shows a schematic illustration of typical potential curves with interatomic distances calculated by the Hartree, HF, and DFT methods. It shows that the DFT, finally covering all energy terms, results in the lowest curve and approaches closely to the exact value. Note also that the energy difference between the Hartree and HF methods represents the exchange energy, while the energy difference between the HF and DFT methods represents the correlation energy.

5.3.2 *Functional derivatives*

We now have all constituting energy functionals for the total energy functional:

$$
\begin{aligned}
E[\rho(\boldsymbol{r})] &= E_{\text{kin}}^{\text{non}}[\phi(\boldsymbol{r})] + E_{\text{ext}}[\rho(\boldsymbol{r})] + E_H[\rho(\boldsymbol{r})] + E_{xc}[\rho(\boldsymbol{r})] \\
&= E_{\text{kin}}^{\text{non}}[\phi(\boldsymbol{r})] + E_{\text{eff}}[\rho(\boldsymbol{r})]
\end{aligned} \tag{5.20}
$$

All terms except kinetic energy are written exclusively in terms of electron density, and kinetic energy is written as a functional of the noninteracting orbitals. What we have to do now is to minimize this energy to find the ground-state electron density that requires differentiation of functionals:

$$\delta F\left[f(x)\right] = F\left[f(x) + \delta f(x)\right] - F\left[f(x)\right] = \int \frac{\delta F\left[f(x)\right]}{\delta f(x)} \delta f(x) dx \tag{5.21}$$

Here, the notation δ indicates that the above equation is a derivative of a functional that is roughly, but not exactly, the same as the usual derivative of a normal function. However, as far as the scope of this book is concerned, the functional derivative with respect to $f(x)$ is simply the regular derivative:

$$\frac{\delta F\left[f(x)\right]}{\delta f(x)} = \frac{\partial F\left[f(x)\right]}{\partial f(x)} = F'\left[f(x)\right] \tag{5.22}$$

Therefore, we can apply the usual product and chain rules such as

$$\frac{\delta E}{\delta \phi_i^*(\mathbf{r})} = \frac{\delta E}{\delta \rho(\mathbf{r})} \frac{\delta \rho(\mathbf{r})}{\delta \phi_i^*(\mathbf{r})} = \frac{\delta E}{\delta \rho(\mathbf{r})} 2\phi_i(\mathbf{r}) \tag{5.23}$$

The actual energy functional must have some degree of a nonlocal nature since it depends on electron density and electron position simultaneously. Especially, the XC functional must be nonlocal since it has no specific coordinate dependence and is involved in all changes throughout the whole system. However, in solids, functionals are often conveniently assumed to be local or semilocal, after which we can express them as a simple bilinear integral form like the external energy functional:

$$E_{\text{ext}}[\rho(\mathbf{r})] = \int U_{\text{ext}}[\rho(\mathbf{r})]\rho(\mathbf{r}) d\mathbf{r} \tag{5.24}$$

Here, the locality of a functional is roughly defined as the extent of the largest energy contribution in the integration. If the shape of a functional looks like the Eiffel tower, meaning that it pops up (positively or negatively) from a rather small part of a system and does not depend upon the gradients, we can safely assume it is local. Then, taking its functional derivatives and obtaining the potential term are rather straightforward as shown in Equation 5.21. Since the differential and integral signs may be interchanged, the functional derivative of the external energy, for example, is

$$\delta E_{ext}[\rho(\boldsymbol{r})] = \int \frac{\delta E_{ext}[\rho(\boldsymbol{r})]}{\delta\rho(\boldsymbol{r})}\delta\rho(\boldsymbol{r})d\boldsymbol{r} = \int \frac{\partial E_{ext}[\rho(\boldsymbol{r})]}{\partial\rho(\boldsymbol{r})}\delta\rho(\boldsymbol{r})d\boldsymbol{r} \tag{5.25}$$

That will lead to

$$\frac{\partial E_{ext}[\rho(\boldsymbol{r})]}{\partial\rho(\boldsymbol{r})} = U_{ext}[\rho(\boldsymbol{r})] \tag{5.26}$$

Thus, the corresponding potential expressions for energy functionals can be obtained by taking the functional derivatives. In the following section, we will use this relationship to derive the KS equations.

5.3.3 *Kohn–Sham equations*

In Chapter 4, by solving the Schrödinger wave equation, we obtained the energy expression as an expectation value of $\hat{H}$:

$$E = \int \Psi^* \hat{H} \Psi d\boldsymbol{r} \tag{5.27}$$

Further, the wave functions are subject to the orthonormal constraint:

$$\int \Psi_i^*(\boldsymbol{r})\Psi_j(\boldsymbol{r})d\boldsymbol{r} = 1 \quad \text{at} \quad i = j, 0 \text{ at } i \neq j \tag{5.28}$$

If we apply the variational principle with a Lagrange multiplier λ with normality constraint on the energy equation, the energy equation returns to the original Schrödinger wave equation:

$$0 = \delta\left(\int \Psi^* \hat{H}\Psi d\boldsymbol{r}\right) = \delta\left[\int \Psi^* \hat{H}\Psi d\boldsymbol{r} - \lambda\left(\int \Psi_i^*\Psi_i d\boldsymbol{r} - 1\right)\right] \rightarrow \hat{H}\Psi = E\Psi \tag{5.29}$$

We now use the same procedure to obtain the KS equations, starting with the energy functionals of DFT. At the minimum energy, the variation of energy functional is zero with respect to wave function or electron density

$$0 = \frac{\delta}{\delta\rho(\boldsymbol{r})}\left(E[\rho(\boldsymbol{r})] - \lambda\left[\int \rho(\boldsymbol{r})\right]d\boldsymbol{r}\right) \rightarrow \frac{\delta E[\rho(\boldsymbol{r})]}{\delta\rho(\boldsymbol{r})} = \lambda \quad \text{or}$$
$$0 = \frac{\delta E[\rho(\boldsymbol{r})]}{\delta\phi_i^*(\boldsymbol{r})} = \frac{\delta}{\delta\phi_i^*(\boldsymbol{r})}\left(E[\rho(\boldsymbol{r})] - \sum_{ij}\lambda_{ij}\left[\int \phi_i^*(\boldsymbol{r})\phi_j(\boldsymbol{r})d\boldsymbol{r}\right]\right) \tag{5.30}$$

Here, the Lagrange multiplier λ_{ij} ensures the orthonormality constraint for orbitals. By rearranging and substituting potentials for the corresponding energy functionals, we now have

$$
\begin{aligned}
0 &= \frac{\delta E_{\text{kin}}^{\text{non}}}{\delta \phi_i^*(\boldsymbol{r})} + \left[\frac{\delta E_{\text{ext}}}{\delta \rho(\boldsymbol{r})} + \frac{\delta E_H}{\delta \rho(\boldsymbol{r})} + \frac{\delta E_{xc}}{\delta \rho(\boldsymbol{r})} \right] \frac{\delta \rho(\boldsymbol{r})}{\delta \phi_i^*(\boldsymbol{r})} - \sum_j \lambda_{ij} \phi_j(r) \\
&= \left(-\frac{1}{2} \nabla^2 + U_{\text{ext}} + U_H + U_{xc} - \lambda_i \right) \phi_i(\boldsymbol{r}) \\
&= \left(-\frac{1}{2} \nabla^2 + U_{\text{eff}} - \varepsilon_i \right) \phi_i(\boldsymbol{r})
\end{aligned}
\tag{5.31}
$$

To derive the above KS equations, the following equalities were used:

$$
\frac{\delta E_{\text{kin}}^{\text{non}}}{\delta \phi_i^*(\boldsymbol{r})} = -\nabla^2 \phi_i(\boldsymbol{r}), \ \frac{\delta \rho(\boldsymbol{r})}{\delta \phi_i^*(\boldsymbol{r})} = 2\phi_i(\boldsymbol{r}), \ U_{xc}[\rho(\boldsymbol{r})] = \frac{\delta E_{xc}[\rho(\boldsymbol{r})]}{\delta \rho(\boldsymbol{r})} \tag{5.32}
$$

The Lagrange multiplier λ_{ij} is identified as energies of each orbital by diagonalization with unitary transformation of $\phi_i(\boldsymbol{r})$. Or we can just replace it with ε_i intuitively because the last line of Equation 5.31 can fit into a wave equation only if $\lambda_{ij} = \varepsilon_i$. The three energy functionals (E_{ext}, E_H, and E_{xc}) are now transformed to their corresponding potentials by these derivative operations, and this leads to a set of coupled KS equations in the form of the familiar Schrödinger equation with the corresponding KS Hamiltonian, $\hat{H}_{\text{KS}}$:

$$
\left[-\frac{1}{2} \nabla^2 + U_{\text{eff}}(\boldsymbol{r}) \right] \phi_i(\boldsymbol{r}) = \varepsilon_i \phi_i(\boldsymbol{r}) \rightarrow \hat{H}_{\text{KS}} \phi_i(\boldsymbol{r}) = \varepsilon_i \phi_i(\boldsymbol{r}) \tag{5.33}
$$

With this equation, the density-governing paradigm of the DFT is completed. For a given system, therefore, actual calculations are performed on this auxiliary Hamiltonian, $\hat{H}_{\text{KS}}$, and it is required to solve this set of KS equations simultaneously. Note that the KS Hamiltonian depends only on $\boldsymbol{r}$, and not on the index of the electron.

Over the years, some doubts about the fictional nature of the KS formalism have been cleared up by actual data that were highly accurate compared to the corresponding experimental values. It appears that the trick of one-electron reformulation is in fact an excellent approximation of the real n-electron world. Note that, assuming the XC functional is exactly known, the electron density and the total energy will be exact by construction, but the KS orbitals and its energies only correspond to a fictitious set of independent electrons.

5.3.3.1 KS orbitals

The orbitals, $\phi_i(\boldsymbol{r})$, are the solutions of the KS equations and are now named specifically as the KS orbitals. For a single atom with n electrons, for example, a set of $n/2$ KS orbitals will result in a solution if we consider all filled-orbitals with two-electrons. During the solution process, they are constructed from electron density and thus do not have the same interpretation as the HF orbitals. However, they calculate the noninteracting kinetic energy exactly and generate the electron shell structure of atoms, all in the framework of DFT.

The antisymmetry is ensured by placing the KS orbitals in a Slater determinant, and the swapping of any pair of rows or columns simply causes the determinant to switch the signs. Note that, under the noninteracting condition, we only need a single Slater determinant of orbitals. The ground-state orbitals, referring to the variational procedure we went through, are those that minimize the total energy while complying with the orthonormality constraints. Contrary to the HF method, however, the KS formalism calculates ϕ exactly and approximates $\hat{H}$, thus not offering any systematical way of improving accuracy.

At this point, it is appropriate to review the concept of orthogonality one more time. It is one of the postulates enforced on orbitals (wave functions or eigenfunctions) in a quantum system. It states that orbitals should be mutually exclusive (independent and unique) without any overlapping nature.

$$\int \phi_i^*(\boldsymbol{r})\phi_j(\boldsymbol{r})d\boldsymbol{r} = 0 \text{ if } i \neq j \tag{5.34}$$

As humans have their own unique characteristics, orbitals (wave functions) have their own identities in nature. Otherwise, they do not fit into the wave equation and do not exist in the quantum world.

5.3.3.2 KS eigenvalues

Eigenvalues, ε_i, could be considered as Lagrange multipliers with no particular physical meaning. Formally, however, ε_i is equal to $\partial E/\partial\rho(\boldsymbol{r})$, which is equivalent to the chemical potential, and the highest occupied state is equivalent to the exact ionization potential. Furthermore, referring to the accumulated data through the years, they are proven to be much more than Lagrange multipliers; they represent energies of each KS electron their sum becomes the total energy after the double counted errors are removed and qualitatively describe band structure, density of states, bonding character, and so on. One remark is that, by calculating KS eigenvalues one by one in terms of electron density, the double counting error is involved in energy terms. The sum of KS eigenvalues thus becomes the actual total energy after this error is removed.

5.4 Exchange-correlation functionals

We now have a practical method for solving the electronic ground-state problem and, if electron density is given, any desired property can be calculated via the KS equations. This has become possible owing to the KS one-electron model. Now, we will face the consequences of adopting this fictitious model instead of the true *n*-electron model in an evaluation of the unknown XC energy.

In fact, the KS approach to the unknown XC energy seems to be no different from an approach taken by a person when faced with a difficult task. When taking an examination, for example, we usually solve the easy problems first and save the hard ones for later. If the hard ones remain unanswered near the end of the exam, we just make our best guess and hope it is the correct answer or close to it. This is the approach Kohn and Sham took when dealing with the unknown XC energy that alone carries the entire burden of the true *n*-electron effects.

In Chapter 2, where we dealt with MD, we were very much concerned about how to generate good interatomic potentials since they determine the accuracy of an MD run. In the DFT, we face a similar situation: how to construct or choose good XC functionals as they determine how much error will be involved in a DFT calculation. Furthermore, if an XC functional is improperly prepared, the connection between the true *n*- and one-electron systems will be completely lost.

In the DFT formulation, all terms are exact, with a sound basis in quantum mechanics, except for the XC energy. In this energy, the troublesome and the unknown terms are cast. Normally, this energy is less than approximately 10% of the total energy, but it actively involves determining materials properties, such as bonding, spin-polarization, and band gap formation. As the name indicates, the XC energy represents the lively activities of electrons among one another. Thus, we have to approximate this energy as exactly as possible. The quality of a DFT run is critically determined by how closely the approximate XC energy reproduces the exact value.

In this section, we will review how the various XC functionals have evolved, as well as evaluate their advantages and limitations. It is not unusual that manuscripts involved with XC functionals have more equations and mathematical symbols than usual texts. The topic of XC functionals is certainly not one that a materials science student spends too much time on. As many devoted scientists continuously work for better XC functionals, our role is to utilize them smartly in our calculations. Therefore, the content of this section is limited within the scope of general and essential topics. For an in-depth understanding of the subject, refer to the excellent reviews and comparative studies on various XC functionals (Paier et al. 2005, Marsman et al. 2008, Haas et al. 2010, Yang et al. 2010).

5.4.1 *Exchange-correlation hole*

The XC energy can be viewed easily with the introduction of the XC hole in an electronic system shown as a rough sketch in Figure 5.10 for two

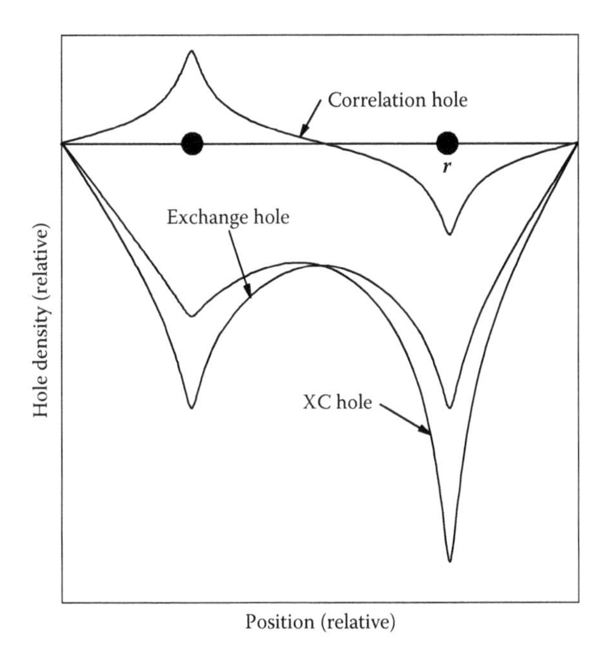

Figure 5.10 Schematic illustration of the exchange and correlation holes forming the XC hole.

electrons at equilibrium distance (Baerends and Gritsenko 2005). Note that the reference is the right electron at $\mathbf{r}$. We expect that the presence of an electron at $\mathbf{r}$ discourages the approach of another electron around it. As a result, there is an effective depletion of electron density, namely, the XC hole, which has two components: exchange and correlation. This concept of the XC hole not only helps us to understand the XC energy but also provides guidelines and rules for the approximation and construction of the largely unknown XC functionals. In addition, as $F[\rho(\mathbf{r})]$ is universal and works for all electronic systems, we may start with a highly simple system, treat it analytically, and extend it to the actual system. Thus, DFT is in principle exact, although it relies in practice on approximations.

5.4.1.1 Exchange hole

The antisymmetry of orbitals requires electrons with the same spin to occupy distinct orthogonal orbitals, and thus forces a spatial separation between those electrons. This reduced electron density is called the *exchange hole* (see Figure 5.10), which will lead to less repulsion with the lowering of the system energy. Because electrons move away from one another, it effectively removes the self-interaction included in the Hartree energy. The exchange energy, therefore, represents the interaction between an exchange hole and electron density over some distance.

5.4.1.2 Correlation hole

Two electrons with different spins can occupy the same orbital, but they avoid each other because of their same negative charges. This electronic correlation also creates a reduced electron density around the electron, thus generating a small attractive energy. We view this effect as a *correlation hole* (see Figure 5.10), and together with the exchange hole, it forms the *XC hole*. Unlike the exchange hole, the correlation hole forms both positive and negative directions because it comes from a correlation between two electrons with opposite spins. Consequently, it does not contribute to the total size of the XC hole but instead changes its shape.

5.4.1.3 Exchange-correlation hole

The XC hole is shared mainly by the exchange part at high electron densities because its origin is rooted in the Pauli exclusion principle that becomes more prominent when electrons are closer to each other. At lower electron densities, however, the correlation part becomes relatively important and comparable with the exchange part. Because most parts of the kinetic energy and the long-range Hartree energy are treated separately, the remaining XC energy can be reasonably assumed to be local or semilocal functionals of electron density and, in addition, the shape of the XC hole is conveniently assumed to be spherical in three-dimensions.

The local XC energy per electron is then the electrostatic interaction energy of an electron at $\boldsymbol{r}$ with XC hole density at $\boldsymbol{r}'$:

$$\varepsilon_{xc}[\rho(\boldsymbol{r})] = \frac{1}{2}\int \frac{\rho_{xc}^{\text{hole}}(\boldsymbol{r},\boldsymbol{r}')}{|\boldsymbol{r}-\boldsymbol{r}'|}d\boldsymbol{r}' \quad (5.35)$$

Then the XC energy functional is the integral over the whole space of the density multiplied by the local XC energy per electron:

$$\begin{aligned} E_{xc}[\rho(\boldsymbol{r})] &= \int \rho(\boldsymbol{r})\varepsilon_{xc}(\boldsymbol{r},\rho)d\boldsymbol{r} \\ &= \frac{1}{2}\iint \frac{\rho(\boldsymbol{r})\rho_{xc}^{\text{hole}}(\boldsymbol{r},\boldsymbol{r}')}{|\boldsymbol{r}-\boldsymbol{r}'|}d\boldsymbol{r}d\boldsymbol{r}' \end{aligned} \quad (5.36)$$

Thus, the treatment of the XC hole is basically the same as the Hartree interaction.

The total amount of the XC hole is subject to the sum rule, which equals exactly one electron, as expected:

$$\int \rho_{xc}^{\text{hole}}(\boldsymbol{r},\boldsymbol{r}')d\boldsymbol{r}' = -1 \quad (5.37)$$

Therefore, a deep exchange hole will be highly localized, and vice versa. However, referring to Figure 5.10, it is apparent that the sum of the correlation hole is zero.

In the Hartree method, the formation of the XC hole is completely ignored, whereas in the HF method, only the exchange hole is fully considered. In the DFT method, both holes are fully accounted for in the XC functional but only in the approximated formulations. The important point is that, by assuming the XC to be potential local or semilocal during the approximation process, the calculation becomes much easier compared to the nonlocal HF approach. The concept of the XC hole, the sum rule, and others will provide guidelines for the construction of XC functionals. Many functionals were reported, possibly more than a hundred, with various accuracies and computational costs. The three groups that are most popular and generally used are functionals of the local density approximation (LDA), the generalized gradient approximation (GGA), and the hybrids.

5.4.2 Local density approximation

When considering approximations for the XC energy, one simple way of accounting for the varying electron densities in a system is by assuming that electrons see the overall landscape in the same way as they see locally. In many cases, this simple picture of the localized XC hole is not a bad assumption to begin with. Then, the complex system can be transformed into many pieces of uniform electron density with different values. It is now possible to calculate the XC energy for each electron with the electron density that is assumed to be constant in that piece. And the energies associated with these local elements can be summed up to equal the total XC energy. This is the local density approximation (LDA), as schematically shown in Figure 5.11.

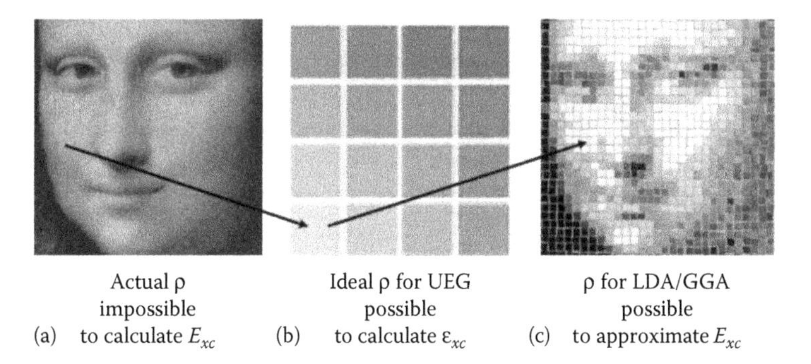

Figure 5.11 Schematic illustration of the actual (a) and approximated XC energy (c) from using a uniform electron gas (UEG) systems (b) as references.

5.4.2.1 Homogeneous electron gas

In a homogeneous electron gas system, electrons are distributed evenly with a uniform positive external potential, and the overall charge neutrality is preserved. In this highly simplified system, all energy terms can be identified conveniently and rather accurately, using the quantum Monte Carlo simulations (Ceperley and Alder 1980). The exchange energies per electron with the electron density, $\rho(\boldsymbol{r})$, can be specified in a simple analytical form:

$$\varepsilon_x^{\mathrm{hom}}(\rho) = -C\rho^{1/3}(\boldsymbol{r}) \tag{5.38}$$

The LDA adopts this simple formula for the construction of the XC functional.

5.4.2.2 Exchange energy

The exact exchange energy can be calculated using the HF method, but it is computationally more expensive to evaluate than Hartree or kinetic energies. It is, therefore, a common practice to approximate exchange energy along with correlation energy, assuming the homogeneous electron gas. Its energy and potential functionals are

$$E_x^{\mathrm{LDA}}\left[\rho(\boldsymbol{r})\right] = -C\int \rho^{4/3}(\boldsymbol{r})d\boldsymbol{r} \tag{5.39}$$

$$U_x^{\mathrm{LDA}}\left[\rho(\boldsymbol{r})\right] = -C\varepsilon_x^{\mathrm{hom}}\left[\rho(\boldsymbol{r})\right] = -C\rho^{1/3}(\boldsymbol{r}) \tag{5.40}$$

where C represents different constants. Remember that, with this approximated potential, the self-interaction portion of the Hartree energy will be largely but not completely removed. And this incomplete removal of the self-interaction will cause some problems and errors such as the underestimation of band gaps (see Section 5.4.5).

5.4.2.3 Correlation energy

Along with the same quantum Monte Carlo simulation, the exact correlation energy for the homogeneous electron gas was calculated with varying electron density, point by point. However, it is difficult to express it in an analytic form even for the uniform electron gas. Here is an example of the correlation energy for a simple homogeneous electron gas at the high-density limit (at $r_s < 1$; Perdew and Zunger 1981):

$$E_c^{\mathrm{LDA}} = C_1 + C_2 \ln r_s + r_s(C_3 + C_4 \ln r_s) \tag{5.41}$$

where:

All C_i are constants

r_s is the Wigner–Seitz radius related to electron density as $\left[3/4\pi\rho(\boldsymbol{r})\right]^{1/3}$

It is the radius of a sphere containing exactly one electron, so that the larger the r_s, the lower the electron density. Based on these data, the unknown correlation energy of the real system is approximated by interpolation. Fortunately, however, the contribution of correlation energy is typically much smaller than that of exchange energy.

5.4.2.4 XC energy

As both exchange and correlation energies are functionals of electron density, we expect that the summed XC energy will also be functionals of electron density:

$$E_{xc}^{\text{LDA}}\left[\rho(\boldsymbol{r})\right] = E_x^{\text{LDA}}\left[\rho(\boldsymbol{r})\right] + E_c^{\text{LDA}}\left[\rho(\boldsymbol{r})\right] \tag{5.42}$$

The LDA becomes exact only in the limit of homogeneous electron gas or very slowly varying densities. In practical calculations, however, the known numerical results from the above quantum Monte Carlo calculations are interpolated, and the values of $\varepsilon_{xc}^{\text{hom}}\left[\rho(\boldsymbol{r})\right]$ per volume of constant ρ for all densities are parameterized. Then, the XC energy of the LDA can be calculated by multiplying $\varepsilon_{xc}^{\text{hom}}\left[\rho(\boldsymbol{r})\right]$ by the local electron density and integrating it over the space:

$$\begin{aligned} E_{xc}^{\text{LDA}}\left[\rho(\boldsymbol{r})\right] &= \int \rho(\boldsymbol{r})\varepsilon_{xc}^{\text{hom}}\left[\rho(\boldsymbol{r})\right]d\boldsymbol{r} \\ &= \int \rho(\boldsymbol{r})\left[\varepsilon_x^{\text{hom}}\left[\rho(\boldsymbol{r})\right] + \varepsilon_c^{\text{hom}}\left[\rho(\boldsymbol{r})\right]\right]d\boldsymbol{r} \end{aligned} \tag{5.43}$$

By taking the derivative of the energy functional, the corresponding XC potential of the LDA is

$$U_{xc}^{\text{LDA}}[\rho(\boldsymbol{r})] = \frac{\delta E_{xc}^{\text{LDA}}}{\delta\rho(\boldsymbol{r})} \tag{5.44}$$

The LDA has been extended to handle spin-polarized systems as the local spin-polarized density approximation (LSDA):

$$E_{xc}^{\text{LSDA}}[\rho_\uparrow(\boldsymbol{r}),\rho_\downarrow(\boldsymbol{r})] = \int \rho(\boldsymbol{r})\varepsilon_{xc}^{\text{hom}}[\rho_\uparrow(\boldsymbol{r}),\rho_\downarrow(\boldsymbol{r})]d\boldsymbol{r} \tag{5.45}$$

The LDA/LSDA description for XC energy in a parameterized format such as the functionals by Perdew and Zunger (1981) and Perdew and Wang (1992) has been one of the most commonly used functionals in the implementations of the DFT. However, they have been becoming less popular and are often considered old-fashioned since the advent of GGA, which generally describes the XC energy more accurately.

5.4.2.5 Remarks

Considering that actual systems are far from the homogeneous electron gas, the LDA works fairly well where the charge density varies relatively slowly, such as in covalent systems and simple metals. Its success was a surprise even to Kohn and Sham, who first proposed it. This is partly due to the cancellation of errors where LDA typically overestimates E_x while it underestimates E_c. This implies that the LDA satisfies the sum rule of the XC hole rather well, although each exchange or correlation hole does not. In other words, the shape of an actual XC hole is different from that of the LDA, but its spherically averaged value required for the energy calculation is well represented.

However, whenever the situation deviates from the LDA model, it causes some errors and problems. The actual XC hole may have local variations such as degenerated energies and long-range tails (for example, in the case of metal surface). Also, the XC potential generally does not accurately follow the asymptotic behavior ($-1/\mathbf{r}$ dependence) of the actual potential. The XC potential becomes too shallow near the nucleus and decays too quickly at a long distance from the nucleus.

Typical drawbacks of LDA functionals include that it

- Overbinds (underestimates the lattice parameters) and thus overestimates the cohesive energy and bulk modulus of solids.
- Calculates the adsorption energies too high and the diffusion barriers too low.
- Underestimates the spin and orbital moments.
- Calculates the band gaps ~50% smaller (e.g., Si) or even no gap (e.g., Ge).
- Cannot describe transition metals or strongly correlated systems (strong electron localization with narrow *d*- and *f*-bands) such as transition metal oxides. It predicts, for example, incorrectly that the ground states of Fe (ferromagnetic) and Cr (antiferromagnetic) are both nonmagnetic.
- Does not work well for materials that involve weak hydrogen bonds or van der Waals attraction (mainly in atoms and molecules).

5.4.3 Generalized gradient approximation

Real systems are evidently not homogeneous and have varying density landscape around electrons. To generate more accurate XC functionals, the generalized gradient approximation (GGA) captures both the local and semilocal information: the electron density and its gradient at a given point. Figure 5.12 illustrates very schematically how LDA and GGA work. In principle, therefore, GGA should return better results with the general formula with density gradient as an additional variable:

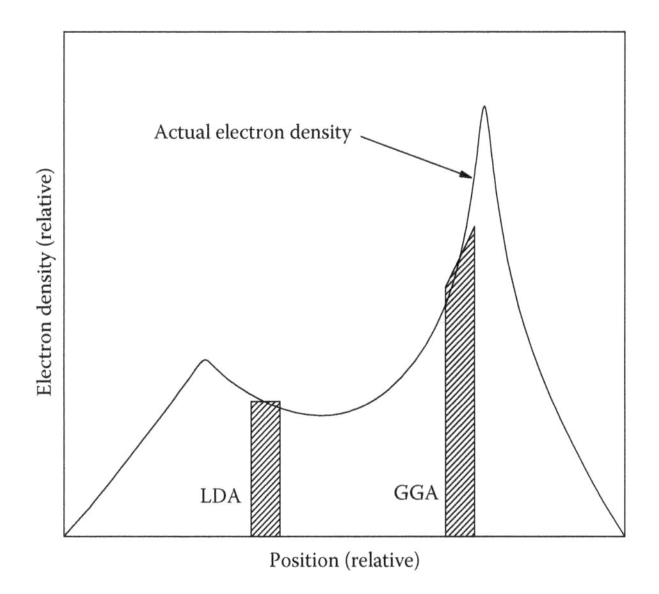

Figure 5.12 Schematic illustration of local and semilocal density approximation by LDA and GGA.

$$E_{xc}^{\mathrm{GGA}}[\rho(\boldsymbol{r})] = \int \rho(\boldsymbol{r})\varepsilon_{xc}^{\mathrm{GGA}}\left[\rho(\boldsymbol{r}), \nabla\rho(\boldsymbol{r})\right]d\boldsymbol{r} \tag{5.46}$$

When the energy functional is in this integral form, the general form of $U_{xc}^{\mathrm{GGA}}[\rho(\boldsymbol{r})]$ is

$$U_{xc}^{\mathrm{GGA}}[\rho(\boldsymbol{r})] = \frac{\delta E_{xc}^{\mathrm{GGA}}}{\delta\rho(\boldsymbol{r})} = \rho(\boldsymbol{r})\frac{d\varepsilon_{xc}^{\mathrm{GGA}}[\rho(\boldsymbol{r})]}{d\rho(\boldsymbol{r})} + \varepsilon_{xc}^{\mathrm{GGA}}[\rho(\boldsymbol{r})] \tag{5.47}$$

If we consider the spin variable, the expression for the energy functional by GGA becomes

$$E_{xc}^{\mathrm{GGA}}[\rho_\uparrow, \rho_\downarrow] = \int \rho(\boldsymbol{r})\varepsilon_{xc}^{\mathrm{GGA}}[\rho_\uparrow, \rho_\downarrow, \nabla\rho_\uparrow, \nabla\rho_\downarrow]d\boldsymbol{r} \tag{5.48}$$

Unlike $E_{xc}^{\mathrm{LDA}}[\rho(\boldsymbol{r})]$, there is no simple functional form that correctly represents the GGA data, and thus the function $E_{xc}^{\mathrm{GGA}}[\rho(\boldsymbol{r})]$ is often expressed with a properly chosen form and is fitted to satisfy various physical constraints. Therefore, the general form of GGA in practice is expressed based on the LDA with an additional enhancement factor $F(s)$ that directly modifies the LDA energy:

$$E_{xc}^{\mathrm{GGA}}[\rho(\boldsymbol{r}), s] = \int \varepsilon_{xc}^{\mathrm{LDA}}[\rho(\boldsymbol{r})]\rho(\boldsymbol{r})F(s)d\boldsymbol{r} \tag{5.49}$$

Here, the s depends on both electron density and its gradient:

$$s = C\frac{|\nabla\rho(\boldsymbol{r})|}{\rho^{4/3}(\boldsymbol{r})} \tag{5.50}$$

In solids, the typical range of s is = 0–3. In this range, the enhancement factors, $F(s)$, of various GGAs for the exchange typically vary from 1.0 to 1.6. Many different forms of GGA have been proposed, and the popular choices are PW91 (Perdew and Wang 1992, Perdew et al. 1992) and PBE (Perdew et al. 1996). These GGA or LDA functionals are normally well tabulated and incorporated with the pseudo-potential files.

5.4.3.1 PW91

PW91 has been very popular in applications for a wide range of materials due to its reasonable accuracy and general applicability. It is firmly based on the known physics and constraints of the XC hole using the data from the uniform electron gas and thus keeps a nonempirical nature. It, however, has a tendency to give spurious wiggles in the XC potential both at high and low electron densities.

5.4.3.2 PBE

Based on PW91, PBE presents a simplified and improved GGA version with no empirical elements. It includes features such as local electron density and its gradient and second-order gradient in the enhancement factors, $F(x)$ and $F(c)$. This functional has been proven to be highly accurate and computationally efficient, and is now the most frequently used functional. PBE does not show the spurious wiggles in the XC potential found for PW91, and thus is better off with pseudopotentials. Table 5.1 shows a typical set of data (Paier et al. 2006) calculated with the PAW/PBE method in comparison with experimental data (Heyd and Scuseria 2004).

A revised version of PBE (Zhang and Yang 1998) is also available, namely, revPBE. It differs only slightly from the PBE functional as a simpler mathematical form for ε_x. This functional is known to give better results when adsorption or hydrogen bonding is involved at no additional computational cost. Another version, RPBE (Hammer et al. 1999), is almost the same as revPBE, but it satisfies the Lieb–Oxford criterion (Lieb and Oxford 1981), which defines a theoretical bound for the maximum value of the E_{xc}. It also reduces adsorption energies, but at the same time has a tendency to have higher energy barriers. It also has a tendency to calculate the total energy several percentages higher than that of PBE.

Table 5.1 Lattice constants and bulk moduli by PAW/PBE calculations and their comparison with experimental data

Materials	Lattice constants (Å) (cal./exp.)	Bulk moduli (GPa) (cal./exp.)
Al	4.040/4.032	77/79
BN	3.626/3.616	370/400
C	3.574/3.567	431/443
Si	5.469/5.430	88/99
SiC	4.380/4.358	210/225
β-GaN	4.546/4.520	169/210
GaP	5.506/5.451	75/89
GaAs	5.752/5.648	60/76
MgO	4.258/4.207	149/165
Cu	3.635/3.603	136/142
Rh	3.830/3.798	254/269
Pd	3.943/3.881	166/195
Ag	4.147/4.069	89/109

Source: Paier, J. et al., *J. Chem. Phys.*, 124(15), 154709–154721, 2006.

5.4.4 Other XC functionals

If we can afford at least a 10-fold increase in computational cost, we may go for one of the advanced functionals beyond common GGAs. Advance functionals (PBE0, meta-GGA, hyper-GGA, HSE, hybrid functional such as B3LYP, etc.) claim better accuracies, reporting successive improvements in energetics. They incorporate with additional variables (for example, higher-order density gradient) or mixing a certain amount of nonlocal HF exchange energy.

The hybrid functionals are, as the name suggests, GGA-type functionals combined with some (~25%) of the accurate exchange energy from the HF method. Remember that the HF method does not account for the correlation energy that causes spatially closer electrons, smaller bond length, larger binding energy, wider band gap, and thus higher energy. In the DFT, on the other hand, systematic errors occur in the opposite way. Thus, it is expected that the DFT exchange energies can reduce the errors if one introduces the HF exchange energy into E_{xc}, typically in the form of

$$E_{xc} = CE_x^{\text{HF}} + (1-C)E_x^{\text{GGA}} + E_c^{\text{GGA}} \tag{5.51}$$

The B3LYP functional has been developed and upgraded by a group of scientists for many years (Stephens et al. 1994). It is a three-parameter functional fitted to atomization energies, ionization potentials, and so on. It is now very popular, especially in the field of molecular chemistry.

Table 5.2 Typical XC functionals commonly used in DFT calculations

Classification	Examples	Remarks
Local	LDA	$\rho_\uparrow, \rho_\downarrow$
Semilocal	GGA	$\rho_\uparrow, \rho_\downarrow, \nabla\rho_\uparrow, \nabla\rho_\downarrow$
Seminonlocal	Meta-GGA	$\rho_\uparrow, \rho_\downarrow, \nabla\rho_\uparrow, \nabla\rho_\downarrow, \nabla^2\rho_\uparrow, \nabla^2\rho_\downarrow$, etc.
Hybrid	B3LYP	GGA + HF

This hybrid functional can describe systems even with rapid variations in electron density or with long-range interaction such as the van der Waals type. It is generally best suited for calculations of bond energies, chemical transition-state barriers, band gap, and so on.

These advanced functionals are often contaminated with empirical elements, losing the first-principles origin. And the claimed improvement could be a system-dependent result. In addition, since so many XC functionals are available, it is sometimes a concern that someone may choose the XC functional that gives the result that one wants to have. One should not just jump into a new functional even if it produces more accurate numbers for a system, unless all aspects are carefully considered and justified. Table 5.2 summarizes the typical XC functionals in DFT in terms of each's locality and considered variables. When it goes toward nonlocality and more added variables, the functional is supposed to give better accuracy at the cost of more computational load.

5.4.5 *Remarks*

This subsection will be rather lengthy and not very entertaining at all, especially for newcomers to the DFT. Therefore, it could be skipped and just saved for later reference. Let us summarize the general trends and limitations of GGA and review some critical issues about its use in practice. The hybrid XC functionals are excluded from this discussion since they are generally impractical in materials calculations.

5.4.5.1 *General trends of GGA*

- GGA works very well with almost all systems, giving most structural properties within a 1%–3% error.
- GGA corrects most of the overbinding problems of LDA, producing an increase in lattice constants and a decrease in cohesive energies, and improving activation barriers. In fact, GGA slightly overcorrects bond length, resulting in adsorption energies on the lower side and diffusion barriers on the higher side.
- GGA also calculates the band gaps approximately 50% smaller, or even with no gap (see the following).

5.4.5.2 Limitations of GGA: Strongly correlated systems

Even with GGA, many features of the actual XC hole can easily be missed. When a system has strongly localized orbitals, an extra repulsive interaction takes place between two electrons on that site that was first recognized by Hubbard (1965). Such cases happen in rare-earth elements and compounds, as well as in transition metal oxides on their narrow and strongly correlated d- and f-orbitals. In these cases, the electron is transferred from the delocalized state to the localized state and becomes more prone to unphysical self-interaction.

This additional onsite repulsive energy U (the Coulomb energy needed to place two electrons at the same site, typically 4–5 eV) is treated by an HF-like approach and added to the LDA/GGA KS Hamiltonian. Normally, the U is fitted to the experimental data, and this additional orbital-dependent interaction will shift the localized orbitals. With the use of a U-approach, an improved description has been reported of band gaps, magnetic moments, and bulk and surface structure, adsorption, and so on.

5.4.5.3 Limitations of GGA: Band gap underestimation

The DFT is inaccurate in the calculation of the band gap, which is not surprising as the method is not designed to describe the non-local nature of electron correctly. The band gap underestimation commonly reaches up to 50% when compared with experimental data. Primarily, it is inherent in the noninteracting one-electron KS scheme because the KS energies are not represented as removal/addition energy for the electron, which defines the true band gaps as $E_g = \varepsilon_{n+1}(n+1) - \varepsilon_n(n)$. The KS version of a band gap is, rather, $E_g^{KS} = \varepsilon_{n+1}(n) - \varepsilon_n(n)$ with the same number of electrons. Remember that the XC potential constructed by the LDA or GGA functional changes continuously with respect to electron density, which is contrary to the discontinuous behavior by the actual XC potential.

We can also blame this problem on another source: the partial removal of self-interaction in the DFT. In an HF calculation, the self-interaction in the Hartree term is exactly canceled by the exchange energy by setting the wave function in a Slater determinant. In fact, HF calculations normally give slightly bigger band gaps. On the contrary, in the DFT, it is not completely removed due to the approximated nature of the XC energy.

The incomplete removal of self-interaction causes the band positions to be dependent on occupation numbers (0–1) and the valence band to be pushed up at $f_i = 1$. In contrast, the conduction band position will be pulled down at $f_i = 0$, and the gap will narrow down accordingly. A remedy, not a fundamental solution, for the band gap problem includes

- The use of hybrid functionals gives a better band gap because the mixing of HF exchange energies certainly provides a better cancelation of self-interaction. All these functionals, however, require

much higher computational cost and thus may not be suitable for materials studies.
- The use of Green function techniques and GW approximation (expansion of the self-energy in terms of the single-particle Green function, *G*, and the screened interaction, *W*) based on the many-body perturbation theory which corrects the self-interaction and the KS eigenvalues, thus giving a better band gap. Note that energy expressed in terms of the Green function contains much more information, becomes more precise, and increases computational demand.
- The use of other self-interaction correction (SIC) methods.

The remedies described here work for both the band gap problem and strongly correlated systems in both the LDA and GGA approaches.

5.5 Solving Kohn–Sham equations

If we write out all the descriptive words for the KS equations, the list will be rather lengthy: ground-state, nonlinear and coupled, self-consistent, noninteracting one-electron, easier-to-solve, and Schrödinger-like wave equations. Indeed, we have a viable method at hand to access any n-electron system by the equations. We just pick a well-tested XC functional and treat the system as if it were made of robot-like noninteracting electrons. In this section, we will review various methods implemented for the actual computations in solving the KS equations. However, the treatment will be brief since the subject will be further discussed in the next chapter in relation to bulk solids.

5.5.1 Introduction

There are three keywords that must be followed in order to solve the KS equations: self-consistency, variational principle, and constraints. In other words, minimizing the energy can be achieved by finding a self-consistent solution to a set of one-electron KS equations whose orbitals are subject to constraints of the orthonormality or fixed number of electrons.

5.5.1.1 Self-consistency

In the DFT scheme, KS orbitals, electron densities, and Hamiltonian are all interrelated during the course of calculation: the KS orbitals calculate the electron densities, the electron densities calculate the KS Hamiltonian, the KS Hamiltonian calculates the new electron densities (and the new KS orbitals), and so on. Therefore, to achieve meaningful results, we have to find a set of KS orbitals that leads to a KS Hamiltonian whose solutions are the KS orbitals we started with. This is called *self-consistency.*

5.5.1.2 Variational principle

Assuming we have a proper XC functional in terms of electron density, the expression for the total energy consists of four terms that are all unique functionals of electron density at the ground state, except the noninteracting kinetic energy:

$$E = -\frac{1}{2}\sum_{i}^{n}\int \phi_i^*(\boldsymbol{r})\nabla^2\phi_i(\boldsymbol{r})d\boldsymbol{r} + \int U_{\text{ext}}(\boldsymbol{r})\rho(\boldsymbol{r})d\boldsymbol{r}$$
$$+\frac{1}{2}\iint \frac{\rho(\boldsymbol{r})\rho(\boldsymbol{r}')}{|\boldsymbol{r}-\boldsymbol{r}'|}d\boldsymbol{r}d\boldsymbol{r}' + E_{xc}[\rho(\boldsymbol{r})] \tag{5.52}$$

At the ground state, the density minimizes variational energy, and the energy becomes stationary with respect to small changes in density everywhere:

$$\frac{\delta E[\rho(\boldsymbol{r})]}{\delta\rho(\boldsymbol{r})} = 0 \tag{5.53}$$

The variational process for a functional is generally no different from the normal minimization process for functions: finding a minimum at $dE/d\rho = 0$. The change in density $\delta\rho(\boldsymbol{r})$ is restricted so that the total number of electrons remains fixed:

$$\int \delta\rho(\boldsymbol{r})d\boldsymbol{r} = 0 \tag{5.54}$$

Note that the general variational principle for the total energy can be applied with respect to either electron density, KS eigenvalues, or norm of the residual vector to an eigenstate.

5.5.1.3 Constraints

Finding the minimum value of the total energy must be carried out under the constraints of a fixed total number of electrons or orthonormality of orbitals:

$$n = \int \rho(\boldsymbol{r})d\boldsymbol{r} \tag{5.55}$$

$$\int \phi_i^*(\boldsymbol{r})\phi_j(\boldsymbol{r})d\boldsymbol{r} = \delta_{ij} \tag{5.56}$$

where δ_{ij} is the Kronecker delta (0 if $i \neq j$, 1 if $i = j$). Otherwise, the density can be any number, or the energy might get lower than the ground-state energy, which is unacceptable and unphysical.

The KS orbitals built in the Slater determinant are initially unspecified but will improve following the above constraints to be antisymmetric and unique, thus fitting into the quantum world. We can use any of these constraints when we solve the KS equations and keep the process on track. We are now fully briefed about the rules of the game and ready for the search of the ground state of a given quantum system.

5.5.2 Direct diagonalization

The straightforward way to solve the KS equations is by the direct and full diagonalization of the KS Hamiltonian matrix. This method is best suited especially when an atom-centered basis set is used for orbitals. Since the localized basis set needs only a small number of basis, direct diagonalization is relatively easy and efficient. For large systems expanded with a large number of plane waves (see Section 6.4), however, the direct method becomes very inefficient because 10^4–10^5 plane waves may be needed for diagonalization of a typical DFT run to have only the lowest ~$n/2$ eigenvalues. Apparently, this method is not suited for materials calculations.

5.5.3 Iterative diagonalization

For solids and materials, the actual DFT calculation involves two energy minimizations in the series: electronic and ionic minimizations as shown in Figure 5.13. The iterative variational approach (Payne et al. 1992) is

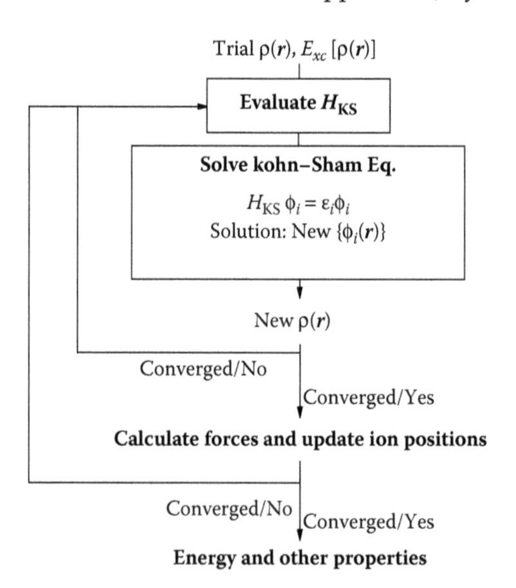

Figure 5.13 Schematic illustration of a typical DFT procedure by iterative self-consistent loop.

known to be the most efficient for electronic minimization, and the classical-mechanics treatment is just sufficient for ionic minimization. As this subject is the mainstream of DFT runs, it will be discussed in more detail in Chapter 6, and only a brief outline for iterative minimization will be presented here.

In order to avoid unnecessary calculations, iterative diagonalization is in general used instead of the direct diagonalization. This method uses only a small number of plane waves (slightly more than $n/2$) to start with and proceeds by iteration:

- Construct the initial electron density of solids approximated by superimposing the electron densities of each atom in its isolated state as provided by the pseudopotential.
- Calculate U_{xc} and other terms to evaluate $\hat{H}_{KS}$.
- Solve a set of coupled KS equations by direct diagonalization in plane-wave-subset and obtain the solutions, a set of KS orbitals. The calculation must cover solutions for all one-electron KS equations (for all occupied $n/2$ levels in case of all filled shells) simultaneously.
- Use the new $\phi_i(\boldsymbol{r})$ and calculate the new $\rho(\boldsymbol{r})$, and with mixing of previous and current densities, repeat the above steps. Then, successive improvements of $\hat{H}_{KS}$ and $\rho(\boldsymbol{r})$ will take place.
- Stop iteration when the energy change (or density change) of the system becomes less than a preset stopping criterion (normally 10^{-4} to 10^{-5} eV).
- Then we declare that self-consistency is reached and the energy calculated is the ground-state energy of the system at the very ground-state electron density.

5.5.3.1 Total energy and other properties

Knowing the electronic ground-state energy, therefore, can lead us to the total energy of a system.

Once the total energy of a system is given (e.g., in terms of lattice parameters), various fundamental properties come out naturally, for example, stable structures, bond lengths and angles, cohesive energies from the energy minima, bulk modulus, elastic constants, surface reconstructions, defect formation energies, pressure-driven phase transitions, vibrational features from the energy curvature.

The first partial derivatives of energy with respect to volume, atom position, and strain (ε_{ij}) give bulk modulus, pressures, forces, and stresses, respectively:

$$B = V\frac{\partial^2 E(V)}{\partial V^2}, P = -\frac{\partial E}{\partial V}, \boldsymbol{F}_I = -\frac{\partial E}{\partial \boldsymbol{r}_I}, \sigma_{ij} = \frac{\partial E}{\partial \varepsilon_{ij}} \tag{5.57}$$

Furthermore, the ground-state calculation serves as a starting point for advanced calculations such as minimum-energy path, barrier energies, density of states (DOS), and band structures. In addition, with further computational efforts, the second derivatives of energy give a force constant matrix (forces acting on all other atoms when each atom is displaced in each direction), STM simulations, phonon spectrums, thermodynamic quantities, reaction rates, and so on.

5.6 DFT extensions and limitations

Quantum mechanics was founded and extended to the DFT by many prominent scientists, including Bohr, Heisenberg, Schrödinger, and Kohn (see Figure 5.14). If we could enter into an atom and observe how the electrons really play, we might come up with a better picture than the DFT scheme. Meanwhile, there is every reason to believe that the DFT is the best tool to comprehend electrons, especially in materials science. It becomes more evident if we once list the remarkable achievements in extensions of the DFT since its inception. We consider in this section how successful the DFT has been, how far it has progressed, and what its limitations are.

5.6.1 DFT extensions

We started this chapter by recognizing two major difficulties in the application of the first-principles methods: the complexities of the n-electron system and the high computational cost with the increasing number of electrons in the system. DFT has succeeded on both issues by achieving favorable scale of $O(N^{2\sim3})$ and comparable accuracy as the HF method. It is now recognized as an indispensable tool in many fields of science, not

Figure 5.14 Scientists who have contributed to the development of DFT.

only in physics, chemistry, and materials science but also in bioscience, geology, soil science, astrophysics, and more.

The original KS method was designed to be valid primarily for local, spin-independent, nondegenerate ground states. The modern KS technique was extended successfully to semilocal, spin-dependent, and degenerate states. Further extension to excited, finite-temperature, relativistic, and even time-dependent states has been progressing steadily. Remember that all these beyond-DFT schemes demand computational resources much more complex than standard DFT methods. The following are some of the extended DFT methods in use and development that are worth mentioning.

5.6.1.1 Spin-polarized DFT

We have neglected the spin degree of freedom up to this point, assuming $\rho_{\uparrow}(\mathbf{r}) = \rho_{\downarrow}(\mathbf{r}) = \rho(\mathbf{r})/2$, which is generally the case in many solids. But, in cases of open-shell systems with an odd number of electrons or magnetic solids, spin-polarization should be considered. The local spin density approximation (LSDA) that immediately followed the LDA is now routinely applied to account for spin density.

The calculation adds a spin variable to electron density and the XC potential so that the energy functionals depend on both the electron density, $\rho(\mathbf{r}) = \rho_{\uparrow}(\mathbf{r}) + \rho_{\downarrow}(\mathbf{r})$ and the spin density, $s(\mathbf{r}) = s_{\uparrow}(\mathbf{r}) - s_{\downarrow}(\mathbf{r})$. Then the calculation considers the total number of electrons as $n = n_{\uparrow} + n_{\downarrow}$ and the XC potential, for example, as $U_{xc}(\mathbf{r}) = U_{\uparrow xc}(\mathbf{r}) + U_{\downarrow xc}(\mathbf{r})$. The spin-polarized DFT with GGA works the same way. Thus, dealing with spin is straightforward only if we accept the doubled calculation time due to the two simultaneous iterative loops for each spin variable.

5.6.1.2 DFT with fractional occupancies

The standard KS formalism can be extended to cases of fractional occupancies. It requires the inclusion of all eigenstates and the assignment of occupancy (0–1) for each orbital. The electron density is now written with this fractional occupancy, f_i, as

$$\rho(\mathbf{r}) = \sum_i f_i \left|\phi_i(\mathbf{r})\right|^2 \tag{5.58}$$

5.6.1.3 DFT for excited states

Extension of DFT to excited states is in principle possible since the diagonalization of the KS Hamiltonian can return an energy spectrum including the full excited states. Therefore, the excited states can also be considered as functionals of the ground-state density under the constraint that their orbitals are orthogonal to orbitals underneath. Further, its properties are often conveniently interpreted in terms of ground-state properties. Current methods to deal with excited states include the many-body perturbation theory within the Green's function approximation, time-dependent DFT, ΔSCF scheme, and so on.

5.6.1.4 Finite-temperature DFT

Most DFT calculations are carried out at the ground state at 0 K, but it is highly desirable to deal with finite-temperature systems that are more realistic in practice. Creating a finite temperature in a system in the framework of DFT is by no means a trivial task. Not like the case of MD which requires a simple velocity scaling or a setup of NVT ensemble, one must think of the energy minimization in terms of free energy of the system and thus deal with the entropy term that cannot be directly obtained from a DFT run. Note also that, at nonzero temperatures, the occupancies of each orbital are no longer discontinuous. In addition, all three types of electronic, ionic, and configurational contributions to temperature should be accounted. The most relevant one in materials is manipulating temperature by structural excitation, commonly carried out by calculating lattice (ionic) vibrations called phonon.

The DFT for the phonon spectra (lattice vibration frequency vs. $\boldsymbol{k}$-vector) can evaluate various properties at elevated temperatures that the conventional DFT cannot. By displacing atoms from the ideal lattice position and calculating the forces on all atoms in a dynamical matrix, various properties can be generated: entropy that leads to free energy, heat capacity, thermal expansion coefficient, phase transition and diagram, prefactor for atomic diffusion, solubility, and so on. Most DFT codes (VASP 5.2, ABINIT, Quantum Espresso, etc.) these days provide efficient ways of calculating the phonon spectrum, for example, by the frozen-phonon technique under harmonic approximation (see Section 7.10).

5.6.1.5 Time-dependent DFT

Standard DFT assumes that electronic relaxation occurs quickly, and thus treats the process time-independently. However, events such as electronic excitation (for example, under an intense laser pulse) require time-dependent treatment. This subject has become a very active field of study recently. In a time-dependent DFT (TDDFT), the system evolves under the presence of time-dependent orbitals and potentials.

Practically, however, TDDFT normally simplifies the problem; the potential changes slowly so that the XC energy of the time-independent can be used and the correspondence between the time-dependent external potential and one-electron density is assumed to be still valid. Thus, the time evolution of a system with noninteracting density and the effective potential is assumed to be equivalent to that of an interacting system. Within this framework, if a system in the ground state is perturbed, a set of time-dependent KS equations is generated, and various properties are calculated.

5.6.1.6 *Linear scaling of DFT*

Linear scaling of DFT is another topic of high interest. Note that the time span covered by today's *ab initio* MD is only in the picosecond (ps) range. Therefore, one needs to be very careful when interpreting the results of *ab initio* MD, statistical sampling of which may be too small.

There are more activities for developing new schemes to perform the DFT calculations, which scales only linearly with the system size, that is, $O(n)$. This $O(n)$ scaling is especially important when the system size is over 1000 atoms. The key to $O(n)$ scaling is to use the nearsightedness of the terms in $\hat{H}$. Therefore, the wave functions are often described in highly localized forms which is why this approach normally applies to insulators or semiconductors.

5.6.2 *DFT limitations*

Knowing the various approximations used in DFT calculations is important for setting up the calculation and interpreting the outcomes. It often happens that the data generated can turn out to be useless by neglecting the sources of errors and the corresponding limitations.

In the first-principles calculations, anything that deviated from the true n-electron picture has potential to cause errors. Here is a list of the approximations often adopted in electronic structure calculations:

- Born–Oppenheimer approximation
- Nonrelativity approximation
- Mean-field approximation
- One-electron DFT approximation
- Single Slater determinant approximation for wave functions
- XC energy approximation

In Chapter 6, when we deal with solids in the framework of the DFT, we will use additional approximations:

- Supercell and PBC (periodic boundary condition) approximations
- PP (pseudopotential) approximation
- Basis expansion and energy cutoff
- $\mathbf{k}$-points sampling, FFT (fast Fourier transformation) grid, smearing, summation replacing integration, numerical truncations, and so on.

All these simplifications are well justified and, in most cases, cause no significant error if we carefully set up the system and run conditions. However, the XC functional (LDA, GGA, or others) is the main source of errors in all DFT calculations, which is not surprising since it contains several approximated energy terms. We already know that the KS scheme is exact only when we have the exact XC energy, but we also know that we may never obtain the exact XC energy. The XC functional describes the general picture of electronic systems in a reasonable degree but cannot collect the fine features of the true landscape in subatomic systems. Here are two main sources of errors coming from the XC functional:

- The XC hole for the generation of an XC functional is normally assumed to be symmetric and spherical centered at electrons although the actual XC hole is most likely asymmetric and centered on nuclei.
- All Coulomb potential, including the XC potential, have a long tail as proportional to $-1/\boldsymbol{r}$ (the so-called asymptotic behavior). The XC functionals decay far too fast and miss this feature. Note that no single functional form can capture both electron density and electron density–dependent potential in an asymptotic curve at the same time unless we divide the potential and treat them separately. This problem, however, mainly occurs for atoms and molecules.

The development of better XC functionals is a long-term task for scientists. Fortunately, however, many materials-related problems require generally standard XC functionals.

Homework

5.1 Find out if the following is a functional or not.
The departing time of an airplane, in terms of the arriving time of the train if the train starts whenever the bus arrives and the airplane departs whenever that train arrives. The number of electrons, n, in Equation 5.2.

5.2 The exchange energy under the LDA can be expressed as a simple integral equation in terms of electron density:

$$E_x^{\mathrm{LDA}}[\rho(\boldsymbol{r})] = A\int \rho(\boldsymbol{r})^{4/3}\,d\boldsymbol{r}$$

Find its functional derivative form, U_x^{LDA} $(= \delta E_x/\delta\rho(\boldsymbol{r}))$.

5.3 Considering the fact that the GGA functionals correct the overbinding problem of the LDA functionals, compare the relative magnitudes of estimation for the following terms and properties between the GGA and the LDA functionals as shown in the examples:
Exchange energy: LDA (under), GGA (proper)
Binding: LDA (over), GGA (slightly under)
Correlation energy: LDA (), GGA (slightly under)
Repulsion between electrons: LDA (), GGA (slightly over)
Cohesive energy and bulk modulus: LDA (), GGA ()

Equilibrium lattice parameter: LDA (), GGA ()
Band gab: LDA (), GGA ()

5.4 It can be proven mathematically that the correlating part of the kinetic energy included in E_{xc} is a positive value. Explain intuitively why it is. And show that, with the example of a plane wave, the kinetic energy cannot simply be calculated from the charge density.

5.5 Hydrogen atom has only one electron and thus has no chance to have any self-interaction. However, in the framework of DFT, it also generates a certain amount of self-interaction energy. Explain why it is.

In general, DFT works much better for solids compared to atoms or molecules. Explain why it is.

References

Baerends, E. J. and O. V. Gritsenko. 2005. Away from generalized gradient approximation: Orbital-dependent exchange-correlation functionals. *J. Chem. Phys.* 123:062202–062218.

Ceperley, D. M. and B. J. Alder. 1980. The homogeneous electron gas: Ground state of the electron gas by a stochastic method. *Phys. Rev. Lett.* 45(7):566–569.

Haas, P., F. Tran, P. Blaha, L. S. Pedroza, A. J. R. Silva, M. M. Odashima, and K. Capelle. 2010. Systematic investigation of a family of gradient-dependent functionals for solids. *Phys. Rev. B* 81:125136–125145.

Hammer, B., L. B. Hansen, and J. K. Norskov. 1999. Improved adsorption energies within DFT using revised PBE functionals. *Phys. Rev. B* 59:7413–7421.

Heyd, J. and G. E. Scuseria. 2004. Efficient hybrid density functional calculations in solids: Assessment of the Heyd-Scuseria-Ernzerhof screened Coulomb hybrid functional. *J. Chem. Phys.* 121(3):1187–1192.

Hohenberg, P. and W. Kohn. 1964. Inhomogeneous electron gas. *Phys. Rev.* 136:B864–B871.

Huang, C.-J. and C. J. Umrigar. 1997. Local correlation energies of two-electron atoms and model systems. *Phys. Rev. A* 56:290–296.

Hubbard, J. 1965. Electron correlations in narrow energy bands. IV. The atomic representation. *Proc. R. Soc. Lond. A* 285:542–560.

Kohn, W. and L. J. Sham. 1965. Self-consistent equations including exchange and correlation effects. *Phys. Rev. A* 140:1133–1138.

Lieb, E. H. and S. Oxford. 1981. Thomas-Fermi and related theories of atoms and molecules. *Int. J. Quantum Chem.* 19:427–439.

Marsman, M., J. Paier, A. Stroppa, and G. Kresse. 2008. Hybrid functionals applied to extended systems. *J. Phys. Condens. Matter* 20:064201–064209.

Paier, J., R. Hirschl, M. Marsman, and G. Kresse. 2005. The Perdew-Burke-Ernzerhof exchange-correlation functional applied to the G2-1 test set using a plane-wave basis set. *J. Chem. Phys.* 122(23):234102–234114.

Paier, J., M. Marsman, K. Hummer, G. Kresse, I. C. Gerber, and J. G. Ángyán. 2006. Screened hybrid density functionals applied to solids. *J. Chem. Phys.* 124(15):154709–154721.

Payne, M. C., M. P. Teter, D. C. Allan, T. A. Arias, and J. D. Joannopoulos. 1992. Iterative minimization techniques for *ab initio* total-energy calculations: Molecular dynamics and conjugate gradients. *Rev. Mod. Phys.* 64:1045–1097.

Perdew, J. P. et al. 1992. Atoms, molecules, solids, and surfaces: Applications of the generalized gradient approximation for exchange and correlation. *Phys. Rev. B* 46:6671–6687.

Perdew, J. P., K. Burke, and M. Ernzerhof. 1996. Generalized gradient approximation made simple. *Phys. Rev. Lett.* 77:3865–3868.

Perdew, J. P. and Y. Wang. 1992. Accurate and simple analytic representation of the electron-gas correlation energy. *Phys. Rev. B* 45:13244–13249.

Perdew, J. P. and A. Zunger. 1981. Self-interaction correction to density-functional approximations for many-electron systems. *Phys. Rev. B* 23:5048–5079.

Stephens, P. J. F. J., Devlin, C. F. Chabalowski, and M. J. Frisch, J. 1994. *Ab initio* calculation of vibrational absorption and circular dichroism spectra using density functional force fields. *Phys. Chem.* 98:11623–11626.

Yang, K, J. Zheng, Y. Zhao, and D. G. Truhlar. 2010. Tests of the RPBE, revPBE, τ-HCTHhyb, ωB97X-D, and MOHLYP density functional approximations and 29 others against representative databases for diverse bond energies and barrier heights in catalysis. *J. Chem. Phys.* 132:164117–164126.

Zhang, Y. and W. Yang. 1998. Comment on Generalized gradient approximation made simple. *Phys. Rev. Lett.* 80:890–890.

Further reading

Kohanoff, J. 2006. *Electronic Structure Calculations for Solids and Molecules.* Cambridge, UK: Cambridge University Press.

Nogueira, F. and M. Marques (Eds.). 2003. *A Primer in Density Functional Theory.* Berlin, Germany: Springer.

Sholl, D. J. and A. Steckel. 2009. *Density Functional Theory: A Practical Introduction.* New York: Wiley-Interscience.

chapter six

Treating solids

> When I started to think about it, I felt that the main problem was to explain how the electrons could sneak by all the ions in a metal. ... By straight Fourier analysis I found to my delight that the wave differed from the plane wave of free electrons only by a periodic modulation.
>
> **F. Bloch (1905–1983)**

Density functional theory (DFT) has extended not only to crystalline solids but also to glasses and minerals such as zeolites, polymers, DNAs, ice, and chemical solutions. Our main concern is, of course, solids. In this chapter, we will introduce electrons into a vast terrain of varying potentials in a solid and try to identify the consequences. At the end, we will have a new formulation of the KS equations in terms of Fourier coefficients and witness the formation of band structures, a very important property of solids. To reach that end, however, two obvious issues have to be resolved: the number of electrons becomes infinite, and so does the number of atoms in solids.

We will first eliminate a large number of electrons from calculation by the pseudopotential (PP) approach. Next, we will see how the data from a small number of atoms become physically relevant to real material, using the periodic nature of solids. Two schemes will be involved in this maneuver: the periodic boundary conditions (PBC) and the supercell approach. The size of the system for calculation will be further reduced and transferred to the reciprocal lattice by the use of Brillouin zone (BZ) construction and *k*-point sampling. Then, at the end of Sections 6.1 and 6.2, our problem for the treatment of a solid will be a small task, such as handling a handful of *k*-points.

Sections 6.3 and 6.4 involve the way quantities on each *k*-point can be represented and calculated efficiently. Three different-looking, but in essence all equivalent, treatments are called for to do this task: the Bloch theorem, Fourier transformation, and plane wave (PW) expansion. With these methods, calculation is carried out back and forth between the real (direct) and reciprocal spaces, depending on the calculation efficiency in each space.

Section 6.5 will discuss some practical topics directly related to the actual DFT runs: energy cutoff, *k*-points integration, and smearing.

The section also deals with topics such as ionic relaxation, *ab initio* MD, and multiscale approaches, all in very brief summaries. After going through these approaches, theorems, and techniques, we become confident in dealing with materials and will be in a position to move to the next chapter for the actual DFT runs. Let us be reminded again of several omissions made for simplicity's sake throughout this chapter:

- The word *operator* and the operator notation (the hat, ∧) are omitted for all operators except $\hat{H}$.
- The coordinate notation $\boldsymbol{r}$ is defined to include both 3-dimensional position (x, y, z) and spin ($\uparrow$,$\downarrow$) coordinates although the spin part is largely ignored and is not explicitly written down.
- Factors resulting from normalizations or summations over the orbitals and $\boldsymbol{k}$-points are often omitted from the derived formulas.

6.1 Pseudopotential approach

This could be a made-up story, but Sir Charlie Chaplin (1889–1977) allegedly participated in a Charlie Chaplin look-alike contest in Monte Carlo and finished second or third. After the contest, the real Chaplin murmured to himself, "He looks like me more than I do to me." It seems that, although that pseudo-Chaplin was definitely a different person, he was as good as the real Chaplin as far as the famous tramp act was concerned (see Figure 6.1).

Figure 6.1 Sir Charlie Chaplin and the pseudo-Chaplin in the tramp act.

The PP approach (Heine 1970), like the act of the pseudo-Chaplin, mimics the true characteristics of the actual potential. The key points are dividing electrons into two groups in terms of their contributing significances, effectively freezing the nucleus and the core electrons together, and pseudizing the remaining valence wave functions. This leads to a significant reduction in the number of electrons in a system to be calculated and to a much easier description and computation of the valence wave functions.

In this section, we will classify electrons in a solid and treat them differently. Then, we will follow how the PPs are generated and how they represent the true potentials effectively for the materials that may easily consist of thousands of electrons. One can often find that any manuscript involved in the generation of a PP is generally full of equations and notations, scaring us off just by looking at it. Fortunately, the subject does not belong to us but to physicists or chemists. We are just the users of PPs—hopefully intelligent users who understand the underlying concept. This section, therefore, will address only the key points of PP generation and its applications in DFT.

6.1.1 *Freezing the core electrons*

6.1.1.1 *Core electrons*

One may say that electrons in an atomic system are all the same as one another in the sense that they normally have the same mass, charge, spin-ups and downs, and so on. However, their role is very much different, depending on where they are. When atoms get together to form a solid, the core electrons (normally the electrons in the inner closed shells) stick tightly to their nucleus in a deep potential well and remain unchanged under most circumstances. Like small children around their mother, the core electrons stay in that well and rarely participate in any change of the system.

In other words, they are so localized that they do not notice whether they are in an atom or in a solid, and only oscillate rapidly due to the strong Coulomb potential by nuclei. And, at the same time, they neutralize the nuclei's charges as much as −1 per electron (in the same way that small children get the mother's immediate attention and love). Thus, their contribution to bonding is minimal when isolated atoms are brought together to form a molecule or a crystal.

6.1.1.2 *Valence electrons*

On the contrary, like big brothers away from their mother, the valence electrons far from their nucleus and high above the potential well are rather independent and quite active in everything. They are the ones forming bonds, being ionized, conducting electricity in metals, forming bands, and performing other atomic activities. In metals, they can even

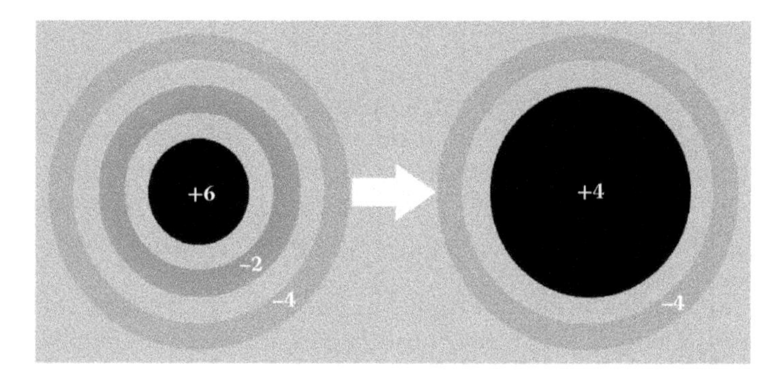

Figure 6.2 Atomic system of carbon showing the frozen core and valence electrons for the construction of a pseudopotential.

travel the whole solid almost like a PW. In covalent or ionic solids, they are not as free as PW but roughly maintain the general picture described above.

6.1.1.3 Frozen-core approximation

From the computational viewpoint, we may simply remove the core (nucleus plus core electrons) from the picture and deal with only the active valence electrons. This is called frozen-core approximation, which is schematically shown in Figure 6.2. The nuclear charge is now largely screened by the core electrons and has much less effect (weaker and smoother attractive force) on the valence electrons.

Note that, in this example in Figure 6.2, the calculation load is already cut by one third. It is more impressive when we go down on the periodic table. For example, the frozen-core model for Pt only accounts for 10 valence electrons ($5d^9$ and $6s^1$) out of a total of 78 electrons ($[Xe]4f^{14}5d^96s^1$). The actual benefit is much more than what these numbers indicate as we further adopt the pseudization scheme. To make the atomic system of frozen-core physically relevant and applicable in practice, several missing details and requirements have to be added. That subject follows next.

6.1.2 Pseudizing the valence electrons

When a valence wave function passes by the highly localized core region, it oscillates rapidly with many wiggles to be orthogonal to the core states as schematically shown in Figure 6.3 (the upper curve with two nodes). Remember that the orthogonal criterion guarantees each wave function to be unique and independent and thus to obey the Pauli exclusion principle.

This kind of wave function with many nodes is simply a headache. It is neither convenient to be expressed in a simple formula nor easy to

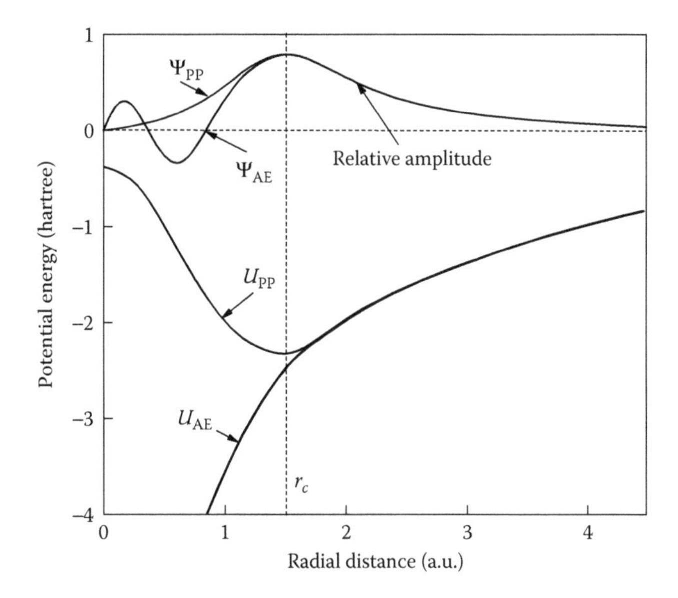

Figure 6.3 Schematic illustration of a pseudo wave function pseudized from a 3s wave function (showing the relative amplitude in arbitrary unit) and the corresponding pseudo- and all-electron (AE) potentials.

be solved computationally. It would be highly appropriate if we could modulate the function to a featureless curve without any of those useless nodes. With the frozen-core approximation, the situation is just right to do that because the valence electrons are now the lowest-lying states, and there is no core electron underneath with which to be orthogonal. Thus, we can "soften" both the wave functions of the valence electrons and their potentials with ions. This procedure is specifically termed as *pseudization*.

6.1.2.1 Pseudizing procedure

Figure 6.3 shows an example of the all-electron (AE) and pseudized wave functions and the corresponding potentials. The standard pseudization steps are as follows:

- Select an atom as the reference state so that the formulated atomic PP can have both transferable and additive properties in application to different environments or many-atom systems.
- Calculate the exact AE potential, wave function, and energy by the DFT calculation with the use of the convenient spherical symmetry of atom; perform a precalculation for the core electrons and keep them frozen for the rest of the calculations.

- Choose a proper r_c of the atom and make the core part ($r < r_c$) of the wave function nodeless and smoother.
- At $r = r_c$, make the first and second derivatives of pseudo- and AE wave functions equal to ensure right scattering characteristics for incoming waves under the different atomic environments.
- At $r > r_c$, make the pseudo and AE wave functions exactly the same since the AE wave function in this part largely decides the behavior of the atom.
- Make the eigenvalues (energies) of the smooth pseudo and original AE wave functions the same.
- Generate a PP from the pseudo wave function and the valence electron density, and parameterize it in spherical Bessel or Gaussian functions for immediate use.

If one takes more wave functions from the core and includes them as a valence state, a more accurate PP can be generated at an increased computational cost in its applications. Note that all PP energies refer to the reference states of each atom for which the potentials were generated, and these are not the same as the true ground states of the atoms. Most DFT codes provide PPs for all isolated atoms in the periodic table in a tabulated format. Since the atomic PP is generated to have both transferable and additive properties, its use in a solid is as simple as just adding them up, regardless of whether it is used for a single atom, molecule, or bulk solid.

6.1.2.2 *Benefits*

In addition to the immediate benefit of removing the core electrons from calculation, other benefits are also recognized:

- The number of PWs needed for the expansion of a pseudo wave function is markedly reduced, and the calculations becomes much faster accordingly (see Section 6.4). When the DFT is armed with a PP and orbitals expanded with PWs, it becomes powerful enough to deal with thousands of electrons and opens up a wide range of problems to first-principles calculation.
- Because we normally calculate energy changes about 0.1–1.0 eV per system in a typical DFT run, any energy change calculated by DFT becomes more noticeable since a large portion of the unchanging energy is taken out as the core energy from calculation. For example, aluminum (Al) has a core potential of about −1700 eV that will be removed by pseudization, and a valence ($3s^2$ and $3p^1$ electrons) potential of about −10 eV will remain, and is subjected to change.

- The errors involved by using PPs are normally less than a couple of percentages.
- The PP approach eliminates the relativistic effect from the system as the core electrons in heavier atoms are most prone to relativity.

After this simplification, the KS equations can be rewritten with the PP (replacing the U_{eff}) and pseudized wave functions, which leads to a different charge density:

$$\left[-\frac{1}{2}\nabla^2 + U_{PP}[\rho(\boldsymbol{r})]\right]\psi_i^{PP}(\boldsymbol{r}) = \varepsilon_i \psi_i^{PP}(\boldsymbol{r}) \tag{6.1}$$

$$\rho(\boldsymbol{r}) = \sum_i \left|\psi_i^{PP}(\boldsymbol{r})\right|^2 \tag{6.2}$$

Therefore, the wave function and the charge density at the core, resister only the pseudized values.

6.1.3 Various pseudopotentials

In the following, three common types of PPs are briefly mentioned: norm-conserving PPs, ultrasoft PPs (USPPs), and projector-augmented wave (PAW) PPs.

6.1.3.1 Norm-conserving PPs

If the pseudo- and AE charge densities within the core are constructed to be equal, the type of PP is called the norm-conserving PP (Hamann et al. 1979; Troullier and Martins 1991). Many PPs are generated to meet this criterion:

$$\int_0^{r_c} \left|\psi_{PP}(\boldsymbol{r})\right|^2 d\boldsymbol{r} = \int_0^{r_c} \left|\psi_{AE}(\boldsymbol{r})\right|^2 d\boldsymbol{r} \tag{6.3}$$

With this scheme, nothing is noticed differently for the valence electrons since the net charge from the core remains the same. Compared to AE methods, however, these PPs give only the valence charge densities, not the total charge densities.

6.1.3.2 Ultrasoft PPs

If we forget about the norm-conserving condition and, in addition to the elimination of radial nodes, shift the peak position of a wave function further to a bigger r_c with reduced peak height, we can in fact make the pseudo wave function as flat as an upside-down bowl. The potentials so generated are called ultrasoft PPs (USPPs; Vanderbilt 1990). As will be discussed in Section 6.4, this type of pseudo wave function with reduced amplitude can be easily expanded with a smaller number of PWs (smaller cutoff energy) that gives a great benefit in computation (up to ~10 times faster). USPPs also give only valence charge densities, not total charge densities.

6.1.3.3 *PAW potentials*

Projector-augmented wave (PAW) potential may be classified as a frozen-core AE potential. This type, first proposed by Blöchl (1994) and adopted by Kresse and Joubert (1999), aims for both the efficiency of the PP and the accuracy of the AE potential. It maps both core and parts of valence wave functions with two separate descriptions as shown in Figure 6.4.

The ψ_{inter} of the valence part is represented with the PW expansion, whereas the ψ_{core} of the core part is projected on a radial grid at the atom center. After the additive augmentation of these two terms, the overlapping part, ψ_{net}, is trimmed off to make the final wave function, ψ_{PAW}, very close to the AE wave function:

$$\psi_{\text{PAW}} = \psi_{\text{inter}} + \psi_{\text{core}} - \psi_{\text{net}} \tag{6.4}$$

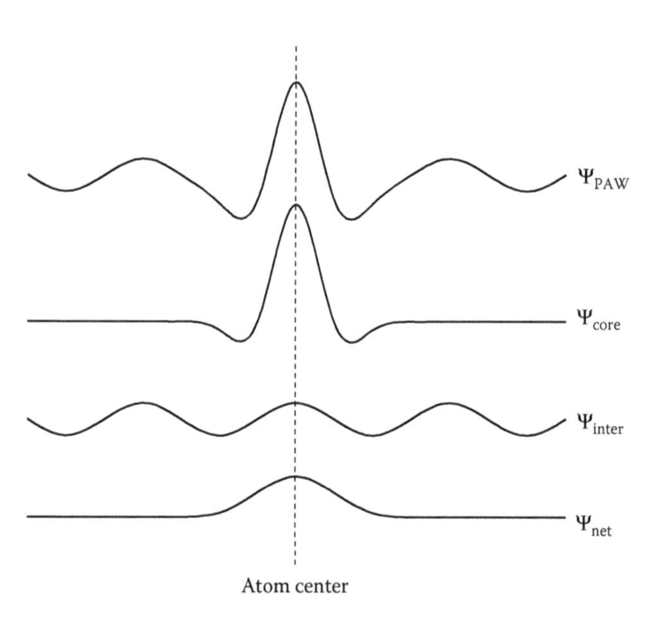

Figure 6.4 Schematic illustration of the wave components used for the construction of PAW.

Owing to the use of ψ_{core}, the core part is well reproduced, and many PWs become unnecessary. Thus, the PAW potential calculates results as accurate as the AE full-potential approach with much less computational effort. Note that this method returns the AE charge density of valence orbitals that cannot be obtained by other standard PPs.

6.2 Reducing the calculation size

Nature displays a variety of periodic events: the sun goes up and down, and the spring comes and summer follows. These temporal repetitions let us prepare for tomorrow and plan for the future. Being a part of nature, most solids are also characterized by the structural periodicity that is absent in the amorphous or liquid phase. Thus, if one crystal cell has two atoms, another one will have two atoms in the same arrangement. In the computational treatment of materials, we try to reduce the system to be calculated as small as possible, and our approach relies heavily on this periodicity of solids. The general flow along this reduction maneuver is outlined in Figure 6.5, which includes the following:

- A solid is first reduced into a supercell made of several unit cells. The constructed supercell is extended to infinity by the PBC.
- The supercell is transformed into reciprocal space and contained in a first BZ. Here, all electronic wave functions are effectively mapped with wave vector $\mathbf{k}$ and reciprocal lattice vector $\mathbf{G}$, and thus all properties are effectively represented by the Bloch equations.

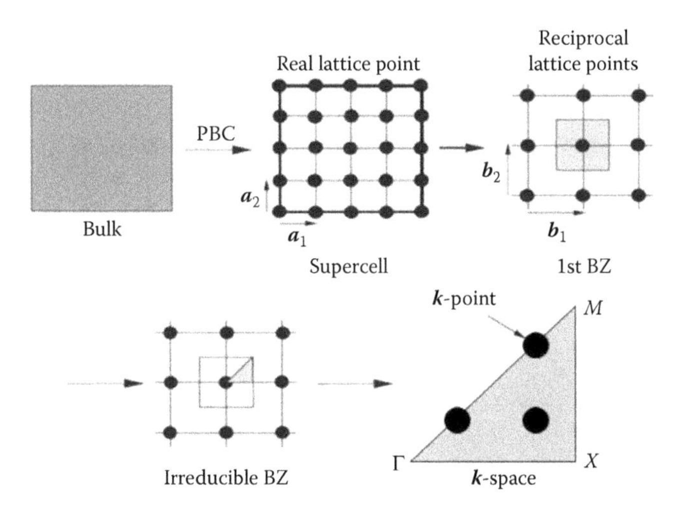

Figure 6.5 Approaches used for the treatment of solids.

- The first BZ is further reduced to a irreducible Brillouin zone (IBZ) by symmetry operations without losing any information.
- The IBZ is mapped with discrete *k*-points, and all necessary quantities are obtained by integration/summation/extrapolation on these points.

Then, dealing with a solid becomes equivalent to treating a handful of *k*-points in a very small volume of IBZ by the periodicity of solids. In this section, the approaches just presented will be discussed one by one in some detail. While the PP approach discussed in the last section reduces the number of electrons to be calculated, the approaches presented in this section will reduce the number of atoms to be calculated from infinity to a small number.

6.2.1 *Supercell approach under periodic boundary conditions*

PBC was discussed in Chapter 2 when we dealt with MD, and it applies here in exactly the same way. Here, the supercell is duplicated periodically throughout the space in all directions. Therefore, even if we simulate a very small box called a supercell (several unit cells to represent the system that one intends to simulate), it can effectively represent a bulk solid. The actual calculation takes place only in the single supercell, and the remaining image supercells (26 in the nearest and more) simply copy it, which causes no significant computational cost.

If the system contains a nonperiodic entity such as a vacancy as schematically shown in Figure 6.6 in two dimensions, we can apply the same approach by including the vacancy into the supercell. Note that,

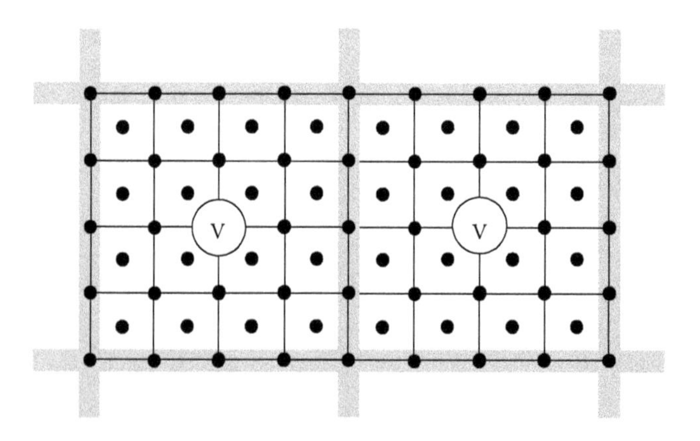

Figure 6.6 Two supercells under the periodic boundary conditions showing the interaction distance of four lattices between vacancies.

however, the supercell must be sufficiently big so that we can safely assume that the interactions between the vacancy and its images in neighboring supercells are negligible.

This supercell approach under PBC can be conveniently extended to any system: bulk, atom, slab, or cluster as shown in Figure 2.10, and thus we can effectively mimic any actual solid behavior. The problem thus becomes one of solving the KS equations only within a single supercell. These imposed tricks can be used only if the supercell is neutral in charge and does not have any dipole moment. If there is any, a correction is necessary.

6.2.2 *First Brillouin zone and irreducible Brillouin zone*

Reciprocal lattice and BZ are where electron waves live and can be folded. We have created these particular coordinate systems to capture the behavior of electrons more easily. Note that we type letters in MS® Word but draw graphs in MS® Excel more conveniently. This will pretty much outline what we are up to in this subsection.

6.2.2.1 *Reciprocal lattice*

Anyone who tried to draw a wave function in real (direct) space immediately realizes that it is not easy. In the reciprocal space, however, the task becomes simply drawing a grid because we construct the unit of the reciprocal lattice matched to that of wave vectors (1/length). Then, the lattice points in that space define the allowed wave vectors, and this is why we are constructing an odd structure called the *reciprocal lattice* out of the real lattice.

Figure 6.7 shows the two corresponding lattices in two dimensions, demonstrating that both magnitude and direction of the real lattice vector are changed. Referring to the primitive vector $\boldsymbol{a}_1$, the corresponding

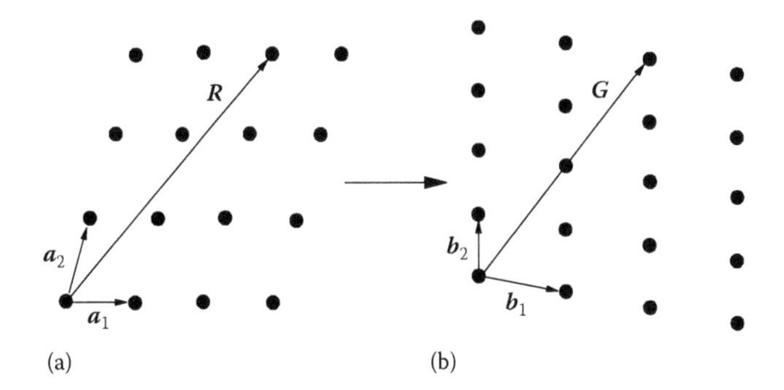

Figure 6.7 Primitive vectors in (a) real (direct) and (b) reciprocal lattices in two dimensions. (A real lattice vector $\boldsymbol{R}$ and a reciprocal lattice vector $\boldsymbol{G}$ are also shown.)

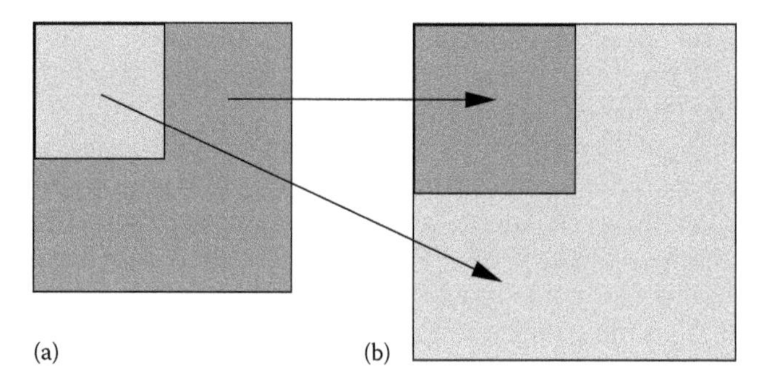

Figure 6.8 Relationship between a supercell in (a) the real space and (b) the corresponding reciprocal lattice in the reciprocal space in two dimensions.

primitive vector, b_1, in the reciprocal lattice is $2\pi/a_1$ (the magnitude is inversed) and its direction is now perpendicular to a_2 (not to a_1). The lattice points in the reciprocal space, therefore, represent the normals of the corresponding set of planes in the real space. Mathematically, the relationship is written as $a_i \cdot b_j = 2\pi\delta_{ij}$, where δ_{ij} is the Kronecker delta (0 if $i \neq j$, 1 if $i = j$). A real lattice vector R and a reciprocal lattice vector G are also shown in Figure 6.7, which are defined as $R = n_1a_1 + n_2a_2 + n_3a_3$ and $G = n_1b_1 + n_2b_2 + n_3b_3$, respectively ($n_i$ are integer numbers). In this illustration in two dimensions, $R = 2a_1 + 3a_2$ and $G = 2b_1 + 4b_2$. Therefore, just as we would mark points for positions in a real lattice, we mark points for wave vectors in a reciprocal lattice.

This is the strange world of electron waves that is in a sense completely opposite to what happens in our world. When we make something big here, it will become something small there. Figure 6.8 shows the relationship between a supercell in real space and the corresponding reciprocal lattice in two dimensions. By doubling the supercell sides in real space (grey → dark grey), the cell sides in the reciprocal space shrink by half (grey → dark grey). Note that all information in the bigger supercell is fully transformed into the smaller reciprocal cell and that much fewer k-points are now needed to achieve the same accuracy by k-point integrations (more details later).

In three dimensions, things are not that simple anymore. Table 6.1 compares the one-to-one relationship between the real and reciprocal lattices.

Note that, for example, an FCC lattice transforms to a BCC lattice in the reciprocal space (see Figure 6.9) with vectors of

$$b_1 = \frac{2\pi}{a_1}(1,1,-1), \quad b_2 = \frac{2\pi}{a_2}(-1,1,1), \quad b_3 = \frac{2\pi}{a_3}(1,-1,1) \tag{6.5}$$

Table 6.1 Summary of characteristic features of real and reciprocal lattices

Characteristics	Real lattice	Reciprocal lattice
Description	Bravais lattice for particles	Fourier transform of the real lattice for waves
Cell shape and size	SC: a BCC: a FCC: a HCP: a, c	SC: $2\pi/a$ FCC: $4\pi/a$ BCC: $4\pi/a$ 30°-rotated HCP $(4\pi/\sqrt{3}a, 2\pi/c)$
Coordinates	x, y, z	k_x, k_y, k_z
Lattice vector	$\boldsymbol{R} = n_1\boldsymbol{a}_1 + n_2\boldsymbol{a}_2 + n_3\boldsymbol{a}_3$	$\boldsymbol{G} = n_1\boldsymbol{b}_1 + n_2\boldsymbol{b}_2 + n_3\boldsymbol{b}_3$
Lattice parameter	$\boldsymbol{a}_i$	$\boldsymbol{b}_i$
Lattice volume	$V = \boldsymbol{a}_1 \cdot (\boldsymbol{a}_2 \times \boldsymbol{a}_3)$	$\boldsymbol{\Omega} = \boldsymbol{b}_1 \cdot (\boldsymbol{b}_2 \times \boldsymbol{b}_3)$
Function	$f(\boldsymbol{r}) = f(\boldsymbol{r} + \boldsymbol{R})$	$f(G) = \sum_{G} c(G)\exp(i\boldsymbol{G}\cdot\boldsymbol{r})$

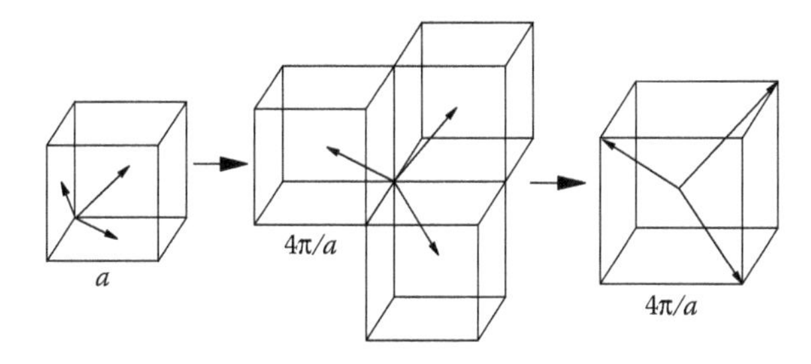

Figure 6.9 A face-centered cubic (FCC) lattice with a in the real space transforms to a body-centered cubic (BCC) lattice with $4\pi/a$ in the reciprocal space.

where $\boldsymbol{b}_1$, $\boldsymbol{b}_2$, and $\boldsymbol{b}_3$ are the primitive reciprocal lattice vectors related to the primitive real lattice vectors $\boldsymbol{a}_1$, $\boldsymbol{a}_2$, and $\boldsymbol{a}_3$ by

$$b_1 = 2\pi\frac{\boldsymbol{a}_2 \times \boldsymbol{a}_3}{V}, \quad b_2 = 2\pi\frac{\boldsymbol{a}_3 \times \boldsymbol{a}_1}{V}, \quad b_3 = 2\pi\frac{\boldsymbol{a}_1 \times \boldsymbol{a}_2}{V} \tag{6.6}$$

Here, V is the volume of the real lattice and is related to the volume of the reciprocal lattice Ω by (again reciprocally)

$$\Omega = \frac{(2\pi)^3}{V}, \quad \Omega_{\mathrm{BZ}} = \frac{(2\pi)^3}{V_{\mathrm{Supercell}}} \text{ for supercell} \tag{6.7}$$

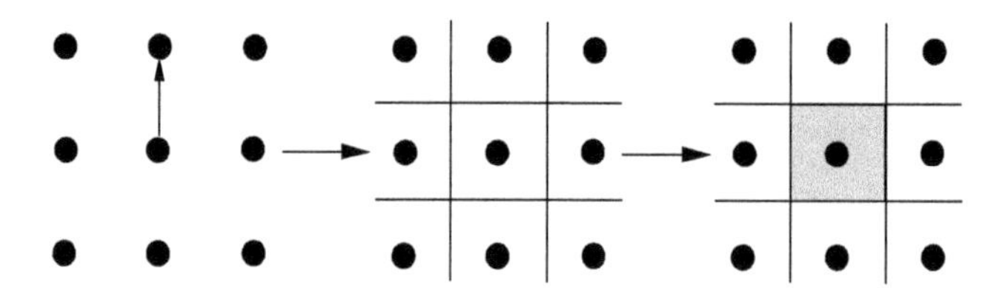

Figure 6.10 Construction steps of the 1st Brillouin zone (grey area) of two-dimensional square lattice.

6.2.2.2 *The first Brillouin zone*

Like a primitive cell in the real space, the first BZ is a primitive cell of the reciprocal lattice. In two dimensions, for example, it is constructed by the following steps (see Figure 6.10):

- Draw a reciprocal lattice.
- Draw lines from a reference lattice point to its nearest neighbors and bisect the lines perpendicularly.
- The formed square (polygon in three dimensions) with these bisecting lines is the first BZ.

The first BZ in three dimensions can be constructed similarly but is much more complicated. Taking the surfaces at the same bisecting distances from one lattice point (Γ-point) to its neighbors, the volume so formed becomes the first BZ. An example is shown in Figure 6.11 for the first BZ of an FCC lattice. It also shows some of the special points and lines of symmetry that become important when we discuss the $\boldsymbol{k}$-points later.

There are also BZs of second, third, and so forth with the same volume as that of the first BZ at increasing distances from the origin. However, these are not our concern at all because, owing to the periodic nature of the reciprocal lattice, any point outside the first BZ can be folded back into the first BZ by the reciprocal lattice vector. More precisely, all $\boldsymbol{k}$-points outside the first BZ can be written as $\boldsymbol{k}' = \boldsymbol{k} + \boldsymbol{G}$, and any $\boldsymbol{k}$-vectors that differ by a reciprocal lattice vector $\boldsymbol{G}$ are all equivalent (see Figure 6.12). In other words, the solutions of KS equations can be completely represented in a single BZ by the folding mechanism in the reciprocal lattice. Thus, we only need to pay attention to the $\boldsymbol{k}$-points inside the first BZ (see Figure 6.5). Table 6.2 shows various points of high symmetry in BZ that we use most often.

6.2.2.3 *Irreducible Brillouin zone*

Symmetry operations by rotation and inversion can further reduce the first BZ to what is called the *irreducible BZ* (IBZ) as shown in Figure 6.5. For example, $10 \times 10 \times 10$ grids of the first BZ in an FCC lattice are made of 1000 $\boldsymbol{k}$-points that can be reduced to only 35 distinctive $\boldsymbol{k}$-points in

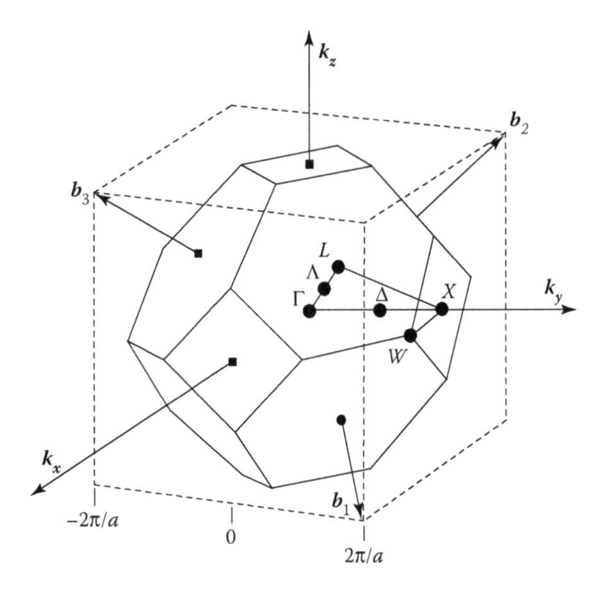

Figure 6.11 First Brillouin zone of FCC lattice showing some of the high symmetry lines and points.

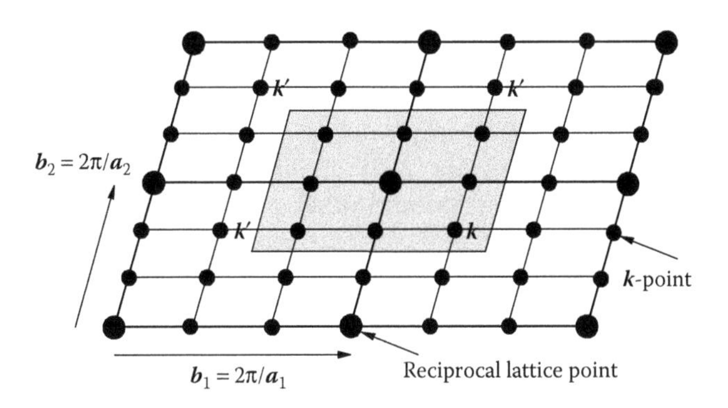

Figure 6.12 Equivalent wave vectors in the reciprocal lattice.

the IBZ. Note that each *k*-point has different importance (weight), which depends on how many times the particular point is folded during symmetry operations. We started with a bulk solid that was far from our reach computationally, but we are now with a small wedge-like piece that seems to be certainly manageable.

6.2.3 k-*points*

Now, we have to account for quantities of concern within this IBZ.

Table 6.2 Points of high symmetry in BZ

System	Symbol	Description
All	Γ	Center of the Brillouin zone
SC (simple cubic)	H	Corner point joining four edges
	N	Center of a face
	P	Corner point joining three edges
FCC (face-centered cubic)	K	Middle of an edge joining two hexagonal faces
	L	Center of a hexagonal face
	U	Middle of an edge joining a hexagonal and a square face
	W	Corner point
	X	Center of a square face
HCP (hexagonal close packed)	A	Center of a hexagonal face
	H	Corner point
	K	Middle of an edge joining two rectangular faces
	L	Middle of an edge joining a hexagonal and a rectangular face
	M	Center of a rectangular face

6.2.3.1 k-*point sampling*

Since any point in an IBZ can represent a ***k***-point, there are an infinite number of discrete ***k***-vectors well qualified to be a wave function. Fortunately, the wave function and other properties in most cases vary smoothly over the IBZ so that we can just sample only a finite number of ***k***-points that represent each small region. Thus, the electronic energy and other properties can be evaluated for the occupied KS eigenstates only on these selected points.

Sampling a set of special ***k***-points in the IBZ is a critical part of the DFT flow and must fulfill two important goals: select as few as possible to reduce the calculation time and, at the same time, select enough so that they can represent the actual quantities adequately. For a large system with hundreds of atoms, for example, the IBZ volume becomes very small, and a few ***k***-points can describe the variation across the zone accurately. For another example, a supercell of a slab often adopted for the study of surface normally has a long dimension in the *z*-direction (perpendicular to the surface), which includes a thick layer of vacuum (see Figure 7.5) in which the wave functions decay to zero. Then only a single ***k***-point is sufficient in the *z*-direction.

Remember that each $\mathbf{k}$-point contains rather rich information about the wave at that point:

- The wave vector $\mathbf{k}$ (wave's magnitude and direction), wave length ($\lambda = 2\pi/k$), and kinetic energy (for example, $E = \mathbf{k}^2/2$ for a plane wave).
- The plane that the particular wave permitted to exist, imposing a particular condition on the wave function to be a solution to the wave equation.
- If we plot all the incoming energies on each $\mathbf{k}$-point, it will form the band structure, the energy dispersion relation (see Section 6.4.3).

The whole bulk material is now narrowed down to a handful of sampled $\mathbf{k}$-points for calculation. A standard method for the generation of a $\mathbf{k}$-point mesh will be discussed next.

6.2.3.2 Monkhorst–Pack method

The Monkhorst–Pack method (Monkhorst and Pack 1976) generates a grid of $\mathbf{k}$-points spaced evenly throughout the IBZ. For example, when a $4 \times 4 \times 4$ grid (64 $\mathbf{k}$-points) is made in an FCC supercell, there will be only 10 $\mathbf{k}$-points in the IBZ by this method. Most DFT codes provide a scheme to generate the special $\mathbf{k}$-points automatically once grid information is given.

6.2.3.3 Γ-point

When a big supercell is used, a single point may be sufficient to describe any property in the IBZ, and the best single point in terms of importance will be the Γ-point that has a high weight factor. At this central $\mathbf{k}$-point ($\mathbf{k} = 0$), the real and reciprocal coordinates coincide, and wave functions will be real so that any consideration for complex numbers is not necessary. The use of high symmetry around this point further increases computational efficiency, reducing the computation time typically to half. Because of these advantages, the Γ-point-only calculation is a popular choice for massive calculations.

6.3 Bloch theorem

In this section, the periodicity in solids will be discussed in relation to the Bloch theorem (Bloch 1928) and the Fourier transformation. This will be extended to plane wave expansion in Section 6.4. These three treatments of wave function and other related quantities in solid are equivalent conceptually but differ in perspectives. At the end, all of them eventually lead to the same conclusion: We can treat a real bulk system efficiently in terms of wave vector $\mathbf{k}$ and reciprocal lattice vector $\mathbf{G}$ either in the real or reciprocal space. Finally, we will see how to rewrite the KS orbitals and equations by Bloch's theorem and thus solve them for each $\boldsymbol{k}$ and band index n in a matrix equation. Note that, throughout this section, the normalization factors (such as $1/\sqrt{\Omega}$, $1/\Omega$, where Ω is volume of the reciprocal box) are ignored for simplicity. However, the meaning that each equation carries is not affected at all by this simplification.

6.3.1 Electrons in solid

Let us look at the solid, perhaps a metallic solid, from the electron's viewpoint. When we say a "solid," we mean something solid. However, the actual "solid" generally is not that solid at all for the electrons, especially for the valence electrons. The larger part of a solid is in fact an open playground for electrons to wave around. Electrons do not even run into each other because of the Pauli exclusion principle, and this will make them more or less independent.

When they pass over each atomic position, they may notice the existence (as an attractive potential) of nuclei, rather remotely, due to the screening by low-lying core electrons. However, it does not disturb them too much, and electrons maintain their near-free nature and move on. This is a very simplified and yet very descriptive picture of valence electrons in metal, which is directly related to what we are up to in this section. We will make this conceptual description materialize into more solid formulations and be combined with the KS equations. Then, we will be able to comprehend the consequences when atoms gather together and form solids and calculate the quantities of interest.

6.3.2 Bloch expression with periodic function

The starting point is recognizing the fact that atoms in most cases are arranged in a periodically repeating pattern in solids. Then, any quantity of interest that depends on position $\boldsymbol{r}$ is naturally periodic, too. For example, the potential acting on electrons is periodic and is invariant under translation with respect to the real lattice vector $\boldsymbol{R}$:

$$U(\boldsymbol{r}) = U(\boldsymbol{r}+\boldsymbol{R}) \tag{6.8}$$

Here, $\boldsymbol{R}$ is the real lattice vector defined by $\boldsymbol{R} = n_1\boldsymbol{a}_1 + n_2\boldsymbol{a}_2 + n_3\boldsymbol{a}_3$ (n_i = any integer number, $\boldsymbol{a}_i$ = unit cell vectors). Similarly, the electron density in a periodic solid is also periodic since it depends on position $\boldsymbol{r}$, too:

$$\rho(\boldsymbol{r}) = \rho(\boldsymbol{r}+\boldsymbol{R}) \tag{6.9}$$

In the case of the wave function, the argument is slightly different. Note that, since it is directly related to $\rho(\boldsymbol{r})$ as $\rho(\boldsymbol{r}) = |\phi(\boldsymbol{r})|^2$, the magnitude of the wave function is definitely periodic, but the wave function itself is not periodic due to the complex number involved in the phase factor. We will see how it can also be treated periodically.

Valence electrons in metals, for example, feel the potential coming from the inner core, but their wave functions normally keep the standard form of a PW:

$$\psi(\mathbf{r}) = C\,\exp(i\mathbf{k}\cdot\mathbf{r}) \tag{6.10}$$

However, under the nonnegligible periodic potentials where the electrons are no longer free, the problem is not that simple. Bloch solved the problem by mapping the same PWs onto the structurally repeating pattern of a solid and made the wave functions quasi-periodic with the introduction of a cell periodic, $u_k(\mathbf{r})$:

$$\psi_k(\mathbf{r}) = u_k(\mathbf{r})\exp(i\mathbf{k}\cdot r) \tag{6.11}$$

Here $u_k(\mathbf{r})$ is a periodic function that has the same periodicity as the potential such that $u_k(\mathbf{r}) = u_k(\mathbf{r}+\mathbf{R})$. The periodic nature of the wave function in a solid expressed by the above equation is illustrated in Figure 6.13 in a highly simplified fashion. In an actual solid, $u_k(\mathbf{r})$ may have more features than this. The wave function now can be written with a new phase factor:

$$\begin{aligned}\psi_k(\mathbf{r}+\mathbf{R}) &= u_k(\mathbf{r}+\mathbf{R})\exp\left[i\mathbf{k}\cdot(\mathbf{r}+\mathbf{R})\right] \\ &= \psi_k(\mathbf{r})\exp(i\mathbf{k}\cdot\mathbf{R})\end{aligned} \tag{6.12}$$

As Bloch noticed, the wave function differs from the PW of free electrons only by a periodic modulation. In other words, the real electrons in an atom can be considered as perturbed free electrons. Referring to the two equivalent Bloch expressions, Equations 6.11 and 6.12, the wave vector $\mathbf{k}$ now has

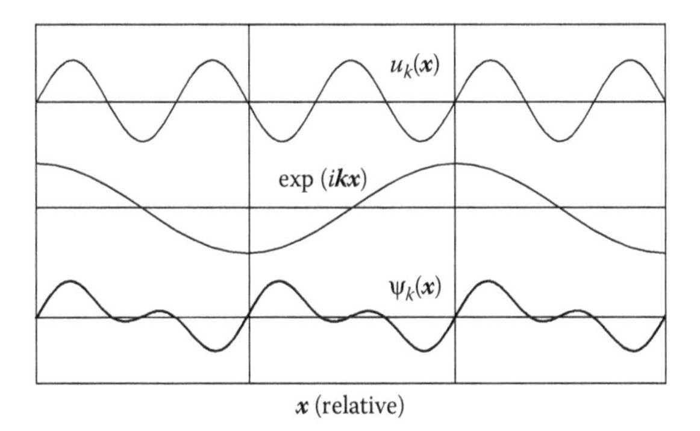

Figure 6.13 Schematic of periodicity of a wave function in one-dimensional solids showing a periodic function (top), a plane wave (middle), and the resulting wave function (bottom).

two roles: It is a wave vector in the PW part of the solution and a quantum index to each wave function, defining a specific condition on any solution (orbitals or wave functions) that must be met under the periodic potential.

Eventually, all relevant entities such as wave functions or KS orbitals, potentials, electron densities (or charge density under the PP approach), or energies can be expressed in periodic forms in real space. This statement is of great importance since, without this theorem, we will be definitely stuck at the end of Chapter 5 with no further progress. By the use of the periodicity of the solids, the Bloch theorem ensures us that a small piece instead of the whole solid is sufficient to have a reliable DFT run. For semiconductors and insulators, there is basically no free electron, but the wave functions of each electron can be treated similarly.

6.3.3 Bloch expression with Fourier expansions

6.3.3.1 Fourier expansions

The Fourier theorem states that, if any function is periodic within a finite real space, it can be expanded in a Fourier series (in PWs in our case) with respect to the reciprocal lattice vector $\boldsymbol{G}$. This means that any periodic function on the interval 2π can be constructed with a superposition of sine and cosine curves of different wavelengths. Note that our complex exponential term can be expressed with sine and cosine terms as $\exp(inx) = i\sin nx + \cos nx$. And if we follow the statement for $u_k(\boldsymbol{r})$ in three dimensions and assign indexes $\boldsymbol{k}$ and $\boldsymbol{G}$, it leads to

$$u_k(\boldsymbol{r}) = \sum_{G} c_k(\boldsymbol{G})\exp(i\boldsymbol{G}\cdot\boldsymbol{r}) \tag{6.13}$$

Here, $c_k(\boldsymbol{G})$ is the Fourier expansion coefficients (complex numbers) of the wave functions that replace the real-space values of wave functions. The phase factor, $\exp(i\boldsymbol{G}\cdot\boldsymbol{r})$, at each $\boldsymbol{G}$ represents a PW traveling in space, perpendicular to the vector $\boldsymbol{G}$. Other quantities such as wave function, charge density, and potential work the same way.

With the use of newly expressed $u_k(\boldsymbol{r})$, the wave function can be written as

$$\begin{aligned}\psi_k(\boldsymbol{r}) &= u_k(\boldsymbol{r})\exp(i\boldsymbol{K}\cdot\boldsymbol{r}) \\ &= \sum_{G} c_k(\boldsymbol{G})\exp(i\boldsymbol{G}\cdot\boldsymbol{r})\exp(i\boldsymbol{K}\cdot\boldsymbol{r}) \\ &= \sum_{G} c_k(\boldsymbol{G})\exp\left[i(\boldsymbol{K}+\boldsymbol{G})\cdot\boldsymbol{r}\right]\end{aligned} \tag{6.14}$$

Here, the first line describes the periodicity of a wave function in the real space with the appearance of $u_k(\boldsymbol{r})$ and the last line expresses it in the reciprocal space by the introduction of **G**. This implies that wave function is now the so-called Bloch wave and is a superposition of many PWs with wave vectors which differ by **G**. There are, of course, an infinite number of electronic states in a bulk material, extending over the entire solid. However, Bloch's expression in cooperation with the Fourier theorem can transform this infinite problem to calculations confined within the IBZ of the system, if the system is structurally periodic.

6.3.3.2 *Fast Fourier transformation*

Referring to Table 6.1, the reciprocal space is the Fourier transformation of the real space. More casually speaking, it is where electrons live, whereas the real space is where *we* live. Thus, the reciprocal space is a little strange for us, and we may guess that the real space is so for electrons. Since we want to describe them, we will map them in their world, the reciprocal space. The key to moving from one space to another and calculating any periodic function efficiently is based on the fast Fourier transformation (FFT). In FFT, continuous functions are mapped on a finite and discrete grid and the integral is replaced by a sum over all grid points. It is conveniently implemented in all DFT programs with the favorable scaling of $O(N_{PW}\log N_{PW})$, where N_{PW} is the number of PWs. Note that the information contained in both spaces is equivalent.

Transforming functions between different coordinate systems offers us a choice to calculate a function at where it is diagonal in matrix or local: for example, kinetic energy and the Hartree potential in the reciprocal space and the XC and external potentials in the real space. Thus, for charge density, we have to calculate it in both spaces since it is required to calculate all energy terms. Note that the accuracy of calculation now depends on how fine the FFT grid is and, as we will find out later, how largely the **G** vector is accounted for calculation.

6.3.3.3 *Matrix expression for the KS equations*

If we expand the orbitals with a basis set, the minimization of an energy functional can proceed with respect to expansion coefficients c_j^*:

$$\phi(\mathbf{r}) = \sum_i c_i \phi_i^{PW}(\mathbf{r}) \tag{6.15}$$

$$\frac{\partial E}{\partial c_j^*} = 0 \tag{6.16}$$

Following the same procedure described in Section 5.3.3, the problem now becomes the general matrix eigenvalue problem:

$$\sum_i \left(\int \phi_j^{*PW} \hat{H}_{KS} \phi_i^{PW} d\mathbf{r} \right) c_i = \lambda_i \sum_i \left(\int \phi_j^{*PW} \phi_i^{PW} d\mathbf{r} \right) c_i \rightarrow \mathbf{Hc} = \lambda \mathbf{Sc} \tag{6.17}$$

Here, each bold letter represents the corresponding matrix. By solving this matrix eigenvalue equation (finding matrix $\boldsymbol{c}$, which diagonalizes $\boldsymbol{H}$ under the orthonormality ensured by the overlap matrix $\boldsymbol{S}$), the diagonal elements of λ return the eigenvalues (energies of each orbital), and the set that returns the lowest total energy will be the ground-state energy.

Note that, if we add or subtract a reciprocal lattice vector $\boldsymbol{G}$ from $\boldsymbol{k}$ in the reciprocal space, it does not make any change of wave function or energy since our system is also periodic in the reciprocal space:

$$\psi_{n,k\pm G}(\boldsymbol{r}) = u_{n,k\pm G}(\boldsymbol{r})\exp\left[i(\boldsymbol{K}\pm\boldsymbol{G})\cdot\boldsymbol{r}\right] \tag{6.18}$$

$$E_n(\boldsymbol{K}) = E_n(\boldsymbol{K}\pm\boldsymbol{G}) \tag{6.19}$$

If we plug the expanded wave function into the KS equations and do the usual quantum manipulation, a new expression for the KS equations becomes

$$\sum_{G'} \hat{H}_{GG'}(\boldsymbol{K})c_{nk}(\boldsymbol{G}') = \varepsilon_{nk}\sum_{G} c_{nk}(\boldsymbol{G}) \tag{6.20}$$

Since there are n wave functions for each $\boldsymbol{k}$-point in a periodic solid, the electronic band index n is introduced as an additional quantum number here. And each $\boldsymbol{k}$ represents a quantum number that indexes the wave vectors of a free electron that is always confined to the first BZ of the reciprocal lattice ($-\pi/a < \boldsymbol{k} \leq \pi/a$). Both n and $\boldsymbol{k}$ can now identify the electronic state in a system completely. Note that n is discrete and $\boldsymbol{k}$ is continuous, although we treat it point by point by sampling. Then we need to be concerned only with the expansion coefficient, c_{nk}, instead of the usual wave function. Here, $\boldsymbol{G}$ and $\boldsymbol{G}'$ represent different vectors for PWs and provide the indexes for the row and column elements in Hamiltonian matrix. Again, remember that $\psi_{nk}(\boldsymbol{r})$, $c_{nk}(\boldsymbol{G})$, and $\rho(\boldsymbol{r})$ are all equivalent in nature to one another.

The above equation is in fact an eigenvalue matrix equation that is far easier to solve than the usual KS equations and can be written in condensed matrix form as follows:

$$\boldsymbol{H}\boldsymbol{c} = \varepsilon\boldsymbol{c} \tag{6.21}$$

where each bold letter represents the corresponding matrices: the Hamiltonian matrix $\boldsymbol{H}$, columnar coefficient matrix $\boldsymbol{c}$, and diagonal eigenvalue matrix ε. Our computer will be ready to handle the above equation since the original differential equation is now transformed into a set of algebraic eigenvalue problems by the use of the Bloch theorem and the Fourier series expansion.

6.4 *Plane wave expansions*

We now understand that, if we expand wave functions, densities, potentials, and energies in conjunction with a PW basis set, they satisfy the Bloch theorem, meet the periodic boundary conditions, and are able to switch between the real and reciprocal spaces via FFT.

In this section, we will fully employ the approaches that we described in the previous section and finalize the practical formulation of the KS equations in a matrix form.

6.4.1 *Basis set*

The KS orbitals could be in any shape, and thus treating them mathematically is not straightforward. If we approximate them as a linear combination of known simple functions such as PWs, the calculations become much easier, especially for computers. Three types of basis sets are commonly used for expansions: local, nonlocal, and augmented.

6.4.1.1 *Local basis set*

A local basis set such as the Gaussian basis set or atom-centered orbitals has its intensity mainly on a local point and is well fitted to orbitals around individual atoms in the real space. Many quantities can be calculated analytically by the use of it, and thus it is a popular choice especially for atoms and molecules whose orbitals are highly localized around each atom. It is claimed that typically 10–20 basis functions of this type per atom are sufficient to achieve reasonable accuracy, whereas the PW-based calculations normally require hundreds of PWs per atom. In addition, they can in principle provide a linear-scaling scheme $O(N)$ for DFT calculations, especially for massive systems (>1000 atoms) of insulators or semiconductors. Due to the local nature of this basis, however, accuracy is often limited, and metals that have long-range interactions cannot be properly described.

6.4.1.2 *Plane wave basis set*

PWs are totally nonlocal and span the whole space equally as

$$\phi_{PW}(\boldsymbol{r}) = C\exp(i\boldsymbol{K}\cdot\boldsymbol{r}) \tag{6.22}$$

PWs are in fact most commonly adopted in DFT codes because of the following advantages:

- PWs are already solutions (orthonormal wave functions) for a free electron satisfying the Bloch condition, and its basis set can represent any smoothly varying functions very well.

- DFT calculations are always involved in taking gradients or curvatures, and PWs have the most convenient exponential forms for that and their derivatives become simple products in reciprocal space (see Equation 4.23).
- The expanded KS orbitals with PWs are already in the transformed form of Fourier, and calculations such as the evaluation of $\hat{H}_{KS\phi k}$ in the reciprocal space are straightforward.
- A systematic improvement of convergence and accuracy is possible with increase in the size of the PW basis set.
- PWs are independent of the ionic (atomic) positions due to their nonlocal nature, and thus the calculated Hellmann–Feynman forces have no Pulay force, which is unavoidable in calculations with other atom-centered basis sets.
- With the PP approach, which is commonly adopted in most DFT codes, the number of PWs needed for expansion is drastically reduced, and a much smaller E_{cut} can be used (see Section 6.5.1).

The use of PWs for expansion also have some disadvantages:

- More than a hundred PWs are normally needed per atom.
- PWs cover all space equally (remember they are nonlocal), and about the same number of basis sets is needed even in a vacuum region.
- The efficiency for parallelization is relatively low due to its nonlocal nature.

However, the overwhelming advantages of the PW basis set may easily nullify these disadvantages.

6.4.2 *Plane wave expansions for KS quantities*

Following are the expanded expressions of various KS quantities in the reciprocal lattice, which are written simply by plugging in the expanded PW expression for the KS orbitals. Fundamentally, observables are calculated as summations over sampled $\boldsymbol{k}$-points within the IBZ since they change rather slowly with varying $\boldsymbol{k}$ within $\boldsymbol{G}$ vectors specified by energy cutoff (see Section 6.5.1).

6.4.2.1 *Charge density*

The charge density in terms of $\boldsymbol{G}$ vectors can be easily obtained by squaring the expanded orbitals:

$$\rho(\boldsymbol{r}) = \sum_{G} \rho(\boldsymbol{G}) \exp(i\boldsymbol{G} \cdot \boldsymbol{r}) \tag{6.23}$$

where $\rho(\boldsymbol{G})$ is the Fourier coefficient for charge density. The charge density is required both in the real and reciprocal spaces, and the calculations of potentials in terms of charge density take place in the following sequence:

$\phi_{nk}(\boldsymbol{G}) \rightarrow \rho(\boldsymbol{G}) \rightarrow E_H[\rho(\boldsymbol{G})]$ since the $E_H[\rho(\boldsymbol{G})]$ is local in the reciprocal space and can be calculated as

$$E_H = C\sum_G \frac{\rho(\boldsymbol{G})^2}{\boldsymbol{G}^2} \tag{6.24}$$

$\phi_{nk}(\boldsymbol{G}) \rightarrow \phi_{nk}(\boldsymbol{r}) \rightarrow \phi_{nk}(\boldsymbol{r}) \times \phi_{nk}(\boldsymbol{r}) = \rho(\boldsymbol{r}) \rightarrow E_{xc}[\rho(\boldsymbol{r})]$ since the $E_{xc}[\rho(\boldsymbol{r})]$ is approximated to be local in the real space. The local external energy, $E_{\text{ext}}[\rho(\boldsymbol{r})]$, can be similarly calculated.

For charge density calculations in the reciprocal space, the Fourier grid must contain all wave vectors up to $2\boldsymbol{G}_{\text{cut}}$ (see Figure 6.14) since it is a square of the expanded orbitals in real space. Otherwise, so-called wrap-around errors may occur.

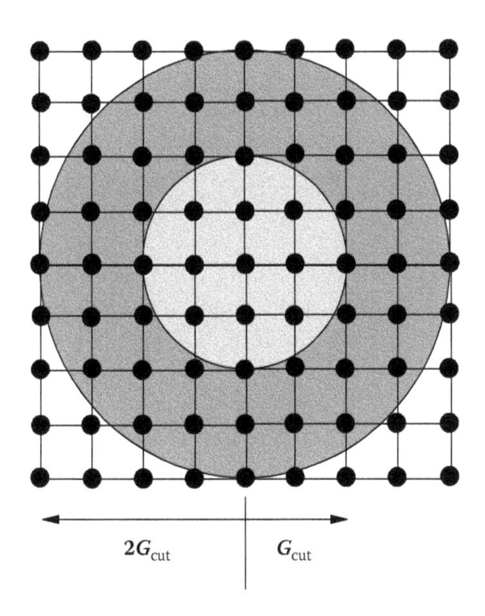

Figure 6.14 The small circle contains all plane waves within $\boldsymbol{G}_{\text{cut}}$ for the finite PW expansion, and the big circle contains all plane waves within $2\boldsymbol{G}_{\text{cut}}$ for electron density calculation in the reciprocal space.

6.4.2.2 Kinetic energy

When electrons are noninteracting and under a static KS potential, the kinetic energy with expanded PWs is local in the reciprocal space (thus diagonal in the matrix expression):

$$\begin{aligned} E_{\text{kin}} &= -\frac{1}{2}\int \phi_{\text{PW}}^{*} \ \nabla^2 \phi_{\text{PW}} d\boldsymbol{r} = \frac{1}{2}\left|\boldsymbol{k}+\boldsymbol{G}\right|^2 \int \phi_{\text{PW}}^{*} \phi_{\text{PW}} d\boldsymbol{r} \\ &= \frac{1}{2}\left|\boldsymbol{k}+\boldsymbol{G}\right|^2 \end{aligned} \tag{6.25}$$

6.4.2.3 Effective potential

The effective potential expanded with a PW basis set is

$$U_{\text{eff}}(\boldsymbol{r}) = \sum_{G} U_{\text{eff}}(\boldsymbol{G}) \exp(i\boldsymbol{G}\cdot\boldsymbol{r}) \tag{6.26}$$

where $U_{\text{eff}}(\boldsymbol{G})$ is the Fourier coefficient for effective potential. And, with respect to different PW vectors of $\boldsymbol{G}$ and $\boldsymbol{G}'$, it becomes:

$$U(\boldsymbol{G},\boldsymbol{G}') = \int U_{\text{eff}} \exp\left[i(\boldsymbol{G}'-\boldsymbol{G})\cdot\boldsymbol{r}\right] d\boldsymbol{r} = U_{\text{eff}}(\boldsymbol{G}-\boldsymbol{G}') \tag{6.27}$$

6.4.2.4 KS equations

This leads to a matrix eigenvalue equation with the potentials in the form of Fourier transforms via the following sequence:

$$\left(-\frac{1}{2}\nabla^2 + U_{\text{eff}}\right)\sum_{G} c_n(\boldsymbol{k}+\boldsymbol{G})\exp\left[i(\boldsymbol{k}+\boldsymbol{G})\cdot\boldsymbol{r}\right] = \varepsilon_{nk}\sum_{G} c_n(\boldsymbol{k}+\boldsymbol{G})\exp\left[i(\boldsymbol{k}+\boldsymbol{G})\cdot\boldsymbol{r}\right] \tag{6.28}$$

Multiplying from the left with $\exp[-i(\boldsymbol{k} + \boldsymbol{G}')\cdot \boldsymbol{r}]$ and integrating over the BZ introduces $U_{\text{eff}}(\boldsymbol{G} - \boldsymbol{G}')$, representing the Fourier transform of effective potential (for a local effective potential, $\boldsymbol{k}$ does not appeal).

$$\begin{aligned} &\sum_{G'}\left[\frac{1}{2}\left|\boldsymbol{k}+\boldsymbol{G}\right|^2 \delta_{GG'} + U_{\text{ext}}(\boldsymbol{G}-\boldsymbol{G}') + U_H(\boldsymbol{G}-\boldsymbol{G}') + U_{xc}(\boldsymbol{G}-\boldsymbol{G}')\right] c_n(\boldsymbol{k}+\boldsymbol{G}') \\ &= \varepsilon_{nk} c_n(\boldsymbol{k}+\boldsymbol{G}) \end{aligned} \tag{6.29}$$

$$\sum_{G'}\left[\frac{1}{2}\left|\boldsymbol{k}+\boldsymbol{G}\right|^2 + U_{\text{eff}}(\boldsymbol{G}-\boldsymbol{G}')\right] c_n(\boldsymbol{k}+\boldsymbol{G}') = \varepsilon_{nk} c_n(\boldsymbol{k}+\boldsymbol{G}) \tag{6.30}$$

If we express the above algebraic equation in a condensed matrix form, it is written as (naturally, the same as Equation 6.21.)

$$Hc = \varepsilon c \tag{6.31}$$

For example, a simple matrix equation with a 3 × 3 square KS Hamiltonian at a given band index n is written as

$$\begin{bmatrix} \frac{1}{2}(\boldsymbol{k}-\boldsymbol{G})^2 & U_{\text{eff}} & 0 \\ U_{\text{eff}} & \frac{1}{2}(\boldsymbol{k})^2 & U_{\text{eff}} \\ 0 & U_{\text{eff}} & \frac{1}{2}(\boldsymbol{k}+\boldsymbol{G})^2 \end{bmatrix} \begin{bmatrix} c_{k-G} \\ c_k \\ c_{k+G} \end{bmatrix} = \varepsilon_k \begin{bmatrix} c_{k-G} \\ c_k \\ c_{k+G} \end{bmatrix} \tag{6.32}$$

Diagonalizing this matrix returns the vectors $c_n(\boldsymbol{k},\boldsymbol{G})$, allowing us to calculate ϕ_{nk} and the corresponding energies, ε_{nk}. Based on the calculated ϕ_{nk}, a new charge density and KS Hamiltonian are calculated, and the process repeats until self-consistency is reached. Note that all unique values of ε_{nk} occur for each band index n and wave vector $\boldsymbol{k}$, yielding a spectrum of energy curves called the *band structure*.

6.4.3 KS orbitals and bands

We are now at the final stage of discussion for solids in relation to the DFT. As schematically shown in Figure 6.15, we reduced a solid into a wedge-shaped IBZ from which a set of KS orbital energies, $\{\varepsilon_{nk}\}$, is calculated. In this subsection, we will follow the process of electrons forming bands in solids.

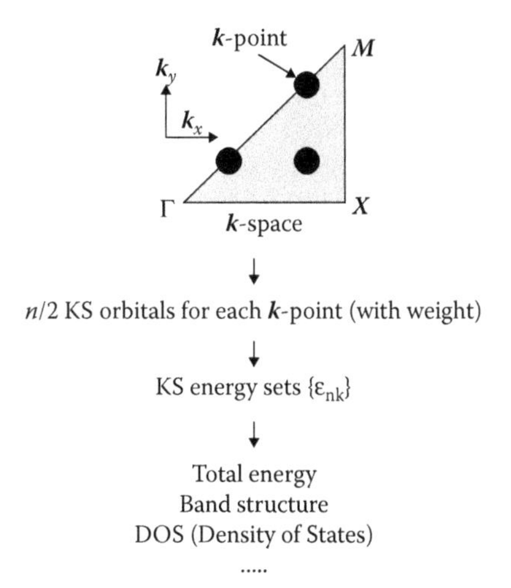

Figure 6.15 Data production from the solutions of KS equations.

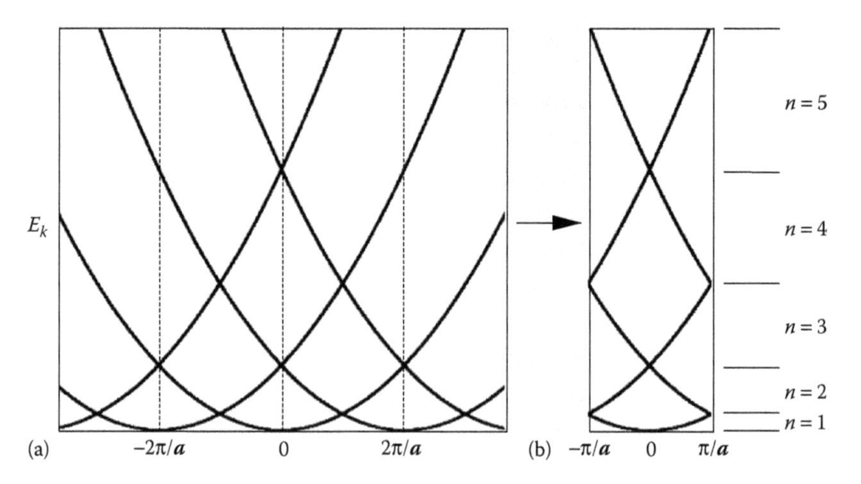

Figure 6.16 Band structures of a free-electron system showing energy curves in terms of $\mathbf{k}$-vector in extended (a) and reduced (b) presentations.

6.4.3.1 Band structure of free electron

Let us first draw a band structure for a free electron. A free electron has only kinetic energy as given in Equation 6.25. Figure 6.16 illustrates the band structures of a free electron showing energy curves in an extended (a) and a reduced (b) presentation. Note that the energy curves are continuous without any gap and any state with $\mathbf{k}$ beyond the first BZ (shown up to $\pm 2G$ in Figure 6.16) is the same to a state inside the first BZ with a different band index n.

6.4.3.2 Band structure of electrons in solids

The first thing happening on the electrons in solids is the formation of bands. When atoms are isolated, each electron occupies specific and discrete orbitals: 1s, 2p, 2d, and so on. When they form a solid, the core electrons remain as they are, sticking to their nuclei in deep potential wells ($U \propto -1/r$). The valence electrons, however, meet each other in solids and try to occupy the same energy level. For example, in an N-atom system, there will be N 3d-orbitals trying to occupy the 3d-energy level. Eventually, they settle down by sharing the 3d-energy level together and form the energy band as shown in Figure 6.17. In this energy band, all the N energy levels from each atom are separated by almost indistinguishable differences. Among the valence bands, the band with higher energy is wider since its electrons interact more actively.

Another change happening on the electrons in solids is the formation of band gaps because the valence electrons in a semiconductor and insulator feel the presence of periodic nuclei and respond to the resulting

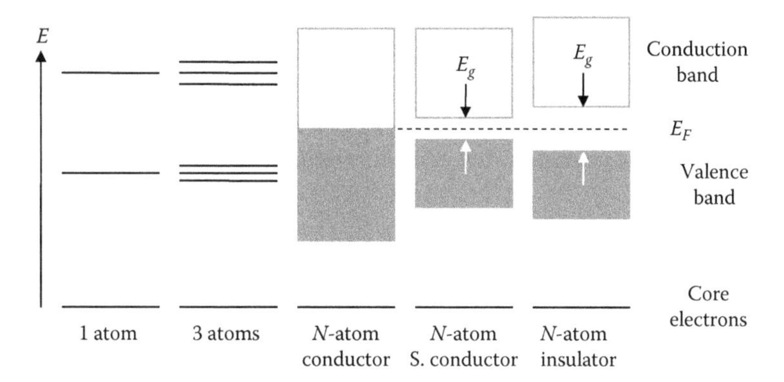

Figure 6.17 Formation of bands and band gaps when isolated atoms become various solids.

potentials. Let us assume that the potentials are very weak for the sake of convenience, and we thus treat the electrons as near-free electrons, meaning that the electrons maintain the general characteristics of free electrons. Then, the situation is not much different from the case of free electrons except when some electrons meet the BZ boundaries with a wave vector of $\mathbf{k} = n\pi/a$. Here, orbitals with vectors that end on the BZ boundaries meet the Bragg condition and thus are diffracted. This electron diffraction at the lattices is the underlying principle of XRD (X-ray diffraction) and the cause of band gap formation in semiconductors. The electrons are perturbed by lattice potentials, reflected (scattered) back, and become standing waves. Note that these standing waves will build up charge densities differently, and there is on average no net propagation of energy at the BZ boundaries.

The band gap width and its relative position with respect to the Fermi level decide a very important criterion for materials: conductor, semiconductor, and insulator. Remember that the Fermi level is the energy of the highest occupied state in a system. The valence bands of typical conductors become wider due to the odd-numbered free electrons and overlap each other, positioning the Fermi level in the conduction band. In semiconductors and insulators, there is no overlap, and the Fermi levels are at the top of the valence band or inside band gaps, which are relatively narrower and wider, respectively. Note that the dark areas represent the occupied energy levels. Conductors have the conduction band partially occupied, which provides no barrier to electron transport, while semiconductors and insulators have their conduction bands completely unoccupied. Thus, the band structure of a solid exhibits all ε_{nk} points at each band and $\mathbf{k}$-point, identifying each energy level of electrons, and describes the band gap formed by interactions between electrons and the periodic potentials coming from the core.

6.4.3.3 Density of states

The electronic structure of a solid can be characterized in another way, that is the density of states (DOS) diagram (see Section 7.8). The DOS defines the number of electronic states per unit energy range. For a free electron in one-dimension, the energy relation is

$$\varepsilon_k = \frac{\hbar^2}{2m}k^2 \tag{6.33}$$

And, the total number of states, $n(\varepsilon)$, is

$$n(\varepsilon) = \frac{V}{3\pi^2}k_F^3 = \frac{V}{3\pi^2}\left(\frac{2m\varepsilon}{\hbar^2}\right)^{3/2} \tag{6.34}$$

The DOS is then

$$D(\varepsilon) = \frac{dn}{d\varepsilon} = \frac{V}{2\pi^2}\left(\frac{2m}{\hbar^2}\right)^{3/2}\varepsilon^{1/2} \propto \sqrt{\varepsilon} \tag{6.35}$$

The DOS of a free-electron system at 0 K is illustrated in Figure 6.18 as the solid curve, indicating the maximum DOS that an energy level can possibly have.

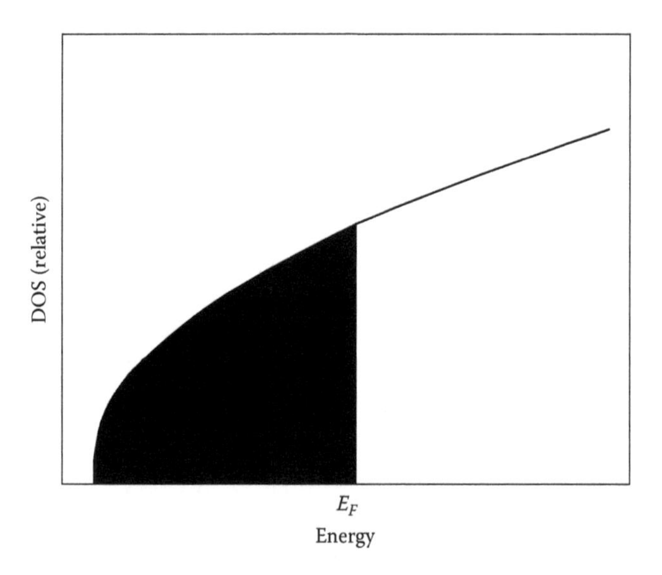

Figure 6.18 DOS of a free-electron system at 0 K. (The dark area represents occupied energy levels.)

Furthermore, the integral of DOS up to the Fermi level gives the total number of electrons in the system:

$$\int_0^{E_F} D(\varepsilon)d\varepsilon = n \tag{6.36}$$

Whether any particular state is occupied by electron or not is decided by the Fermi–Dirac distribution function, $f(\varepsilon)$ (Dirac 1926; Fermi 1926) at nonzero temperatures:

$$\int_0^{\infty} D(\varepsilon)f(\varepsilon) = n \tag{6.37}$$

At a finite temperature, there will be a change of the DOS line at the Fermi energy, that is, some electrons are thermally excited and cross over the Fermi line, changing the line to a reversed "S" curve. The actual band structure and DOS are much different from this simple case but have similar features (see Section 7.8).

6.5 *Some practical topics*

6.5.1 *Energy cutoff*

The PW basis set for the expansion is in principle infinite to present the wave function precisely. But we can make the size of the expansion finite by introducing a proper cutoff energy, E_{cut}, since the $c_n(\boldsymbol{k}+\boldsymbol{G})$ with higher kinetic energies have a negligible contribution to the total energy and can be neglected.

6.5.1.1 *Cutoff energy*

In a solid system, the unbound free electrons occupy the higher energy levels. As the solutions with lower energies are more important than those with higher energies (remember we are aiming to have the lowest ground-state energy), we would rather cut the energy at some upper limit. Because the kinetic energy of the free electron calculated previously is the highest energy in a system, E_{cut} is determined referring to this kinetic energy. Then, the cutoff energy E_{cut} is defined as

$$E_{cut} \geq \frac{1}{2}(\boldsymbol{k}+\boldsymbol{G})^2 \tag{6.38}$$

So we only include PWs with kinetic energies smaller than this cutoff energy for calculation. This cutoff energy is system dependent, and thus it is necessary to ensure that the cutoff energy is high enough to give accurate results. We repeat the calculations with higher and higher energy cutoffs until properties such as energy change no more (see Section 7.4.1). This is also the reason that the PP approach described in Section 6.1.1 improves the calculation efficiency greatly since it effectively removes very large **G** components at the core from calculation.

The Fourier expansion for the wave function now becomes

$$\psi_k(\boldsymbol{r}) = \sum_{|k+G|\text{cut}} c_{k+G} \exp\left[i(\boldsymbol{k}+\boldsymbol{G})\cdot \boldsymbol{r}\right] \tag{6.39}$$

With this energy cutoff, the calculation becomes finite, and its precision within the approximations of DFT can be controlled by the E_{cut}. Therefore, at a given volume V of a simulation box, the number of plane waves N_{pw} for bulk materials increases with this energy cutoff:

$$N_{\text{pw}} \approx \frac{1}{2\pi^2} V E_{\text{cut}}^{3/2} \tag{6.40}$$

As shown in Figure 6.14, the N_{pw} increases discontinuously with the E_{cut}. Usually, more than a hundred PWs are needed per atom, and this implies that a typical calculation may require 10^3–10^5 PWs. Note that ***k***-vectors become finite by ***k***-points sampling in the IBZ and **G**-vectors become finite by energy cutoff, and thus the PW expansion approach comes in the tractable range.

6.5.2 Smearing

For insulators and semiconductors, charge densities decay smoothly to zero up to the gap, and the integration at ***k***-points is straightforward. Metals, however, display a very sharp drop of occupation ($1 \rightarrow 0$) right at the Fermi level at 0 K. Since these factors multiply the function to be integrated, it makes the resolution very difficult with PW expansions.

Several schemes are available to overcome this problem, generally replacing the delta function with a smearing function that makes the integrand smoother. In these methods, a proper balance is always required: Too large smearing might result in a wrong total energy, whereas too small smearing requires a much finer ***k***-point mesh that slows down the calculation. This is in effect the same as creating a fictitious electronic temperature in the system, and the partial occupancies so introduced will cause some entropy contributions to the system energy. However, after the calculation is done with any smearing scheme, the artificially introduced temperature is brought back to 0 K by extrapolation, and so is the entropy.

6.5.2.1 Gaussian smearing

The Gaussian smearing method (Fu and Mo 1983) creates a finite and fictitious electronic temperature (~0.1 eV) using the Gaussian-type delta function just like heating the system up a little, and thus broadens the energy levels around the Fermi level.

6.5.2.2 Fermi smearing

The Fermi smearing method also creates a finite temperature using the Fermi–Dirac distribution function (Dirac 1926; Fermi 1926) and thus broadens the energy levels around the Fermi level:

$$f(\varepsilon) = \frac{1}{\exp\left[(\varepsilon - E_F)/k_B T\right] + 1} \tag{6.41}$$

Here, E_F is Fermi energy, k_B is Boltzmann's constant, and T is absolute temperature. Now, eigenstates near the Fermi surface are neither full nor empty but partially filled, and singularity is removed at that point during calculation. Note that electronic thermal energy is roughly $k_B T$, which corresponds to about 25 meV at 300 K (see Figure 6.19).

6.5.2.3 Methfessel–Paxton smearing

This method as suggested by Methfessel and Paxton (1989) is one of the most widely used methods in practice with a single parameter σ. It uses step function expanded into polynomials and Gaussian function.

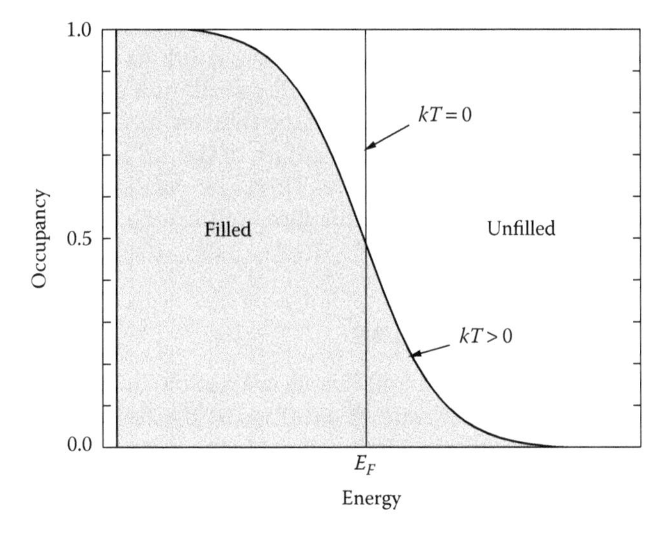

Figure 6.19 Generation of partial occupancies around the Fermi level by smearing.

6.5.2.4 *Tetrahedron method with Blöchl corrections*

This method (Blöchl et al. 1994) is conceptually different from the previously described three methods. It divides the IBZ into many tetrahedrons (for example, a cube into six tetrahedrons), and the KS energy is linearly interpolated and integrated within these tetrahedrons. It is especially well suited for transition metals and rare earths whose delicate Fermi surfaces require a fine resolution. At least four $\boldsymbol{k}$-points including Γ-point are needed. It is also known that this method is preferred for electronic structure evaluations such as band structure or DOS calculations of semiconductors.

6.6 *Practical algorithms for DFT runs*

After going through the topics covered so far, we can make the conclusion that the level of a DFT run in terms of accuracy and efficiency crucially depends on three choices:

- The choice of an XC functional.
- The choice of a PP.
- The choice of a basis set for the expansion of the KS orbitals.

Let us assume that we have made excellent decisions for all choices: for example, the PBE XC functional for the calculation of XC energy, the PAW PP for the calculation of orbitals and potentials, and the KS orbitals expanded with PWs. In this section, we will add one more element: the algorithms adopted for solving the KS equations. Remember that a DFT calculation must repeatedly treat large nonlinear eigenvalue problems to reach the electronic ground state of a solid. At a given computer power, therefore, the rest depends on the calculation algorithms implemented. Note that calculating the charge density or energy terms is of minor importance, but the diagonalization of the KS Hamiltonian takes about 70% of overall computer time. Thus, the consideration here will be only on the diagonalization algorithms within the limit of materials interest.

6.6.1 *Electronic minimizations*

The electronic minimization (electronic relaxation or static relaxation) means finding the ground state of electrons at the fixed atom positions and it can be achieved by diagonalization for the H matrix either directly or iteratively. For the full minimization of the system, an ionic minimization must follow after each electronic minimization.

6.6.1.1 *Direct diagonalization*

The conjugate gradient and CP algorithms belong to this category. Note, however, that we only need those of the lowest occupied orbitals (~$n/2$) for the total energy at the end. Therefore, the direct method is not suitable for materials calculations which normally require large matrix elements of PWs (10^4–10^5).

6.6.1.2 *Iterative Davidson method*

Rarely with normal matrix–matrix operations do we find the eigenvalues of large matrices. Rather, the methods always involve a variety of algorithms, manipulations, and tricks to speed up the diagonalization: the Hamiltonian matrix is projected, approximated to smaller sizes, preconditioned, or spanned in subspace. In this iterative diagonalization (normally in conjunction with the charge density mixing), matrix elements are preconditioned and thus the needed number of orbitals can be reduced to slightly more than $n/2$. Furthermore, the orbitals other than the ground states will be removed along the iteration process. The most notable iterative algorithms along this line are the blocked Davidson (1975) (Kresse and Furthmüller 1996) and the residual minimization/direct inversion in the iterative subspace (RMM-DIIS) methods. All these are algebraic manipulations to solve the matrix problem efficiently. Their details, however, are not our concern at all but rather that of mathematicians and physicists. In fact, they are normally imbedded in the DFT codes so that we do not notice any of their actions during a DFT run.

The Davidson method provides an efficient way for solving eigenvalue matrix problems of a very large size, iteratively and self-consistently. The Davidson algorithm is considered to be slower, but it is very reliable in a crude convergence of the wave function at the initial stage of iteration. Normally, by combining sets of bands into blocks, energy minimization is performed iteratively in parallel.

6.6.1.3 *RMM-DIIS method*

The RMM-DIIS method (Pulay 1980; Wood and Zunger 1985) also provides an efficient and fast way of solving eigenvalue matrix problems iteratively and self-consistently. The diagonalization is carried out iteratively with respect to PW coefficients, $c_{nk}(\boldsymbol{G})$, improving the orbitals, ϕ_{nk} (and thus charge densities). The actual calculation, however, does not handle the orbitals directly, but instead minimizes the residual vector, $\boldsymbol{R}$, which is generally written as follows:

$$\boldsymbol{R} = (\hat{\boldsymbol{H}} - \varepsilon)\phi \tag{6.42}$$

In this residuum minimization approach, $\boldsymbol{R}$ becomes smaller and smaller as iteration proceeds and eventually vanishes to zero at convergence. It also involves subspace diagonalization and plays with residuum (actually the norm of the residual vector), the difference between input and calculated wave functions. The convergence is achieved when the residual becomes zero. Note that working with the residual vector (the norm of the residual vector is clearly positive) removes the orthogonality requirement because the norm of the residual vector has an unconstrained local minimum at each eigenvector, which can speed up the calculation significantly. After updating all wave functions, orthogonalization is then carried out conveniently by subspace diagonalization.

Its parallel implementation is also possible based on a band-by-band optimization of wave functions (information about other bands is not required). Thus, the RMM-DIIS algorithm is considered to be fast and efficient for large systems but sometimes is apt to miss the right wave function at the initial stage of iteration. (If the initial set of wave functions does not span the real ground state, it might happen that, in the final solution, some eigenvectors will be missing.) This fact naturally suggests the optimal option for us: starting with the Davidson method, followed by the RMM-DIIS method.

The following describes the general iterative procedure in a typical DFT run:

- Choose a proper PP, XC functional, and ENCUT.
- Construct the initial electron density of solids approximated by superimposing the electron densities of each atom in its isolated state as provided by the PP.
- Calculate U_{xc} and other terms to evaluate $\hat{H}_{KS}$.
- Generate one-electron orbitals covering all occupied $n/2$ levels (and some more) and expand into PWs.
- Solve a set of coupled KS equations by diagonalization of the KS matrix and obtain the solutions, which is a set of $c_{nk}(\boldsymbol{G})$ which minimizes the total energy and equivalent to a set of KS orbitals.
- Use the new $\phi_i(\boldsymbol{r})$ and calculate new $\rho(\boldsymbol{r})$, and mix it with the previous densities (see Section 6.6.3).
- Resume for the next self-consistent (SC) loop and, after successive improvements of $\hat{H}_{KS}$ and $\rho(\boldsymbol{r})$, stop iteration when the energy change (or density change) of the system becomes less than a preset stopping criterion (normally 10^{-4}–10^{-5} eV).
- Then the convergence is reached and the energy calculated is the ground-state energy of the system at the very ground-state charge density.

Charge mixing step is an essential part of in SC loop for convergence. Consistent overshooting of charge density causes a problem known as charge sloshing where the charge density drifts around without reducing the residuum. Charge mixing is a standard way to avoid it by updating the density with old and new densities in an optimal proportion (e.g., 60% old ρ plus 40% new ρ). All DFT programs with an iterative approach adopt linear or other elaborate mixing schemes.

6.6.2 Ionic minimizations

The DFT calculation aims for a system fully relaxed both electronically and structurally. All topics up to now have concerned minimizing the electronic energy, with the inner SC loop at fixed ionic positions. After each electronic relaxation, the forces acting on each ion are calculated, and ions are driven to downward directions of forces for the next electronic relaxation. This outer loop, the ionic minimization, includes relaxations of ionic positions and/or unit cell shape/size. After successive minimization of electronic and atomic energies as shown in Figure 5.13, the system will reach the minimum-energy configuration at which the forces at each atom vanish close to zero (normally <0.01–0.05 eV/Å). We first see how the forces at each atom are calculated and then move to three algorithms that are commonly implemented for this ionic minimization: the quasi-Newton method, the conjugate gradient (CG) method, and the damped MD method.

6.6.2.1 Hellmann–Feynman forces

Forces on atoms arise from both atomic and electronic sources, and the Hellmann–Feynman theorem (Hellmann 1937; Feynman 1939) provides an efficient way to account for them, which does not require any additional effort (e.g., calculation of $\partial\phi/\partial r_r I$) other than normal electronic minimization. The theorem states that, if an exact $\hat{H}$ and the corresponding ϕ_i are calculated, the force on an atom is the expectation value of the partial derivative of $\hat{H}$ with respect to atomic position $\boldsymbol{r}_I$. Since only two potential terms are related to $\boldsymbol{r}_I$, the theorem leads to

$$\boldsymbol{F}_I = -\frac{dE}{d\boldsymbol{r}_I} = -\left\langle \phi_i \left| \frac{\partial \hat{H}}{\partial \boldsymbol{r}_I} \right| \phi_i \right\rangle = -\frac{\partial U_{IJ}}{\partial \boldsymbol{r}_I} - \int \frac{\partial U_{\text{ext}}}{\partial \boldsymbol{r}_I} \rho(\boldsymbol{r}) d\boldsymbol{r} \tag{6.43}$$

Therefore, after an electronic iteration, the forces can be calculated accordingly by taking simple derivative operations on two potential terms. This is why the force calculations, which are so fast, are almost unnoticed in a normal DFT run. Based on these calculated forces, we can direct in which direction and how much the atoms should move. One remark should be made here. If the basis set depends on atomic positions, the Hellmann–Feynman equation has an extra term called Pulay forces. The PW approach has no such problem since it is nonlocal.

6.6.2.2 Minimization methods

Three commonly implemented schemes for ionic minimization are the conjugate gradient (CG), the quasi-Newton, and the damped MD methods. The first two are already discussed in Section 2.5.2, which apply in principle the same way here. The damped MD scheme decides search direction based on the previous energy gradient and a force term adjusted with a damping factor. This method for the structural relaxation is most appropriate when the system is rather away from its minimum configuration, for example, the NEB calculations, *ab initio* MD, and so on.

6.6.3 Born–Oppenheimer molecular dynamics

In classical MD, the use of empirical potentials causes inevitable drawbacks: many different situations require different potentials and parameters. Furthermore, the electronic and magnetic properties cannot be determined, and the bonding nature cannot be specifically identified. If we couple the MD and the DFT together and treat atoms and electrons respectively, these drawbacks can be eliminated (Payne et al. 1992).

Two methods are most commonly implemented in practice: the Born–Oppenheimer MD (BOMD) and the Car–Parrinello MD (CPMD). All methodologies used in MD in Chapter 2 basically apply here for *ab initio* MD as far as atomic movement is concerned: Newton's equations of motion, periodic boundary conditions, timestep, and integration algorithms such as velocity Verlet method. The critical issue is how to introduce the electronic dynamics effectively.

BOMD can be considered an extension of the ionic minimization that we have discussed. Here, electrons are allowed full relaxation self-consistently via the DFT each time for a given set of ionic configurations:

- Fix positions of nuclei; solve the KS equations self-consistently for electronic minimization.
- Find electrostatic forces on each atom via the Hellmann–Feynman theorem.
- Move atoms via classical mechanics (forward integration) with a timestep and find new positions of atoms.
- Repeat the process until convergence.

The Lagrangian is given by

$$L_{\text{BO}} = E_{\text{kin}} - E = \frac{1}{2}\sum_I m_I \dot{r}_I^2 - E\left[\rho(\boldsymbol{r}), \{\boldsymbol{r}_I\}\right] \tag{6.44}$$

Here, the first term is the kinetic energies of nuclei, and the second term is the electronic total energy obtained by the DFT, which serves as the potential energy for the nuclei. Computationally, this is a very slow process due to full electronic minimization by standard DFT procedures at each MD step.

6.6.4 Car–Parrinello molecular dynamics

6.6.4.1 CPMD for electronic minimization

Variational minimization for the search of the electronic ground state can be achieved dynamically by the simulated annealing approach pioneered by Car and Parrinello (1985). The method couples both electronic and ionic degrees of freedom on the classical coordinate system. The set of coefficients of the PW basis set is treated classically as a set of coordinates and improved by a series of iterations. The given kinetic energies cool down, and the $\{c_{nk}\}$ (corresponds to ϕ) reaches the ground state that minimizes all energy functionals. The orthonormality constraints on the wave functions are achieved with the Lagrange multiplier method. Generally, standard CPMD only for the electronic ground state is relatively slow and often encounters difficulties when applied to metals due to charge fluctuations of the mobile valence electrons. Note, however, that this method has successfully been extended to an efficient *ab initio* MD.

Figure 6.20 visualizes the final two timesteps of the CPMD, showing iterative diagonalizations for the ground-state energy in a simplified illustration.

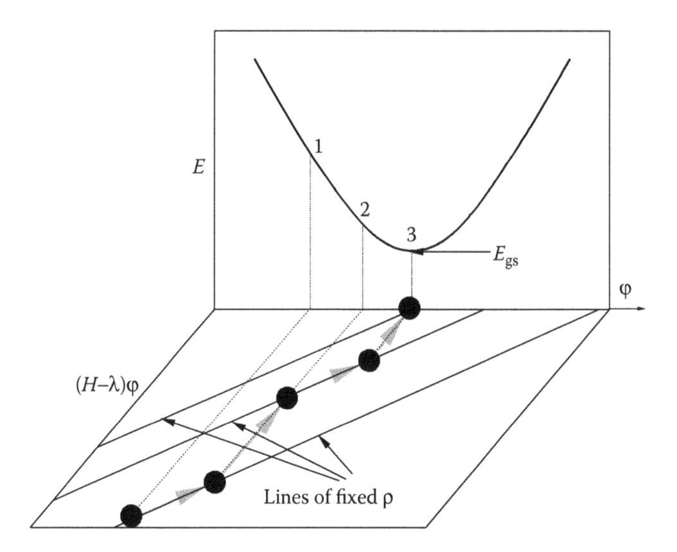

Figure 6.20 Schematic of an iterative minimization in a CPMD run for the search of the ground state where $\delta E/\delta\phi = 0$.

It connects the residuum change (lower part of the figure) with orbital optimizations and drives the energy toward the minimum value (upper part of the figure), which proceeds as follows:

- At the start of iteration, the trial charge density is rather away from the ground state, and the corresponding energy is also off the minimum value (point 1). Note that if a calculation follows along a solid line of fixed charge density all the way to the ϕ axis, this would correspond to the conventional matrix diagonalization of a static KS Hamiltonian, $\hat{H}$.
- With iterative diagonalization at this fixed charge density, the residual vector, $\boldsymbol{R}$, is reduced, the ϕ is improved, and the corresponding energy moves closer to the minimum value (point 2).
- At this point after a timestep, a new and better charge density is calculated, and the new iterative diagonalization starts with a new static $\hat{H}$ along the line of this fixed charge density.
- The residual vector is further reduced to zero, and the corresponding energy moves to the minimum value (point 3 where $\delta E/\delta\phi = 0$).

6.6.4.2 *CPMD for ionic minimization*

If we move ions (or nuclei) in the system toward lower forces, a series of U-type energy curves will result as shown in Figure 6.20. When the ionic and electronic minimizations are achieved together (reaching the minimum point of the lowest U curve), we can declare that the system is at its ground state and the corresponding total energy is the ground-state energy. In BOMD, the process is carried out in series. Car and Parrinello (1985), however, accounted for the motions of nuclei and electrons at the same time in parallel fashion in an extended classical MD.

The CPMD approach assigns a small fictitious mass μ to all orbitals so that they can have coordinates like a classical particle and become time dependent. Here, the mass assigned is small enough that the coordinates of nuclei and electrons are still decoupled. Then, the dynamics are written by the extended Lagrangian for both nuclei and orbitals:

$$L_{CP} = \frac{1}{2}\sum_I m_I \dot{r}_I^2 + \frac{1}{2}\mu_i \sum_i \int \left|\dot{\phi}_i\right|^2 dr - E\left[\rho(r),\{r_I\}\right] + \sum_{ij} \Lambda_{ij}(\langle\phi_i|\phi_j\rangle - \delta_{ij}) \quad (6.45)$$

Here:

m_I is the mass of the nuclei

μ_i is the fictitious electron mass with unit of energy $\times t^2$ and the dot indicates a time derivative

Note that if μ_i is zero, this CPMD equation becomes the BOMD equation (Equation 6.44). The first and second terms are the kinetic energies of nuclei and orbitals, respectively. The fictitious kinetic energy of an orbital serves as a measure for deviations of the system from the exact Born–Oppenheimer surface. The third term is the electronic total energy obtained by the DFT, which serves as the potential energy for the nuclei. The last term is a constraint that is necessary to ensure the orthonormality of orbitals with the Lagrange multipliers, Λ_{ij}.

In this scheme, one needs only a rough estimation on the forces for a given atomic configuration, and thus it is not necessary to achieve full electronic minimizations, especially when the system is far from its ground state. Electronic density, therefore, is allowed to fluctuate around its ground-state value during the on-the-fly optimization. However, the force calculations are 10^{4-5} times more expensive than that of classical MD due to electronic minimization at each time evolution.

This CPMD seems to have an edge over BOMD in speed due to the classical treatment employed for electrons. The method works well for insulators and semiconductors and is especially effective with small timesteps and for simulated annealing. For metals, however, the system may drift away from the potential surface due to the coupling of electrons and ions. To speed up calculations, the Γ-point-only sampling is often used.

6.6.5 Multiscale methods

The DFT has progressed to the size level of about 1000 atoms if one has computational power that is more than moderate, which is a remarkable feat. Considering the actual laboratory experiments, however, we are still far from the scale ranges that we want to have in both size and time. The so-called multiscale simulations have been developed to be the alternative and to extend the size and time scales accessible by computations.

6.6.5.1 Crack propagation in silicon

In an embedding approach (Csanyi et al. 2004) for crack propagation in silicon, for example, a small part that requires a fine DFT treatment was embedded into a large matrix where only the MD treatment could be sufficient. The crack tip (about 300 atoms) was treated quantum mechanically, and the rest (200,000 atoms) classically using the force-field potential. At the interface between two regions, the parameters of the force field were continually provided by quantum mechanical calculations. The study was able to observe the reconstruction of the crack surface with (2 × 1) periodicity and clearly demonstrated the promising potential of this multiscale approach for further developments and extensions.

A similar approach applies to the case of Si oxidation (Nakano et al. 2001): Si bulk by the MD treatment and the oxidation on surface by the *ab initio* MD. Other examples in this category that require the hierarchical coupling of different approaches include the descriptions of diffusion, chemical reaction, catalytic process, and corrosion.

Homework

6.1 Show that any product of real and reciprocal lattice vectors is $2\pi\times$ integer number.

6.2 Show that PW basis is all orthonormal.

Show that electron density is periodic in periodic solid with the Bloch theorem.

6.3 The reciprocal lattice is the set of wave-vectors, $\mathbf{G}$, that satisfies $\exp(i\mathbf{G}\cdot\mathbf{R}) = 1$. Using this relationship, prove that any local quantity in a periodic solid is invariant by the translation with $\mathbf{R}$ as

$$U(\mathbf{r}) = U(\mathbf{r}+\mathbf{R})$$

6.4 Why metals generally require more $\mathbf{k}$-points for a DFT calculation? For electron density calculations in the reciprocal space, the Fourier grid must contain all wave-vectors up to $2G_{\text{cut}}$ (see Figure 6.14). Explain why it is. A given E_{cut} will decide the number of PWs, N_{pw}, required for the finite PW expansion. Show that $N_{\text{pw}} \propto E_{\text{cut}}^{3/2}$.

6.5 The Poisson equation for the Hartree potential is

$$\nabla^2 U_H(\mathbf{r}) = -4\pi\rho(\mathbf{r})$$

Show that the equivalent equation in terms of $\mathbf{G}$ in reciprocal space by the PW expansion.

References

Bloch, F. 1928. Über die Quantenmechanik der Elektronen in Kristallgittern. *Z. Physik* 52:555–600.

Blöchl, P. E. 1994. Projector augmented-wave method. *Phys. Rev. B* 50:17953–17979.

Blöchl, P. E., O. Jepsen, and O. K. Andersen. 1994. Improved tetrahedron method for Brillouin-zone integration. *Phys. Rev. B* 49:16223–16233.

Car, R. and M. Parrinello. 1985. Unified approach for molecular dynamics and density-functional theory. *Phys. Rev. Lett.* 55:2471–2474.

Csanyi, G., T. Albaret, M. C. Payne, and A. De Vita. 2004. "Learn on the fly": A hybrid classical and quantum-mechanical molecular dynamics simulation. *Phys. Rev. Lett.* 93:175503–175506.

Davidson, E. R. 1975. The iterative calculation of a few of the lowest eigenvalues and corresponding eigenvectors of large real-symmetric matrices. *J. Comput. Phys.* 17:87–94.
Dirac, P. A. M. 1926. On the theory of quantum mechanics. *Proc. R. Soc. A* 112:661–677.
Fermi, E. 1926. Sulla quantizzazione del gas perfetto monoatomico. *Rend. Lincei* 3:145–149.
Feynman, R. P. 1939. Forces in molecules. *Phys. Rev.* 56:340–343.
Fu, C. L. and K. M. Ho. 1983. First-principles calculation of the equilibrium ground-state properties of transition metals: Applications to Nb and Mo. *Phys. Rev. B* 28:5480.
Hamann, D. R., M. Schluter, and C. Chiang. 1979. Norm-conserving pseudopotentials. *Phys. Rev. Lett.* 43:1494–1497.
Heine, V. 1970. The pseudopotential concept. *Solid State Phys.* 24:1–36.
Hellmann, H. 1937. *Einfuhrung in die Quantenchemie.* Leipzig, Germany: Deuticke.
Kresse, G. and J. Furthmüller. 1996. Efficient iterative schemes for ab initio total-energy calculations using a plane wave basis set. *Phys. Rev. B* 54:11169–11186.
Kresse, G. and J. Joubert. 1999. From ultrasoft pseudopotentials to the projector augmented-wave method. *Phys. Rev. B* 59:1758–1775.
Methfessel, M. and A. T. Paxton. 1989. High-precision sampling for Brillouin-zone integration in metals. *Phys. Rev. B* 40:3616–3621.
Monkhorst, H. J. and J. D. Pack. 1976. Special points for Brillouin-zone integration. *Phys. Rev. B* 13:5188–5192.
Nakano, A. et al. 2001. Multiscale simulation of nanosystems. *Comp. Sci. Eng.* 3:56–66.
Payne, M. C., M. P. Teter, D. C. Allan, T. A. Arias, and J. D. Joannopoulos. 1992. Iterative minimization techniques for ab initio total-energy calculations: Molecular dynamics and conjugate gradients. *Rev. Mod. Phys.* 64:1045–1097.
Pulay, P. 1980. Convergence acceleration in iterative sequences: The case of SCF iteration. *Chem. Phys. Lett.* 73:393–398.
Troullier, N. and J. L. Martins. 1991. Efficient pseudopotentials for plane wave calculations. *Phys. Rev. B* 43:1993–2006.
Vanderbilt, D. 1990. Soft self-consistent pseudopotentials in a generalized eigenvalue formalism. *Phys. Rev. B* 41:7892–7895.
Wood, D. M. and A. Zunger. 1985. A new method for diagonalizing large matrices. *J. Phys. A: Math. Gen.* 18:1343–1359.

Further reading

Kohanoff, J. 2006. *Electronic Structure Calculations for Solids and Molecules: Theory and Computational Methods.* Cambridge, UK: Cambridge University Press.

chapter seven

DFT exercises with Quantum Espresso

In the following three chapters, we are dealing with Density functional theory (DFT) in actual calculations for real atoms, molecules, solids, surfaces, and interfaces. To do that, we will bring three different packages as shown in Table 7.1. These are undoubtedly the leading packages in the field capturing all features of DFT capabilities built on the same approaches: periodic boundary conditions (PBC), plane wave (PW) expansion of the KS orbitals, and pseudopotential (PP) scheme. Note, however, that all of packages have their own flavors and merits, which we will explore by going through various exercises. In this chapter, we employ our first pick, Quantum Espresso.

7.1 Quantum espresso

7.1.1 General features

Quantum Espresso (QE) is a full *ab initio* package implementing all features of DFT:

- Electronic and ionic minimizations (or relaxations) for total energies, band structures, density of states, etc.) using ultrasoft (US), norm-conserving, and PAW pseudopotentials.
- Linear response methods to calculate phonon spectra, dielectric constants, and Born effective charges.
- Very efficient CPMD (Car-Parrinello MD), FPMD (First-Principles MD), and so on.

Although QE can deal with any level of electronic structure calculation, it is most suited for beginners to quantum world in three respects. First, it is free and open code. And it can be fast in calculation with some sacrifice on accuracy. Finally, the Windows version has been recently available. In short, it is best for anyone who wants to start the first principles approach in no time. And that is exactly what we will do in this chapter with five simple examples, which are very introductory and have been introduced in various tutorials for QE. Yet, we will have a good opportunity to review all basics we have learned in the previous three chapters.

Table 7.1 Description of three packages implemented for the DFT runs

Packages	Description	Remarks
Quantum Espresso	• Best for classrooms • Best for CPMD • Limited built-in PP	Free and open code
VASP	• The leading package • Best for serious calculations	Purchase required
MedeA-VASP	• VASP with GUI • Many convenient tools including flow diagrams • Best for all users for faster results	Purchase required

7.1.2 Installation

We use the Windows version of QE for a start since it requires no additional preparation.

- Download the package from http://www.quantum-espresso.org/; DOWNLOAD > the download page > qe-5.1.2-32bit-serial.exe (54 MB). Path will be set automatically during installation.
- Install in C/Program Files; confirm bin directory where executable codes reside and "pseudo" directory where pseudopotential files reside.
- Confirm codes such as "scf," "relax," "bands," "nscf," "md," and so on. These code names will be written into "&control/calculation=" in the input file.
- Download resources; QE-on-Windows.pdf (331 KB), input file description, tutorial, frequent-errors-during-execution, and so on.

7.2 Si2

7.2.1 Introduction

In this first exercise, we look for the energy of a primitive Si system with two atoms (an FCC lattice or diamond lattice) by a self-consistent field (scf) calculation only at a fixed ion positions. In fact, all the following exercises are also for Si systems to be familiar with DFT easier and faster.

7.2.2 Si2.in

Running with QE, we write all instructions in one input file for a calculation such as Si2.in for this exercise. Note that the file name should always be as ***.in contrary to the LAMMPS case. The file structure consists of many so-called name lists and cards. For this simple run, however, we use only the mandatory ones:

- Three name lists: &control, &system, and &electrons.
- Three cards: ATOMIC_SPECIES, ATOMIC_POSITIONS, K_POINTS.
- See more information on http://www.quantum-espresso.org/wp-content/uploads/Doc/INPUT_PW.html

```
############################################################
&control
   calculation = 'scf',          # self-consistent field
                                 calculation, default
   prefix = 'Si2'                # prefix for output files
   pseudo_dir = QE/pseudo        # if PP is not on the path
/                                # end of &control
&system
   ibrav = 2,                    # 2; fcc, 1; simple cubic,
                                 3; bcc, 4; hcp
   celldm(1) = 10.333,           # lattice parameter (bohr),
                                 5.47 Å, exp. = 5.43 Å
   nat = 2, ntyp = 1,            # 2 atoms, 1 type (by
                                 ATOMIC_SPECIES)
   ecutwfc = 20                  # size of PW basis set to
                                 expand KS orbitals
/
&electrons
   mixing_beta = 0.7             # 0.7 new + 0.3 old
                                 densities
/
ATOMIC_SPECIES
 Si 28.086 Si.pbe-rrkj. UPF      # name, mass, PP
ATOMIC_POSITIONS (alat)          # direct x, y, z
 Si 0.0 0.0 0.0
 Si 0.25 0.25 0.25
K_POINTS (automatic)             # MP method
  6 6 6 1 1 1                    # Nk1, Nk2, Nk3, shift1,
                                 shift2, shift3
                                 # 0; no shift, grid from
                                 0 to Nk
                                 # 1; shift, grids shifted
                                 by half a grid
############################################################
```

Note that a uniform MP grid is generated automatically for K_POINTS with a shift along the three directions by half a grid (Monkhorst and Pack 1976). It is obvious that the sifting can have a better coverage for sampling, and normally reaches convergence faster with less *k*-points. The rest is set as the default values, for example, convergence threshold = 1.0E-06. In the actual run, the remarks with "#" should be removed in the actual Si2.in file to avoid any run failure.

7.2.3 Si.pbe-rrkj.UPF

Confirm that the potential file, Si.pbe-rrkj.UPF, is on the path, and can be read automatically. Normally, The PP directory "pseudo" is automatically set on the path when the QE package is first installed.

```
###############################################################
      <PP_INFO>
Generated using Andrea Dal Corso code (rrkj3)
Author: Andrea Dal Corso   Generation date: unknown
Info: Si PBE 3s2 3p2 RRKJ3
  0    The Pseudo was generated with a Non-Relativistic
       Calculation
  2.50000000000E+00 Local Potential cutoff radius
nl  pn  l  occ   Rcut            Rcut US        E pseu
3S  1   0  2.00  2.50000000000  2.60000000000  0.00000000000
3S  1   0  0.00  2.50000000000  2.60000000000  0.00000000000
3P  2   1  2.00  2.50000000000  2.70000000000  0.00000000000
3D  3   2  0.00  2.50000000000  2.50000000000  0.00000000000
</PP_INFO>

<PP_HEADER>
  0    Version Number
  Si   Element
  NC   Norm - Conserving pseudopotential
  F    Nonlinear Core Correction
 SLA PW PBE PBE          PBE Exchange-Correlation functional
  4.00000000000          Z valence
  -7.47480832270         Total energy
  0.0000000 0.0000000    Suggested cutoff for wfc and rho
  2                      Max angular momentum component
  883                    Number of points in mesh
  2 3                    Number of Wavefunctions, Number of
                         Projectors
Wavefunctions            nl l occ
                         3S 0 2.00
                         3P 1 2.00
</PP_HEADER>

<PP_MESH>
  <PP_R>
  1.77053726905E-04   1.79729551320E-04   1.82445815642E-04
  1.85203131043E-04
…..
###############################################################
```

Confirm that the Si potential with four valence electrons is constructed by the norm-conserving way, and its XC energy is calculated by the PBE approach. The rest is just full of numbers for PP calculations, which is not much concern for us at this stage.

7.2.4 Run

Make a directory such as "Si2" and place the Si2.in file. Open the Windows Command Prompt (CMD window or DOS window) and check that the QE package is on the path by typing "path":

```
############################################################
> path
path=C:\Program files (x86)\Quantum ESPRESSO 64 bit serial\
bin;
############################################################
```

If it is not, you can add the path by typing the following command on the Windows Command Prompt:

```
############################################################
> setx path "%path%; C:\Program files (x86)\Quantum
ESPRESSO 64 bit serial\bin;"
############################################################
```

Then, go to the prepared directory and run the Si2.in script by:

```
############################################################
C:\Windows\System32>D:
D:\>cd QE-runs\Si2
D:\QE-runs\Si2>pw -in Si2.in > Si2.out
############################################################
```

The pw.exe program runs the script in no time, because we have only two atoms of Si in our system.

7.2.5 Si2.out

In the directory, two output files and one directory are generated by the run, and the Si2.out file has everything about the run including the total energy. We can confirm the converged total energy by looking for an exclamation mark (!) in the Si2.out file or by typing "find" command on the Windows Command Prompt:

```
################################################################
D:\QE-runs\Si2>find "total energy" Si2.out

---------- SI2.OUT
      total energy = -15.73887005 Ry
      total energy = -15.74081275 Ry
      total energy = -15.74122417 Ry
      total energy = -15.74125386 Ry
!     total energy = -15.74125494 Ry
      The total energy is the sum of the following terms:

D:\QE-runs\Si2>
################################################################
```

It tells us that the total energy of –15.74 Ry (1 Rydberg [Ry] = 13.6057 eV) is obtained in five iterations of electronic minimization at fixed ions.

- Doing the same way to find the "number of k points," we see that symmetry operations reduced the number of *k*-points from 6 × 6 × 6 = 216 to 28.
- There are eight electrons in this two-atom cell of nonmagnetic semiconductor, and the lowest four valence bands (KS states) are considered for calculation.
- Note that the Linux version of QE on a Linux machine runs with a slightly different syntax as "pw.x < Si2.in > Si2.out."
- Note also that kinetic-energy cutoff = 20.0000 Ry and charge density cutoff = 80.0000 Ry, which is four times of the former.

```
################################################################
D:\QE-runs\Si2>find "cutoff" Si2.out

---------- SI2.OUT
kinetic-energy cutoff = 20.0000 Ry
charge density cutoff = 80.0000 Ry

D:\QE-runs\Si2>
################################################################
```

Because electron density is the square of wave function, it can vary twice as rapidly. This implies that the cutoff for electron density should be four times larger than the corresponding cutoff for wave function. This leads to the conclusion that the PW basis set for electron density normally contains eight times more plane waves than that of wave function expansion, because $N_{pw} \propto E_{cut}^{3/2}$, as shown in Equation 6.40.

7.3 *Si2-convergence test*

7.3.1 *Introduction*

We now run convergence tests (more details later in Section 8.4) to have a proper ground-state energy for the Si system. Three tags of the input file are subjected to the test: energy cut for PW, *k*-point grid, and smearing degree. Because of the variational nature of *ab initio* calculations, this procedure is a must whenever a new system is introduced. In addition, we decide the equilibrium lattice parameter, ground-state energy, and bulk modulus at the converged condition.

7.3.2 *Si2-conE.in*

The kinetic energy cut, ecutwfc, determines how far we can go down for the search of the ground-state energy, and we vary it in the input file "Si2-conE.in" until we see no further change in energy.

- Make a new directory such as "Si2-conE."
- Use the same Si2.in file and change only ecutwfc = 12–28 by 2 Ry to include less or more plane waves in the basis set for wave function.
- Run each of them as before and plot ecutwfc vs. total E.

7.3.3 *Results*

The resulted data looks like:

```
#################################################################
12                     -15.7210
14                     -15.7329
16                     -15.7380
18                     -15.7399
20                     -15.7407
22                     -15.7410
24                     -15.7412
26                     -15.7412
28                     -15.7413
#################################################################
```

Figure 7.1 shows the plot, which indicates a monotonic convergence as a consequence of the variational principle. Note that total energies depend on which PP is used and do not have any physical meaning. But their differences with the same PP as shown in the plot do. And somewhere around ecutwfc = 20 Ry could be the right cutoff as we did in the Si2 exercise before.

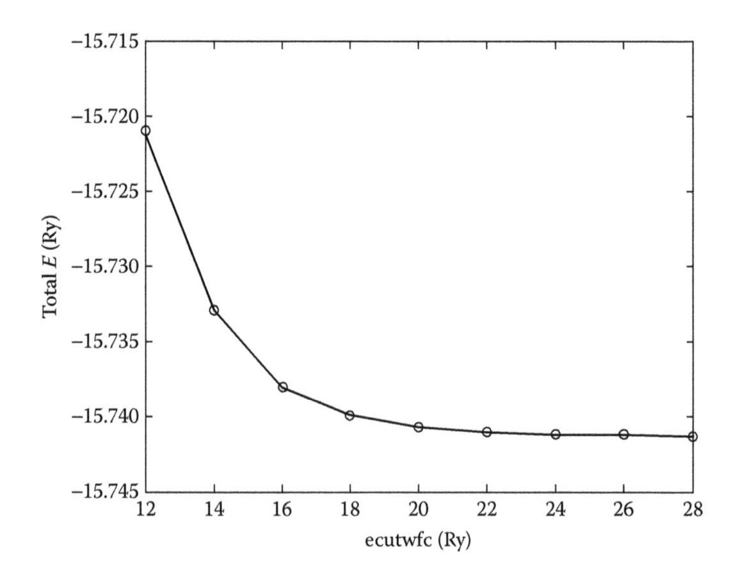

Figure 7.1 Convergence test for cutoff energy.

7.3.4 *Further runs*

Convergence tests for *k*-point grid is the same way as energy convergence by varying the *Nk* (number of *k*-point grids) = 2–9 at a fixed ecutwfc = 20 Ry. Running each of them as before, we have the following data as *Nk* vs. total *E*, which is plotted as shown in Figure 7.2.

```
##############################################################
Nk                  total E
222                 -15.7291
333                 -15.7394
444                 -15.7405
555                 -15.7407
666                 -15.7407
777                 -15.7407
888                 -15.7407
999                 -15.7407
##############################################################
```

We now confirm the converged *k*-points of 6 × 6 × 6. The curve needs not necessarily be monotonic, because *Nk* is not a variational entity. We can test for smearing width by varying the Gaussian/degauss values (normally 0.01–0.05 Ry) at a fixed ecutwfc = 20 Ry and *k*-points of 6 × 6 × 6. Note that we used the following default setting for smearing and its information is skipped in the input file.

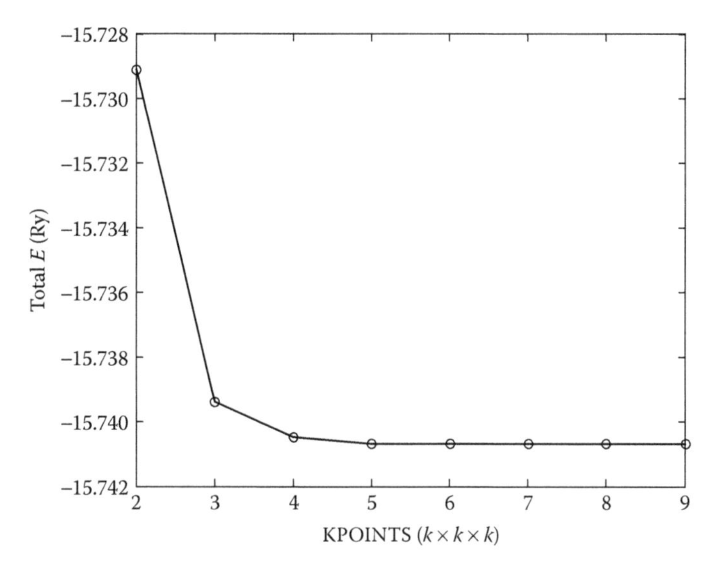

Figure 7.2 Convergence test for k-point grids.

```
################################################################
occupations = 'smearing'
smearing = 'gaussian'              # default, or mp, mv, fd
degauss = 0.01                     # value of the gaussian
                                   broadening (Ry)
################################################################
```

We move ahead to the next stage of this exercise, calculations for lattice constant vs. total energy at fixed ecutwfc = 20 Ry and k-points of $6 \times 6 \times 6$. We prepare Si2-lattice.in files with various celldm(1) numbers from 9.8 to 11.2 bohr (1 bohr = 0.529177 Å) by 0.2 bohr, and run 8 scf calculations. After the run, the following data file, Si2latticeE.dat (celldm(1) vs. total E), can be tabulated:

```
################################################################
 9.8                  -15.7167
10.0                  -15.7320
10.2                  -15.7397
10.4                  -15.7411
10.6                  -15.7372
10.8                  -15.7291
11.0                  -15.7174
11.2                  -15.7031
################################################################
```

Let us plot the data on MATLAB® and find out the minimum energy at the equilibrium lattice parameter graphically:

```
#################################################################
plot(A(:,1).*0.529177249, A(:,2).*13.6056981,'-o');
Tools > Basic Fitting > cubic > Show equations > 4
syms x;
y=-1.846*x^3 + 33.74*x^2 - 203.4*x+191.2;
dy=diff(y)
#################################################################
```

The plotted parabolic curve is symmetric about the minimum, and Figure 7.3 shows where the differentiated potential curve becomes zero. We have –15.74 eV at the equilibrium lattice parameter of 5.47 Å (=10.333 bohr), which is the ground state at the converged condition. The equilibrium lattice parameter is in fair agreement with the experimental value of 5.43 Å, and may indicate the tendency of overestimation by GGA/PBE PP. Note that we can have the same result by using MATLAB's "diff" or "solve" function.

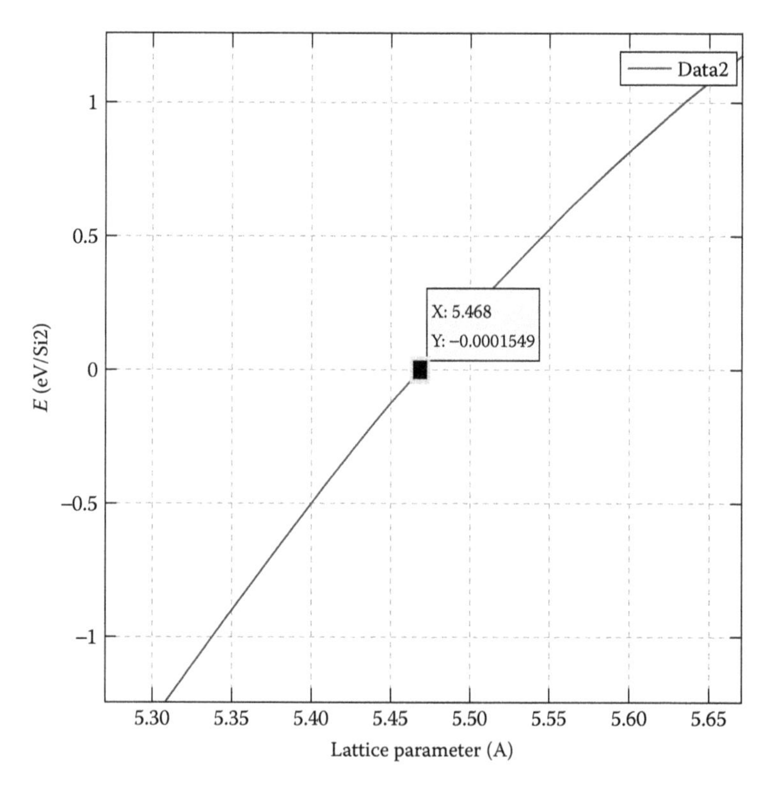

Figure 7.3 Graphical determination of the equilibrium lattice parameter of Si2.

The final stage of this exercise is calculation of bulk modulus by using the Si2latticeE.dat file and the ev.exe program. The program will fit this data to an equation of state with the information provided:

```
############################################################
D:\QE-runs\Si2BulkM>ev
D:\QE-runs\Si2BulkM>ev
     Lattice parameter or Volume are in (au, Ang) > au
     Enter type of bravais lattice (fcc, bcc, sc, noncubic)
     > fcc
     Enter type of equation of state:
     1=birch1, 2=birch2, 3=keane, 4=murnaghan > 2
     Input file > Si2latticeE.dat
     Output file > Si2latticeE.out
############################################################
```

The resulted Si2latticeE.out shows a bulk modulus of 87.2 GPa (k0 value) at a0 = 5.47561 A, which is comparable with the experimental value of 97.6 GPa (Hopcroft et al. 2010):

```
############################################################
# equation of state: birch 3rd order. chisq = 0.5331D-09
# a0 = 10.3474 a.u., k0 = 871 kbar, dk0 = 4.31 d2k0 =
-0.009 emin = -15.74128
# a0 = 5.47561 Ang, k0 = 87.2 GPa, V0 = 276.971 (a.u.)^3, V0
= 41.043 A^3
…. .
############################################################
```

7.4 Si2-band

7.4.1 Introduction

Plotting the band structure of Si, we now need three QE execution programs in a roll: pw.ex for scf and bands calculations, and bands.ex for orbital energy calculation at *k*-points.

7.4.2 Si2-scf

First, make a new directory such as "Si2-scf" and run a scf calculation as we have already done in Section 7.2 with a minor editing for a bands calculation as:

```
############################################################
&control
    calculation = 'scf',
    prefix = 'Si2-bands',
```

```
    verbosity = 'high'
# prints out additional information such as occupation
numbers, etc.
…..
#############################################################

#############################################################
D:\QE-runs\Si2-scf>pw -in Si2-scf.in > Si2-scf.out
#############################################################
```

The run generates three outputs and the next step will read two of them: Si2-bands.save directory and Si2-bands.wfc file.

7.4.3 Si2-bands

To obtain the band structure, we now have to run a non-self-consistent calculation:

- Make a new directory such as "Si2-bands."
- Copy the previous Si2-scf.in and rename it as Si2-bands.in.
- Edit it with revised and new instructions as:

```
#############################################################
&control
  calculation = 'bands',
  prefix = 'Si2-bands',
  verbosity = 'high'
.....
&system
       nbnd = 8
.....
K_POINTS
       16
       0.5 0.5 0.5   1
       0.4 0.4 0.4   2
       0.3 0.3 0.3   3
       0.2 0.2 0.2   4
       0.1 0.1 0.1   5
       0.0 0.0 0.0   6
       0.0 0.0 0.1   7
       0.0 0.0 0.2   8
       0.0 0.0 0.3   9
       0.0 0.0 0.4   10
       0.0 0.0 0.5   11
       0.0 0.0 0.6   12
       0.0 0.0 0.7   13
       0.0 0.0 0.8   14
```

```
      0.0 0.0 0.9   15
      0.0 0.0 1.0   16
############################################################
```

Note that K_POINTS follows the high symmetry points instead of grid points. Run a "bands" calculation with a set of *k*-points on the same number of processors as the first run.

```
############################################################
D:\QE-runs\Si2-bands>pw < Si2-bands.in > Si2-bands.out
############################################################
```

Now, the bands.exe program separates the band values from the data with the following bands.in file:

```
############################################################
&bands
  prefix = 'Si2-bands',
  filband = 'bands.dat'
/
############################################################

############################################################
D:\QE-runs\Si2-bands> bands < bands.in > bands.out
############################################################
```

The bands.dat file is generated with 8 KS energies at each point of 16 *k*-points (total 128 data points), and the bands.dat.gnu (128×2 matrix) is also generated to be used in gnuplot or MATLAB. The most convenient way of plotting it is by loading the Si.bands.out file directly on VNL/Bandstructure Analyzer/Bandstructure, which will plot a nice band structure shown in Figure 7.4. Note that VNL (Virtual Nano Lab)* is now GUI (graphic user interface) for QE and normally provides a 30-day trial license.

7.4.4 *Results and discussion*

Comparing with experimental data, we realize that the calculated value of about 0.6 eV is only half of the experimental value of 1.2 eV. We also can see that a similar run with the VASP program has almost the same result in the next chapter. As discussed in Section 5.6.2, this is due to the inherent inaccuracy for the calculation of XC energy in the frame of DFT. We improve the accuracy later in Section 9.1 by using a hybrid potential, HSE06, instead of the PBE potential.

* VNL (Virtual Nano Lab). http://quantumwise.com/products/download.

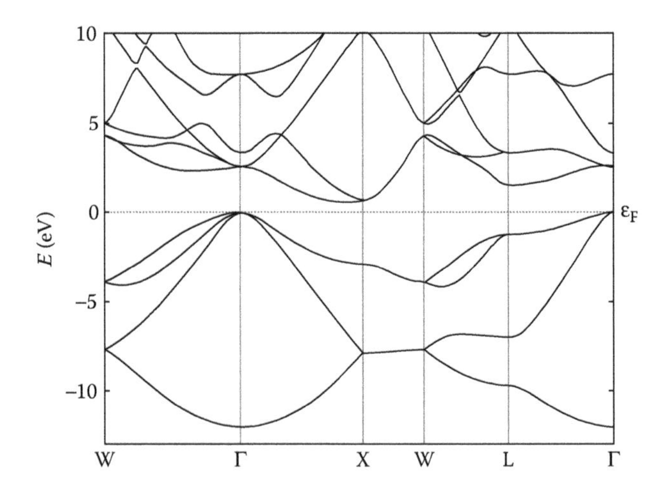

Figure 7.4 Band structure of Si plotted by VNL.

7.5 *Si7-vacancy*

7.5.1 *Introduction*

Perfect solid does not exists because any solid definitely has one of the following defects:

- *Point defects*: 0-dimensional; vacancy, interstitial, substitutional
- *Line defects*: 1-dimensional; dislocation
- *Planar defects*: 2-dimensional; interface, grain boundary, surface
- *Bulk defects*: 3-dimensional; void, inclusions, precipitates, and so on
- *Crystallinity*: poly, amorphous

While defects such as dislocation and interface raise the free energy of a material, the presence of a certain number of point defects to an otherwise perfect crystal reduces its free energy, because of a gain of configurational entropy caused by the many possible sets of places in the crystal in which the point defects can exist as shown in Figure 7.5.

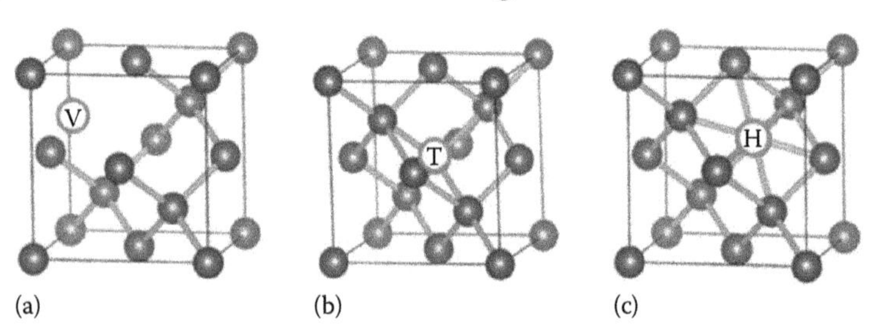

Figure 7.5 Point defects in Si: (a) vacancy, (b) interstitials in tetrahedral, and (c) hexagonal sites.

The goal of this exercise is calculation of vacancy formation energy in Si with an Si8 system. To create a vacancy in Si8, an atom is removed and the atom positions are relaxed, and its energy is compared with that of a perfect Si8 by using two QE execution programs:

- pw.ex/scf
- pw.ex/relax

Undoubtedly, the supercell is too small to represent the actual vacancy in a bulk solid. But remember that our goal here is learning the principle and procedure in a given time. Calculation of vacancy formation energy in a bigger supercell of Si64 is given in Homework 7.2.

7.5.2 Si8-scf

For this run, only thing we need is a minor editing of the Si2.in file to have the Si8-scf.in file as:

```
############################################################
.....
    prefix = 'Si8',
.....
    nat = 8,
.....
ATOMIC_POSITIONS (alat)
 Si 0.0 0.0 0.0
 Si 0.5 0.5 0.0
 Si 0.0 0.5 0.5
 Si 0.5 0.0 0.5
 Si 0.25 0.25 0.25
 Si 0.75 0.75 0.25
 Si 0.75 0.25 0.75
 Si 0.25 0.75 0.75
K_POINTS {automatic}                # KPOINTS
4 4 4 1 1 1
############################################################
```

Running is the same as before and we can have the total energy of a perfect Si8 crystal in five self-consistent calculations without any ionic position change:

```
############################################################
D:\QE-runs\Si8-scf> pw < Si8.scf.in > Si8.scf.out
> find "!" Si8.scf.out
! total energy = -62.9643 Ry
############################################################
```

7.5.3 *Si7v-relax*

From the above Si8-scf.in file, we remove an Si atom at 0.25 0.25 0.25 to create a vacancy, and make the Si7v.relax.in file:

```
##############################################################
&CONTROL
       calculation='relax',                        # INCAR
       prefix='Si7vRelax'
       disk_io='high'                              # INCAR
/
&SYSTEM
       ibrav = 1, celldm(1) =10.333,               # POSCAR
       nat=7, ntyp= 1,                             # POSCAR
       ecutwfc = 20.0                              # INCAR
/
&ELECTRONS
       mixing_beta = 0.7                           # INCAR
       conv_thr = 1.0d-8                           # INCAR
/
&IONS
       ion_dynamics='bfgs'                         # INCAR
/
ATOMIC_SPECIES
Si 28.086 Si.pbe-rrkj. UPF                         # POTCAR
ATOMIC_POSITIONS (alat)                            # POSCAR
Si     0.003081476   0.003081476   0.003081476
Si     0.496918524   0.496918524   0.003081476
Si     0.496918524   0.003081476   0.496918524
Si     0.003081476   0.496918524   0.496918524
Si     0.750000000   0.750000000   0.250000000
Si     0.750000000   0.250000000   0.750000000
Si     0.250000000   0.750000000   0.750000000
K_POINTS {automatic}                               # KPOINTS
4 4 4 1 1 1
##############################################################
```

New card &IONS is needed to control ionic motion for structural relaxation and MD run where atoms move. Note that each line is marked with the corresponding input file names of the VASP program, which will be implemented in the next chapter. In VASP, inputs are divided into four different files, while QE writes all in one file as shown above. Run a "relax" calculation with the above Si7v.relax.in file and check the total energy:

```
##############################################################
> pw -in Si7v.relax.in > Si7v.relax.out
# from now on, we skip the directory path and just use ">"
instead.
##############################################################
```

The pw code "relax" performs eight ionic relaxations to accommodate a vacancy in the supercell as shown in the Si7v.relax.out file below. We can confirm that force on all atoms becomes close to 0, and the final relaxed energy is −54.8673 Ry. We can also notice that four atoms (#1 Si to #4 Si) around the vacancy relaxed toward the vacancy.

```
############################################################
…..
convergence has been achieved in 8 iterations

    Forces acting on atoms (Ry/au):

    atom 1 type 1 force =
    -0.00043565 - 0.00043565 - 0.00043565
    …..
    Total force = 0.001509 Total SCF correction = 0.000084
    …..
    Final energy = -54.8673161655 Ry
Begin final coordinates

ATOMIC_POSITIONS (alat)
Si      0.003081476   0.003081476   0.003081476
Si      0.496918524   0.496918524   0.003081476
Si      0.496918524   0.003081476   0.496918524
Si      0.003081476   0.496918524   0.496918524
Si      0.750000000   0.750000000   0.250000000
Si      0.750000000   0.250000000   0.750000000
Si      0.250000000   0.750000000   0.750000000
End final coordinates
…..
############################################################
```

Let us view Si7v.relax.out on the VNL program* and export the file so that we can visualize nicely on the VESTA program†:

```
############################################################
VNL > Classic View > My Computer > D: > QE > qeEx >
qeEx4SiVacancy > Si7v.relax.out > Viewer to see the
structure and Export > File name: ***, Files of type:
VASP(*)
############################################################
```

Figure 7.6 shows structures of Si7v on VNL, perfect Si8 and Si7v on VESTA. Note that #5 Si at the 0.25 0.25 0.25 site is missing in the Si7v structure.

* VNL (Virtual Nano Lab). http://quantumwise.com/products/download.
† VESTA. http://www.geocities.jp/kmo_mma/index-en.html.

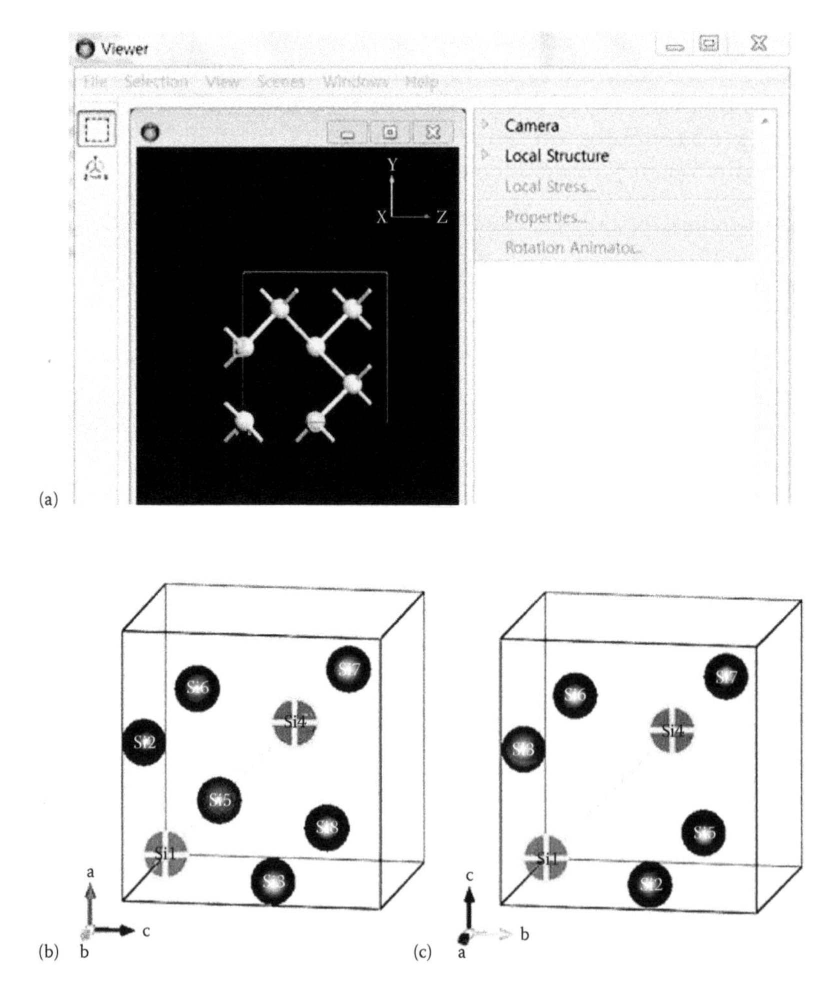

Figure 7.6 Structures of Si7v on VNL (a), perfect Si8 (b), and Si7v on VESTA (c).

We can check the change of interatomic distances around the vacancy before and after vacancy creation (e.g., #1 and #2 Si in the perfect Si8, and the corresponding #1 and #4 Si in the Si7v) on VESTA:

```
#################################################################
   l (Si1-Si2)  =  3.82020(0) Å
    1      Si1 Si 0.00308 0.00308 0.00308 (0, 0, 0)+ x, y, z
    2      Si2 Si 0.49692 0.49692 0.00308 (0, 0, 0)+ x, y, z
   l (Si1-Si4)  =  3.81879(0) Å
    1      Si1 Si 0.00308 0.00308 0.00308 (0, 0, 0)+ x, y, z
    4      Si4 Si 0.00308 0.49692 0.49692 (0, 0, 0)+ x, y, z
#################################################################
```

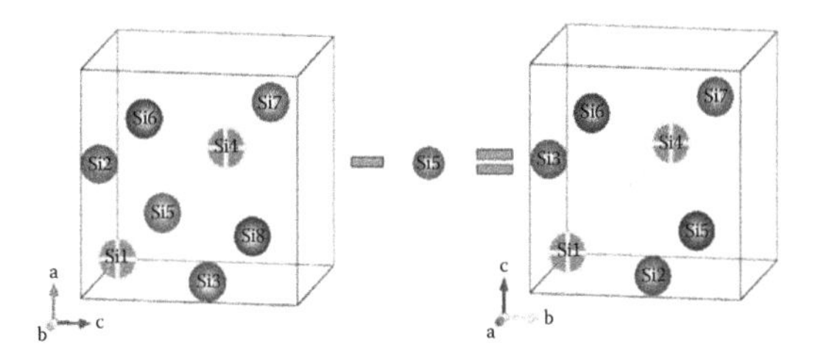

Figure 7.7 Calculation of the vacancy formation energy for Si.

It tells us that the two Si atoms relaxed toward the vacancy site by about 0.001 Å.

Calculation of the vacancy formation energy, E_v^f, is the same as a chemical reaction with the conserved mass balance (see Figure 7.7):

$$E_v^f = E_v - \frac{7}{8}E = (-54.8675) - \frac{7}{8}(-62.9643)$$
$$= 0.2263\,\text{Ry} = 3.08\,\text{eV} \tag{7.1}$$

Other DFT calculation data for the vacancy formation energy of Si show 3.61 eV by GGA/PBE and 4.49 eV by hybrid/HSE06. Experimental values range from 3.4 to 4.0 (Ramprasad et al. 2012). Note that our supercell is a very small one and there is some interaction via PBC making the result a little off from the more accurate values. Calculation with a bigger supercell is assigned in HOMEWORK problems.

7.6 Si7-vacancy diffusion

7.6.1 Introduction

This calculation involves in the diffusion of a vacancy in a bulk Si system, and the barrier energy over which the diffusing vacancy has to pass. To be called a "bulk," we normally imply a system of atoms more than Avogadro's number. In this run, however, we will use a supercell of eight Si atoms as our system. Undoubtedly, the supercell is too small to represent the actual event. But remember again that our goal here is learning the principle and procedure in a given time. In addition, the results of our previous run can be conveniently used.

7.6.2 Calculation method

- *Model*: Conventional Si lattice (diamond structure) with seven atoms and a vacancy
- Vacancy diffusing into the nearest-neighbor site that is identical as an atom in the nearest-neighbor site diffusing into the vacancy site

- We generate the first and last images (atomic configurations) accordingly
- *QE execution programs*: pw.ex/relax, neb.ex (details of the NEB [nudged elastic band] method in Section 8.7)
- *Setup*: ecutwfc = 20.0, Pseudopotential; Si 28.086 Si.pbe-rrkj.UPF, K_POINTS {automatic} 4 4 4 1 1 1

7.6.3 Step 1: First image

We use the relaxed structure of Si7-vacancy (see Section 7.5.3 and Figure 7.6) from the Si7v.relax.out file as the first image:

```
############################################################
Si     0.003081476  0.003081476  0.003081476
Si     0.496918524  0.496918524  0.003081476
Si     0.496918524  0.003081476  0.496918524
Si     0.003081476  0.496918524  0.496918524
Si     0.750000000  0.750000000  0.250000000
Si     0.750000000  0.250000000  0.750000000
Si     0.250000000  0.750000000  0.750000000
############################################################
```

7.6.4 Step 2: Last image

We run another Si7v.relax calculation with a vacancy at the next site. The Si7v.relax. NewV.in file is the same as the Si7v.relax.in file except the position of the #2 Si atom:

```
############################################################
.....
&ions
 ion_dynamics='bfgs'
 /
ATOMIC_SPECIES
 Si 28.086 Si.pbe-rrkj.UPF
ATOMIC_POSITIONS (alat)
Si     0.003081476  0.003081476  0.003081476
Si     0.250000000  0.250000000  0.250000000
Si     0.496918524  0.003081476  0.496918524
Si     0.003081476  0.496918524  0.496918524
Si     0.750000000  0.750000000  0.250000000
Si     0.750000000  0.250000000  0.750000000
Si     0.250000000  0.750000000  0.750000000
K_POINTS {automatic}
4 4 4 1 1 1
############################################################
```

When compared to the Si7v.relax.in file, we can notice that the Si atom at 0.25 0.25 0.25 is back and the Si atom at 0.496918524 0.496918524

0.003081476 is missing now. The Si7v.relax.out file after nine ionic iterations shows the relaxed atomic coordinates for the last image for the next step, a NEB run:

```
############################################################
.....
    Final energy = -54.8675212649 Ry
Begin final coordinates

ATOMIC_POSITIONS (alat)
Si      0.000275647   0.000275647   - 0.000223050
Si      0.254257702   0.254257702     0.245742298
Si      0.500223050   0.000275647     0.499724353
Si      0.000275647   0.500223050     0.499724353
Si      0.747881870   0.747881870     0.247714312
Si      0.747881870   0.252285688     0.752118130
Si      0.252285688   0.747881870     0.752118130
End final coordinates
.....
############################################################
```

7.6.5 Step 3: Si7v.NEB20.in

We now run the last step, calculation of barrier energy for vacancy in Si8. The ***.in file is very different from the previous ones because we have to provide information for NEB including the atomic positions of the first and last images, number of images, and so on. Here is the Si7v.NEB20.in for 20 NEB calculations:

```
############################################################
BEGIN
BEGIN_PATH_INPUT
&PATH
 string_method = 'neb',
 restart_mode = "from_scratch",
 nstep_path = 20,          # number of ionic + electronic
                           steps
 num_of_images = 7,        # number of images including
                           first, last images
 ds = 2.D0,                # step length (Hartree a. u.)
 # a guess for diagonal of Jacobian matrix with
 opt_scheme="broyden"
 opt_scheme = "broyden",   # quasi-Newton Broyden's 2nd
                           method (suggested)
 k_max = 0.6D0,
 k_min = 0.4D0,
 # elastic constants range [k_min, k_max], (Hartree a. u.)
 # useful to rise the resolution around the saddle point
```

```
 # use_freezing =.TRUE.,   # default;.FALSE.
 # images with an error larger than half of the largest are
 optimized
 # the other images are kept frozen
 path_thr = 0.05D0,        # default, stopping force
                           criterion (eV/A)
/
END_PATH_INPUT
BEGIN_ENGINE_INPUT
&CONTROL
 prefix = "Si7vNEB10",
/
&SYSTEM
  ibrav = 1,
  celldm(1) = 10.33,
  nat = 7,
  ntyp = 1,
  nbnd = 44,
  ecutwfc = 20.D0,
  occupations = "smearing",
  smearing = "gauss",
  degauss = 0.01D0,
/
&ELECTRONS
  conv_thr = 1.0D-7,
  mixing_beta = 0.7,
/
ATOMIC_SPECIES
 Si 28.086 Si.pbe-rrkj.UPF
BEGIN_POSITIONS
FIRST_IMAGE
ATOMIC_POSITIONS {alat}
Si     0.003081476   0.003081476   0.003081476
Si     0.496918524   0.496918524   0.003081476
Si     0.496918524   0.003081476   0.496918524
Si     0.003081476   0.496918524   0.496918524
Si     0.750000000   0.750000000   0.250000000
Si     0.750000000   0.250000000   0.750000000
Si     0.250000000   0.750000000   0.750000000
LAST_IMAGE
ATOMIC_POSITIONS {alat}
Si     0.000275647   0.000275647   - 0.000223050
Si     0.254257702   0.254257702     0.245742298
Si     0.500223050   0.000275647     0.499724353
Si     0.000275647   0.500223050     0.499724353
Si     0.747881870   0.747881870     0.247714312
Si     0.747881870   0.252285688     0.752118130
Si     0.252285688   0.747881870     0.752118130
```

```
END_POSITIONS
K_POINTS automatic
4 4 4 1 1 1
END_ENGINE_INPUT
END
############################################################
```

We are now ready to run an NEB run of 20 paths for the Si7v system. Again, you have to remove all remarks with # to run the above file.

```
############################################################
>neb -in Si7vNEB20.in > Si7vNEB20.out
      parsing_file_name: Si7vNEB20.in
############################################################
```

With the flickering prompt, the calculation goes on and several output files are saved along with each progressing path in the working directory:

```
############################################################
prefix.path: data to restart a NEB calculation
prefix.axsf: path in xcrysden format
prefix.xyz: path in xyz format
prefix.dat: reaction coordinate, energy barrier, and error
of each image
prefix.int: a cubic interpolation for the energy profile
############################################################
```

The run converges in about 40 min after 15 paths. If we rename the Si7vNEB20. dat file after 0, 5, 10, and 15 NEB calculations during the run, we can plot the progress of NEB paths. Figure 7.8 shows that the image4 has a barrier energy of about 0.85 eV at the half way between the first and last images after 15 paths. We can also find the change of activation energy (barrier energy) by typing "find 'activation energy (->)' Si7vNEB20.out." This value is of course in the high side due to the small supercell adopted in this run. Calculation by the CPMD code with GGA (BLYP) functional (Kumeda et al. 2001) found a barrier of 0.58 eV with supercells containing 64 and 216 atoms.

We can also plot the last Si7vNEB20.dat on MATLAB (or other plotting programs):

```
############################################################
neb=[Si7vNEB20.dat];        # copy and paste
% to create a neb.mat (7 × 2 double) in Workspace; Table
plot(neb(:,1),neb(:,2),'-ob');
xlabel('Distance (relative)'); ylabel('Eb (eV)');
title('Vacancy Diffusion in Si7v System'); legend('after
NEB15');
############################################################
```

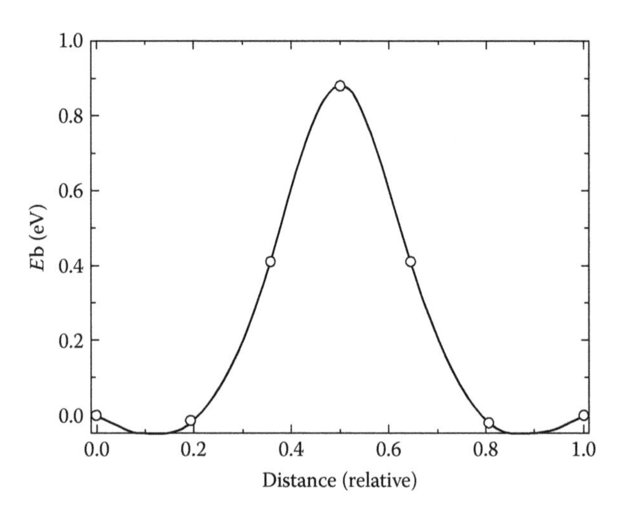

Figure 7.8 Barrier energies by a NEB run for the vacancy diffusion in the Si7v system.

The Si7vNEB20.xyz file has final atomic positions for all seven images, and we can check the position of any Si atom in any image. For example, we edit it and save only the fourth image as Si7vNEB20-image4.xyz, and open it on VESTA. We can tell that the diffusing #2 Si atom experiences the highest barrier for the fourth image when it passes through four Si atoms (#3, #4, #6, and #7). Let us edit that Si7vNEB20.xyz file again and make a new Si7vNEB20-traj.xyz file to see the trajectory of the diffusing #2 Si atom into the vacancy site, which corresponds to the move of the vacancy toward the opposite direction:

```
#############################################################
13
Si 0.0168445821 0.0168445821 0.0168445821
Si 2.7163557002 2.7163557002 0.0168445821 # image1
Si 2.5127545794 2.5127545794 0.2204457030 # image2
Si 2.2548772762 2.2548772762 0.4783230061 # image3
Si 2.0720382916 2.0720382916 0.6611619908 # image4
Si 1.9289334495 1.9289334495 0.8042668329 # image5
Si 1.6086783114 1.6086783114 1.1245219710 # image6
Si 1.3898744458 1.3898744458 1.3433258366 # image7
Si 2.7163557002 0.0168445821 2.7163557002
Si 0.0168445821 2.7163557002 2.7163557002
Si 4.0998004236 4.0998004236 1.3666001412
Si 4.0998004236 1.3666001412 4.0998004236
Si 1.3666001412 4.0998004236 4.0998004236
#############################################################
```

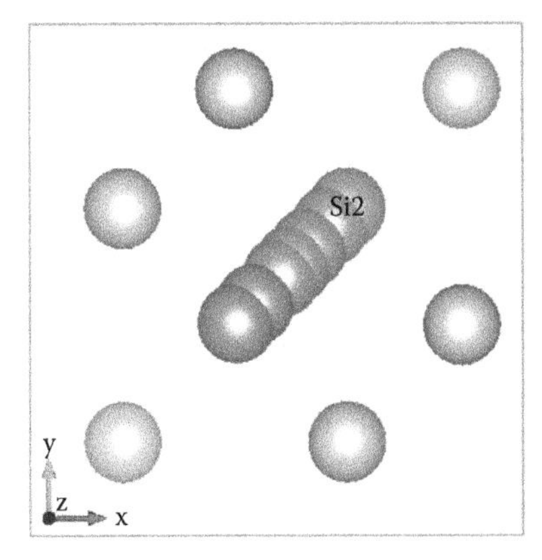

Figure 7.9 Trajectory of the #2 Si atom into the vacancy site in the Si7v system.

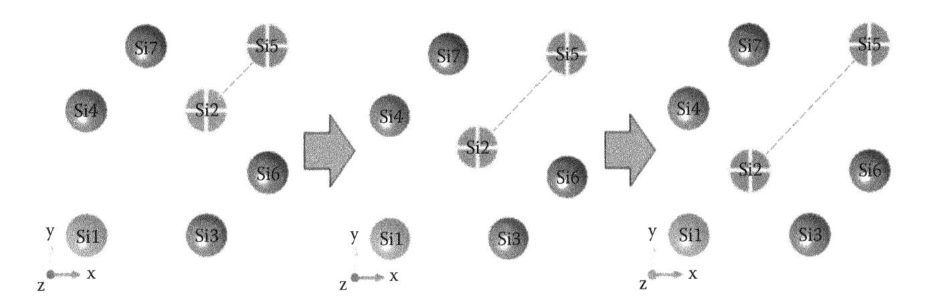

Figure 7.10 Change of distance between the #2 and #5 Si atoms during diffusion of a vacancy in Si7v.

Figures 7.9 and 7.10 show the trajectory of the #2 Si atom into the vacancy site in the Si7v system, and the traveled distance of ~2.3 Å for the #2 Si atom, respectively. A more realistic run can be done by extending this exercise with a bigger system (supercell of >64 Si atoms) if enough run time is given.

Homework

7.1 We define the equilibrium in any system when it is in the minimum energy at its equilibrium lattice parameter. Another definition is that the forces on any atom are all zero. Verify this by setting "tprnfor=. true." in name list "&control" and looking for forces printed at the end of the output file. Use the same setting as Si2-lattice (celldm(1) = 10.333, ecutwfc = 20 Ry, and *k*-point grid = 4 4 4 1 1 1).

Answer

```
##############################################################
.....
Total force = 0.000000
.....
# no movement of atoms from their equilibrium positions
##############################################################
```

7.2 Make a bigger supercell of Si64 and create a vacancy and run a "relax" calculation with the prepared Si63v.relax.in file. The new vacancy formation energy becomes closer to the experimental one?

7.3 Make a bigger supercell of Si64 and make an interstitial and run a "relax" calculation with the prepared Si65i.relax.in file. Is the formation energy bigger or smaller than that of vacancy?

7.4 Referring the runs for Si systems, what changes do we need to run a graphene system (hexagonal structure with a = 2.462 Å, c = 10,000 Å) of two carbon atoms in the graphene.in file? What is the total energy calculated?

7.5 Referring homework 7.4, perform the same procedure to run a CH4 system (molecular in a vacuum box of 16 bohr) of one carbon atom and four hydrogen atoms in the CH4.in file? What is the total energy calculated?

References

Hopcroft, M. A., W. D. Nix, and T. W. J. Kenny. 2010. What is the Young's modulus of silicon? *Micro-Electromech. Syst.* 19(2):229–238.

Kumeda, Y., D. J. Wales, and L. J. Munro. 2001. Transition states and rearrangement mechanisms from hybrid eigenvector-following and density functional theory: Application to $C_{10}H_{10}$ and defect migration in crystalline silicon. *Chem. Phys. Lett.* 341:185–194.

Monkhorst, H. J. and J. D. Pack. 1976. Special points for Brillouin-zone integration. *Phys. Rev. B* 13:5188.

Ramprasad, R., H. Zhu, P. Rinke, and M. Scheffler. 2012. Perspective on formation energies and energy levels of point defects in nonmetals. *Phys. Rev. Lett.* 108:066404 1–5.

chapter eight

DFT exercises with VASP

> DFT is the most detailed "microscope" currently available to glance into the atomic and electronic details of matter.
>
> **J. Hafner, C. Wolverton, and G. Ceder (2007)**

We often predict something in our daily life: football scores, election outcomes, stock prices, and so on. We usually end up with the odds much less than 50/50 and realize that the prediction was either too bold or too foolish. When we are right, however, we feel good even if it is out of sheer luck. We definitely feel very good if it is out of a calculated and deliberate effort. Predicting something and being right is a very satisfying and an exhilarating experience, especially when the object is as foreign as electrons. In this chapter, we will run some actual Density functional theory (DFT) calculations and venture to predict the electron's activities with our scientific knowledge and, hopefully, experience that exceptional feeling.

At present, tens of programs (or codes or packages) are available in the framework of DFT with varying features and capabilities. Table 8.1 shows a list of typical programs that are popular in physics, chemistry, and materials science. As no program is designed to fit all our needs, we have to think hard about why simulation is needed and what can be achieved by doing it. Then a selection process may be followed, which requires a very careful balancing act between accuracy and efficiency.

For the exercises in this chapter, we will use the Vienna *ab initio* simulation package (VASP) developed by Kresse and Hafner (1994) and further upgraded by Kresse and Furthmüller (1996). Over the past decade, VASP has proven to be fast and accurate so that up to a hundred atoms can be treated routinely with only tens of CPUs in the parallel mode. The simple fact that more than 1000 groups are using it now clearly shows its worldwide popularity. Note that the Austrian tradition on computational science in VASP traces its origin directly back to Kohn and Schrödinger.

In this chapter, by using VASP,* we will practice some DFT calculations to understand how the concepts and methodologies studied in the

* VASP. http://cms.mpi.univie.ac.at/vasp/vasp/vasp.html.

Table 8.1 Various DFT-based programs

Program (language)	Potential	Basis	Remarks
VASP (Fortran90)	Pseudo-potential, PAW	Plane waves	Most popular for solids and surfaces $3000 for academia http://cms.mpi.univie.ac.at/vasp/
CASTEP (Fortran90)	Pseudo-potential	Plane waves	Marketed commercially by Accelrys http://www.tcm.phy.cam.ac.uk/castep/
QUANTUM ESPRESSO	US-PP, NC-PP PAW PP	Plane waves	Free for all Best to start *ab initio* calculation http://www.quantum-espresso.org/
ABINIT	Pseudo-potential, PAW	Plane waves	DFT, BIGDFT/Wavelets, etc. Free (GNU) http://www.abinit.org
CPMD	NC-PP, US-PP	Plane waves	Car–Parrinello MD, etc. Free http://www.cpmd.org
WIEN2K (Fortran90)	All-electron	Linear-augmented plane waves	Good for band structure, DOS, etc. €400 for academia http://www.wien2k.at/
GAUSSIAN	All-electron	Local	Most popular for molecules Hartree–Fock, DFT, etc. http://www.gaussian.com

previous chapters are implemented in actual simulations at 0 K. The exercises cover from a very simple system of a single atom to a catalytic system of 46 atoms. They are carefully selected so that a single node with eight CPUs can return the results in a reasonable time, for example, in most cases within 30 steps for both the electronic and ionic iterations. They are also arranged so that one can easily follow them one by one without facing any surprise. If some runs (or homework runs) are expected to take rather extensive computation under the given computer resources and time, for example, the NEB (nudged elastic band) run (see Section 8.7), nearly converged CONTCARs may be provided for students so that they can finish them without missing any essentials involved. Since the aim of these exercises is not accuracy but efficient practice, the precision of calculations is set as moderate. However, their extensions and variations into the more accurate and higher-level calculations are rather straightforward. For a detailed description of VASP calculations, refer to the VASP guide written by Georg Kresse et al. (2016).

8.1 VASP

Let us first assess the program very briefly and see what it can do and how it runs.

8.1.1 General features of VASP

- Performs the electronic and ionic minimizations (relaxations or optimizations) and calculates the ground-state properties (total energies, band structures, density of states, phonon spectra, etc.) using the KS orbitals expanded with the PW basis set.
- Solves the KS equations self-consistently with an iterative matrix diagonalization and a mixing of charge densities.
- Calculates the energy terms of the KS Hamiltonian and the charge densities in real and reciprocal spaces via fast Fourier transformations (FFT).
- Works for any system including atoms, molecules, bulk solids, surfaces, clusters, and so on.
- Provides most PPs including the USPP (Vanderbilt 1990), the PAW potentials (Blöchl 1994), and so on, for all elements.
- Provides the XC functionals of LDA, PW91 (Perdew and Wang 1992), PBE (Perdew et al. 1996), hybrids, and so on.
- Runs in a parallel mode (efficiently up to 32 nodes) and can treat up to 4000 valence electrons.
- Provides tens of flags with default values and is thus well-suited even for a first DFT runner as well as for experts.

As highlighted by Hafner (2007), the role of VASP in the first-principles calculations is becoming vital, and there is no doubt that it will contribute a great deal to the solutions of materials science problems in both properties and processes.

8.1.2 Flow of VASP

Figure 8.1 shows the general flow of a VASP run, which includes both electronic and ionic minimizations with the corresponding input and output files:

- Select a PP and an XC functional and construct a trial electron density.
- Using a PW basis set for the expansion of KS orbitals, calculate the $E_{\text{kin}}^{\text{non}}$ and the U_{ext} at the given atom positions and crystal structure.
- Calculate the U_H and the U_{xc} and construct the KS Hamiltonian, $\hat{H}_{\text{KS}}$.
- Solve the KS equations (a matrix eigenvalue equation) by an iterative diagonalization in the real and reciprocal spaces.
- With the newly calculated orbitals, generate a new electron density and repeat the process until self-consistency is reached.

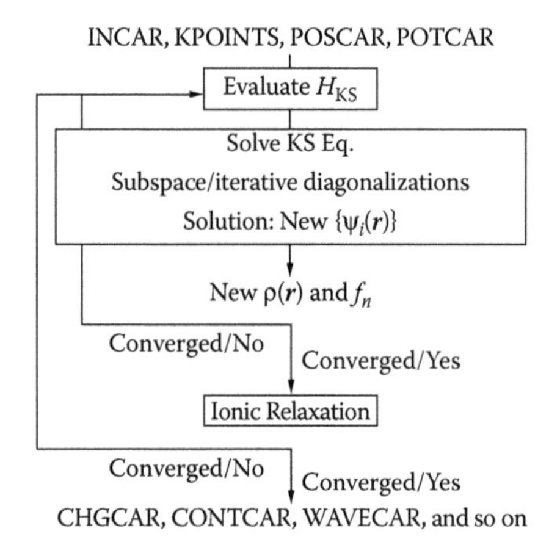

Figure 8.1 General flow of VASP.

- After the electronic minimization is achieved, calculate the forces on atoms and move the atoms to the new positions of lower forces.
- Repeat the electronic and ionic minimizations in series until an energy-convergence criterion is met.

VASP is a very complex program with myriad manipulations, algorithms, schemes, approaches, methodologies, and databases that have been added, patched, upgraded, and modified for many years. Depending on the run setup, a variety of solutions in different calculation times may be possible for a given problem. For a beginner, therefore, running VASP in the right way requires a careful balancing act between accuracy and efficiency, which can be achieved only by good guidance from an expert with some good experience.

8.1.2.1 *Ten things you should not do in a VASP run*

A VASP run on a computer and actual experimental work in a laboratory have a lot in common. Unfortunately, the similarity list also includes, among many other things, the making of mistakes. Indeed, we often make mistakes during the implementations of DFT as we do in experiments. The following is a list of simple mistakes or meaningless moves that we are apt to make, especially at the beginner's stage. Let us be aware of them ahead of time and avoid them in the course of the exercises in this chapter:

- Starting a run with fewer than four input files (except the case of using a Linux shell script)
- Continuing a run with the CONTCAR file without changing its name to POSCAR
- Setting a POSCAR file with the "Selective dynamics" flag on and not specifying "F F F" or "T T T" at the end of each atomic coordinate

- Running with the tetrahedron method for smearing and setting a sigma value
- Setting ENCUT = smaller than the ENMIN
- Using the *k*-points as 4×4×4 for a slab calculation
- Placing two defects at the diagonal corners of a supercell and claiming that the interaction between the two defects is minimized
- Running for a single atom or a molecule in a 16-node parallel mode
- In the POSCAR file, selecting "Direct" and writing the atomic coordinates bigger than 1
- Trying various XC functionals until the want-to-see numbers show up

8.2 Pt-atom

This is a very simple run for the ground-state energy of an isolated Pt atom in a vacuum. To bring electrons to the ground state, we will follow a typical routine of the VASP run for electronic minimization and get familiar with the input and output files of the program. In addition, we will practice a continuous run with the use of the output files of a preceding run, which is often adopted for higher accuracy, post processing, and so on. All step-by-step procedures will be described here for this first run. All Linux commands, its stdouts (standard outputs), and the contents of input/output files are written in a smaller font size to distinguish them from the normal text and are confined between two lines of sharps (#).

8.2.1 Input files

First, confirm that the VASP code is properly installed such as ~/vasp/vasp.5.2 (or later). Under a separate directory properly named, for example, "Pt atom" in this case, prepare the four input files and name them exactly as INCAR, KPOINTS, POSCAR, and POTCAR. You may simply copy the first three files from any source and edit them. The POTCAR file that contains a specific PP and a XC functional can be copied from the PP library of VASP.

8.2.1.1 INCAR

The INCAR file is the central brain and controls the run with the given flags. We will specify here only the four particular flags for this run and leave the rest to the defaults given by the program.

```
############################################################
Pt-atom                  # title
ISMEAR     = 0           # Gaussian smearing
SIGMA      = 0.1         # broadening of Fermi level, 0.1~0.2
                         for transition metals (eV)
ISPIN      = 2           # spin-polarized run for a single atom
ISYM       = 0           # symmetry off
############################################################
```

- *ISMEAR*: This flag sets a smearing method for the Fermi level for an easier convergence. Here, the Gaussian smearing is used to assign some partial occupancies and to broaden the Fermi level with a SIGMA flag. This method is well suited for the case of a single-atom system without causing any negative occupancy.
- *ISPIN*: This run will consider the spin polarization since an isolated atom tends to be spin-polarized, and the Pt atom is a transition metal with a valence state of $5d^9 6s^1$. We now have two electron densities to calculate, spin-up and spin-down, that roughly double the computation time compared to that of a spin-unpolarized case.
- *ISYM*: The symmetry consideration is switched off for the better description of an atomic Pt.

8.2.1.2 KPOINTS

The KPOINTS file decides the ***k***-point sampling in the IBZ. Note that the Bloch theorem does not apply in this case of one atom, and we need only one ***k***-point at the Γ (gamma)-point.

```
#############################################################
k-points    # title, make sure no empty line between entities
0           # automatic k-point mesh generation
G           # Γ (gamma)-point centered
1 1 1       # only a single k-point is needed for atoms,
            molecules, and clusters
0 0 0       # shift
#############################################################
```

8.2.1.3 POSCAR

The POSCAR file provides structural information for the system, and we will place a Pt atom in the middle of a big empty box (10 × 10 × 10 Å) to mimic an isolated atom.

```
#############################################################
Pt-atom                         # title, make sure no empty line
                                between entities
10                              # lattice parameter (Å) of a
                                supercell
1.000000 0.000000 0.000000      # unit lattice vector
0.000000 1.000000 0.000000      # unit lattice vector
0.000000 0.000000 1.000000      # unit lattice vector
1                               # number of atoms
Direct                          # direct coordinates considering
                                each side of supercell as 1
0.500000 0.500000 0.500000      # atom positions
#############################################################
```

8.2.1.4 POTCAR

In this practice, we will use the PAW_PBE file that has been proven to be accurate and computationally efficient. Simply copy the potential file of Pt (potcar.PBE.paw/Pt/POTCAR) from the PP library of VASP and place it in the current directory. The file contains various data used to generate the PP of a Pt atom.

```
############################################################
PAW_PBE Pt 05Jan2001
10.0000000000000000
parameters from PSCTR are:
 VRHFIN =Pt: s1d9
 LEXCH = PE
 EATOM = 729.1171 eV, 53.5886 Ry
 TITEL = PAW_PBE Pt 05Jan2001

 LULTRA = F use ultrasoft PP ?
 IUNSCR = 1 unscreen: 0-lin 1-nonlin 2-no
 RPACOR = 2.330 partial core radius
 POMASS = 195.080; ZVAL = 10.000 mass and valence
 RCORE = 2.500 outmost cutoff radius
 RWIGS = 2.750; RWIGS = 1.455 wigner-seitz radius (a.u.)
 ENMAX = 230.283; ENMIN = 172.712 eV
.....
############################################################
```

Note that the ENMAX value of 230.283 eV is the default for the ENCUT in this case.

8.2.2 Run

A VASP run requires only a few Linux commands (listed in Appendix B), and any Linux command is written following the "$" notation in this book. Check that all necessary files are in order in the directory and begin the calculation by simply typing "vasp.exe." The progress begins to scroll past on the monitor and, after only a few seconds, the run will be over.

```
############################################################################
$ ls
INCAR KPOINTS POSCAR POTCAR
$./vasp.exe                              # start run
running on 8 nodes
distr: one band on 1 nodes, 8 groups
vasp.5.2.2 15Apr09 complex
POSCAR found: 1 types and 1 ions
LDA part: xc-table for Pade appr. of Perdew
POSCAR, INCAR and KPOINTS ok, starting setup
WARNING: small aliasing (wrap around) errors must be expected
FFT: planning...(1)
```

```
reading WAVECAR
entering main loop
   N  E            dE       d eps    ncg    rms       rms(c)
DAV: 1 0.547544841162E+02 0.54754E+02 - 0.10430E+03 64 0.167E+02
DAV: 2 0.938646877437E+01 - 0.45368E+02 - 0.42330E+02 128 0.273E+01
.....
DAV: 19 - 0.576338999258E+00 - 0.33539E-02 - 0.12634E-03 64 0.101E-01
0.111E-01
DAV: 20 - 0.576244709456E+00 0.94290E-04 - 0.65478E-04 64 0.753E-02
 1 F = -.57624471E+00 E0 = -.52677121E+00 d E = -.989470E-01 mag = 2.0046
 writing wavefunctions
##################################################################
```

The "wrap around" error message may frequently show up if not enough FFT-grid points are assigned. The message indicates that certain parts of electron densities are wrapped to the other side of the grid due to insufficient FFT-grid numbers. Check OUTCAR, search *wrap*, and follow the recommended values of NGX, NGY, and NGZ. However, the warning could often be ignored since the error associated with it is normally very small. The last line, "writing wavefunctions," indicates that the run is done.

8.2.3 Results

There will be 12 output files generated by the run in the directory.

```
##################################################################
$ ls
CHG CHGCAR CONTCAR DOSCAR EIGENVAL IBZKPT INCAR KPOINTS
OSZICAR OUTCAR PCDAT POSCAR POTCAR WAVECAR XDATCAR vasprun.xml
##################################################################
```

For the moment, we review only two files: OSZICAR and OUTCAR.

8.2.3.1 OSZICAR

The OSZICAR file contains a summary of the iteration loop. Open OSZICAR and confirm the E0 value (−0.527 eV) on the last line that is the ground-state energy of a single Pt atom in a vacuum under the given setup:

```
##################################################################
$ cat OSZICAR
.....
DAV: 19 - 0.576338999258E+00 - 0.33539E-02 - 0.12634E-03 64
0.101E-01 0.111E-01
DAV: 20 - 0.576244709456E+00 0.94290E-04 - 0.65478E-04 64
0.753E-02
 1F= -.57624471E+00 E0= -.52677121E+00 d E =-.989470E-
 01 mag= 2.0046
##################################################################
```

The first column indicates that the convergence (with the default tolerance of <E-04 eV) is achieved in 20 iterations by the Davidson algorithm. Note that the E0 value (–0.527 eV) is not the true energy but the value in reference to the PAW potential generated for a Pt atom. The magnetic moment per one Pt atom by spin polarization is also written as mag = 2.0046 μ_B (Bohr magneton), coming from the electronic configuration of [Xe]$4f^{14}5d^{9}6s^{1}$. As expected, the two unpaired electrons show magnetic moment of one each:

$$\text{mag} = \mu_B[\rho_\uparrow - \rho_\downarrow] \tag{8.1}$$

8.2.3.2 OUTCAR

The OUTCAR file writes down everything about the run including the four input files, information on lattice, symmetry analysis, the *k*-points and their positions, the PW basis used (the number of PWs on each *k*-point), the nearest-neighbor distances at the start of the run, and every electronic and ionic minimization step. It is, therefore, strongly recommended that the OUTCAR file should be saved even if it was a waste run. It also gives a summary of the computation time spent for the run at the last part of the file.

```
############################################################
.....
Total CPU time used (sec): 13.953
.....
############################################################
```

It shows that the energy of a single atom can be calculated in a matter of seconds under the given computational resource.

8.2.3.3 Continuous run

There are three important output files: CHGCAR, CONTCAR, and WAVECAR. They are files of the calculated charge densities, the final atomic positions, and the final wave functions (in a binary format), respectively. A continuous run can be started with the use of these output files and time can be saved. For example, we can continue our run by simply copying the three files and adding two lines into the new INCAR file.

```
############################################################
Pt-atom-continuous         # title
.....
ISTART = 1                 # 0: new run, 1: continuous run
ICHARG = 1                 # 1: use CHGCAR file, 2: use atomic
                           charge density from PP
.....
############################################################
```

Then the VASP will read six files including the CHGCAR and WAVECAR files and resume a continuous run. After an additional 15 iterations, the result improves further with E0 = −0.528 eV/atom and mag = 2.0001 μ_B. This ground-state energy of a Pt atom will be used when we calculate the cohesive energy of a bulk Pt in Section 8.5.

8.3 Pt-FCC

In this run, we will calculate the ground-state energy of an FCC (face-centered cubic) lattice with four Pt atoms, assuming we already know about the structure and a reasonable lattice parameter (a_0 = 3.975 Å). Both the electronic and ionic minimizations (relaxations) will take place as a slightly displaced Pt atom relaxes to its normal position in an FCC structure.

8.3.1 Input files

For this run, we will prepare three input files that are newly edited to describe the given Pt-FCC system. The POTCAR file is the same as in the previous run.

8.3.1.1 INCAR

Here, we switch off the spin flag since the spin densities spread out rather evenly in a many-atom lattice, and we add six new flags to control the ionic minimization.

```
##############################################################
Pt-FCC

Electronic Relaxation:
 ENCUT = 400             # sets PW cutoff (eV)
 PREC = normal           # default, NGX=1.5Gcut, NGXF=2NGX
 EDIFF = 1E-04           # energy-stopping criterion for
                         electronic minimization (eV)
 ALGO = Fast             # 5 initial steps of blocked Davidson
                         followed by RMM-DIIS algorithms

DOS related values:
 ISMEAR = 1              # smearing by Methfessel/Paxton
                         method for metals
 SIGMA = 0.10            # broadening parameter (eV)
 # ISPIN = 2             # spin-polarization switched off

Ionic Relaxation:
 NSW = 100               # max number of steps for structural
                         relaxation
```

```
IBRION = 1              # ionic relaxation using quasi-Newton
                        method
EDIFFG = -0.02          # stopping criterion by force on
                        atoms (eV/A)
ISIF = 1                # moves ions at fixed shape and
                        volume of lattice, calculate forces
                        only
ISYM = 0                # no symmetry, 2 (on) is default for
                        PAW
POTIM = 0.2             # scaling constant for the forces
############################################################
```

- *ENCUT*: This flag sets a cutoff energy for the PW basis set and must be specified manually, which overrides the value set by the PREC flag. Here, the value is set much higher than the maximum value recommended in the POTCAR file (ENMAX = 230.283). The reason is to make the ENCUTs consistent when we run later a system of Pt atoms with an O atom that has the ENMAX of 400 eV. One must use the highest ENMAX of the constituting atoms whenever one deals with multicomponent systems.
- *PREC*: Once the ENCUT is specified manually, this flag sets the fineness of FFT grids for wave functions and charge densities and thus decides the accuracy of the calculation. The "normal" provides sufficient accuracy in most calculations. If a higher accuracy is needed, we may use PREC = Accurate.
- *EDIFF*: This flag sets a stopping criterion for electronic minimization. The "1E-04" is the default and provides sufficient accuracy in most calculations.
- *ALGO*: This flag sets an electronic minimization algorithm. The "fast" is the most popular choice and provides the stable Davidson algorithm for the initial five iterations, followed by the faster RMM-DIIS algorithm. After the first ionic minimization, however, electronic minimization is carried out with only a single Davidson algorithm followed by the RMM-DIIS algorithm.
- *EDIFFG*: This flag sets a stopping criterion for ionic minimization. Atoms are moved until forces (absolute value) are smaller than this value (0.02 eV/Å) on any of the atoms in the supercell.

For any new run, the initial wave functions are calculated from the given atomic PP, which usually makes an excellent starting point. By using wave functions from the previous calculations even with a different cutoff energy or slightly different geometry, calculation time can be reduced. We will leave the rest of the flags at the moment and explain them in later exercises. To compare the total energies of two systems, we need to set all parameters the same in both calculations: ENCUT, PREC, EDIFF, EDIFFG, and ISMEAR, which directly determine accuracy.

8.3.1.2 KPOINTS

The **k**-points mesh of 9×9×9 is generated by the Monkhorst–Pack method, which is most preferred for metals.

```
##############################################################
k-points
0                        # auto mesh
M                        # MP method
9 9 9                    # for a single FCC cell
0 0 0                    # shift
##############################################################
```

8.3.1.3 POSCAR

We construct a POSCAR file of the FCC lattice with four Pt atoms positioned at the origin and at the centers of three faces of the cube sharing the origin.

```
##############################################################
Pt-FCC
3.975                                # FCC lattice parameter
                                     (Å) with unit lattice
                                     vectors in x, y, z
1.000000 0.000000 0.000000
0.000000 1.000000 0.000000
0.000000 0.000000 1.000000
4                                    # no. of atoms
Selective dynamics                   # atoms are allowed to
                                     relax selectively
Direct                               # direct lattice with atom
                                     positions in x, y, z
0.000000 0.000000 0.000000 F F F     # F: false (fixed)
0.510000 0.500000 0.000000 T T T     # T: true (free to move)
0.500000 0.000000 0.500000 T T T
0.000000 0.500000 0.500000 T T T
##############################################################
```

Note that the first Pt atom is fixed for the stability of structure and the rest of the atoms are set to relax freely. Note also that the second Pt atom is slightly out of position by 0.01 in direct coordinate so that we can confirm its move to the normal position after the run.

8.3.2 Run

8.3.2.1 run.vasp

This time, we will make a very convenient shell script, "run.vasp," for the run command and use it for the rest of the exercises.

```
############################################################
nohup mpirun -machinefile ~/mf1 -np 8 /usr/local/vasp/
vasp5.2 &
############################################################
```

The "nohup [commands] &" makes the command continue in the background even after we log out so that we can do something else while the calculation goes on by the computer. Note that the "nohup" is coined from "no hang-up." The "mpirun -machinefile ~/mf1 -np 8" is all related to parallel computation, meaning that the calculation will use eight CPUs listed in a machinefile "mf1." The rest is the path where the program is installed. Save this run file along with the four input files and execute it by the following line.

```
############################################################
./run.vasp
############################################################
```

It calculates the forces on each atom using the Hellman–Feynman theorem and updates the positions of the atoms in accordance with these forces. The whole procedure repeats until the magnitude of the forces on all atoms is less than the setting of EDIFFG (0.02 eV/A). During the course of the run, we can monitor the forces on atoms by looking at the lines in the OUTCAR file.

```
############################################################
$ grep "drift" OUTCAR
.....
total drift:  0.013264 0.000002 0.000002
total drift:  0.013792 0.000001 0.000008
total drift:  0.051897 0.000029 0.000006
############################################################
```

8.3.2.2 nohup.out

The whole process will be summarized and saved in a new output file, nohup.out.

```
#################################################################
.....
entering main loop
    N           E                    dE             d eps       ncg      rms        rms(c)
DAV: 1    0.398585504866E+03   0.39859E+03  -0.15408E+04  23568  0.102E+03
DAV: 2   -0.114673743508E+02  -0.41005E+03  -0.37702E+03  23360  0.379E+02
DAV: 3   -0.254580381254E+02  -0.13991E+02  -0.13559E+02  28432  0.803E+01
DAV: 4   -0.256918719273E+02  -0.23383E+00  -0.23351E+00  27264  0.932E+00
DAV: 5   -0.256999084327E+02  -0.80365E-02  -0.80356E-02  28672  0.167E+00  0.726E+00
RMM: 6   -0.243541401231E+02   0.13458E+01  -0.21381E+00  23360  0.122E+01  0.224E+00
RMM: 7   -0.242316526128E+02   0.12249E+00  -0.46990E-01  23360  0.432E+00  0.584E-01
RMM: 8   -0.242168946061E+02   0.14758E-01  -0.18260E-02  23385  0.132E+00  0.148E-01
RMM: 9   -0.242165915581E+02   0.30305E-03  -0.17472E-03  23561  0.232E-01  0.594E-02
RMM: 10  -0.242165890846E+02   0.24735E-05  -0.37905E-04  19049  0.160E-01
 1 F= -.24216589E+02 E0= -.24216576E+02 d E =-.242166E+02
 BRION: g(F)= 0.518E-02 g(S)= 0.000E+00
 bond charge predicted
.....
DAV: 1   -0.242253782297E+02   0.91002E-04  -0.17742E-01  24040  0.178E+00  0.223E-02
RMM: 2   -0.242255961696E+02  -0.21794E-03  -0.21741E-03  22764  0.348E-01  0.213E-02
RMM: 3   -0.242256123515E+02  -0.16182E-04  -0.19612E-04  13158  0.816E-02
 11 F= -.24225612E+02 E0= -.24225718E+02 d E =-.139917E-03
 BRION: g(F)= 0.790E-05 g(S)= 0.000E+00 retain N= 2 mean eig= 5.33
 eig: 5.332 5.332
 reached required accuracy - stopping structural energy minimization
 writing wavefunctions
#################################################################
```

As set by the ALGO = fast, the first five electronic minimizations were carried out by the Davidson method followed by the RMM-DIIS method. After solving for the electronic ground states, atoms are moved to the new positions based on the calculated forces on each atom. Then the next electronic minimization starts with the new atomic configuration. This series of minimizations continues until the preset criteria (EDIFF and EDIFFG) are met. The last line, "writing wavefunctions," indicates that the job is done, and the WAVECAR file is written for a further run, if necessary.

8.3.3 Results

8.3.3.1 CONTCAR

```
#################################################################
Pt-FCC
3.975
1.000000 0.000000 0.000000
0.000000 1.000000 0.000000
0.000000 0.000000 1.000000
Pt
4
Selective dynamics
Direct
0.0000000000000000 0.0000000000000000 0.0000000000000000 F F F
0.5004141680668154 0.4999999997119582 0.0000000877225392 T T T
```

```
0.5004600291443179 0.0000000864829462 0.4999999260091121 T T T
0.0005052936914971 0.4999999752016114 0.4999999462990823 T T T
.....
###############################################################
```

Note that the second Pt atom initially positioned at (0.510000, 0.500000, and 0.000000) has moved to the stable FCC position.

8.3.3.2 OUTCAR

Let us confirm the calculated total energies for each ionic step on the lines of "energy without entropy" in the OUTCAR file.

```
###############################################################
$ grep "energy without entropy" OUTCAR
.....
energy without entropy = -24.22569487 energy(sigma->0) = -24.22548378
energy without entropy = -24.22591274 energy(sigma->0) = -24.22570169
energy without entropy = -24.22592894 energy(sigma->0) = -24.22571788
###############################################################
```

Confirm the E0 value of –24.226 in the last iteration that corresponds to the ground-state energy of –24.226/4 = –6.056 eV/atom for a Pt atom in the FCC structure at the assumed lattice parameter of 3.975 Å. In the next section, we will optimize the lattice and determine the equilibrium lattice parameter.

8.4 Convergence tests

This section treats the convergence issue that was briefly outlined in Section 7.3. During a well-set DFT run, the system energy decreases steadily with a little fluctuation and eventually reaches the ground-state value. This is called *convergence,* and we must have it to achieve a meaningful run. Whenever a new system is about to be calculated, therefore, it is essential to perform a proper convergence test to avoid any chance of divergence and to have the ground-state value at the end. Remember that the DFT runs are based on the variational theorem, and thus we have to make sure that all setups for the run are good enough to drive the system all the way down to the ground state. Otherwise, we may be ending up with wrong results. Two most important tags are the ENCUT in the INCAR file, and the ***k***-points mesh in the KPOINTS file. Both parameters should be increased until the result stops changing.

8.4.1 Encut convergence

We already defined the cutoff energy, E_{cut}, in Section 6.5.1 as

$$E_{cut} \geq \frac{1}{2}\left|\boldsymbol{k}+\boldsymbol{G}\right|^2 \tag{8.2}$$

Thus, we only include PWs with energies less than this cutoff energy. The ENCUT is normally within the range of 150–400 eV, and its default values (maximum and minimum) are given as ENMAX and ENMIN in each POTCAR file of elements. However, the proper ENCUT varies depending on the system and the run setups, and we must find out the right value by convergence test. In this exercise, we will use a shell script that contains all three input files (INCAR, KPOINTS, and POSCAR) and will scan lattice parameters at four different ENCUTs (200, 225, 250, and 350 eV) at fixed *k*-points (9×9×9). Thus, we will decide the ENCUT for convergence and the equilibrium lattice parameter at the same time.

8.4.1.1 *Shell script run.lattice*

A shell script, "run.lattice," is prepared for the run, for example, for the ENCUT = 250 eV.

```
##############################################################
for a in 3.90 3.92 3.94 3.96 3.98 4.00 4.02 4.04 4.06
do
rm WAVECAR                # removes WAVECAR to start a fresh run
echo "a = $a"

cat >INCAR << end         # INCAR from here to 'end'
SYSTEM = Pt-FCC-lattice
ENCUT = 250               # set ENCUT at 250 eV
ISMEAR = 1
SIGMA = 0.1
PREC = normal
ALGO = FAST
end

cat >KPOINTS << kend      # KPOINTS from here to 'kend'
k-points
0
M
9 9 9
0 0 0
kend

cat >POSCAR << !          # POSCAR from here to '!'
Pt-FCC-lattice
$a
1.0000000000000000 0.0000000000000000 0.0000000000000000
0.0000000000000000 1.0000000000000000 0.0000000000000000
0.0000000000000000 0.0000000000000000 1.0000000000000000
4
```

```
Direct
0.0 0.0 0.0
0.5 0.5 0.0
0.5 0.0 0.5
0.0 0.5 0.5
!

mpirun -machinefile ~/mf1 -np 8 /usr/local/vasp/vasp-5.2
E='tail -1 OSZICAR'                    # take the last line of
                                       the OSZICAR file
echo $a $E >> Pt-lattice-999-E.dat     # write a and E in the
                                       Pt-lattice-999-E.dat
                                       file
done
############################################################
```

Here, we vary the variable for lattice parameter, $a, from 3.90 Å to 4.06 Å, and the run will go on continuously, changing the lattice parameter automatically. For a convergence test, we can skip the spin flag because the PW convergence does not depend on the spin variable. One remark is that, when we type in the sections of INCAR, KPOINTS, and POSCAR, we have to make sure that each command line is not intervened with a comment or an empty line.

8.4.1.2 Run

The prepared script, "run.lattice," should be first changed to an execution file and then executed.

```
############################################################
$ chmod +x./run.lattice    # change the file mode to
                           execution mode
$./run.lattice
rm: cannot remove 'WAVECAR': No such file or directory
a = 3.90
 running on 8 nodes
 distr: one band on 1 nodes, 8 groups
 vasp.5.2.2 15Apr09 complex
 POSCAR found: 1 types and 4 ions
.....
############################################################
```

8.4.1.3 Results

The Pt-lattice-999-E.dat is written and has the following energies at different lattice parameters of Pt-FCC, for example, for the ENCUT = 250 eV.

```
###############################################################
$ cat Pt-lattice-999-E.dat
3.90 1 F= -.24048778E+02 E0= -.24049106E+02  d E =0.983607E-03
3.92 1 F= -.24131354E+02 E0= -.24131599E+02  d E =0.736199E-03
3.94 1 F= -.24187299E+02 E0= -.24187475E+02  d E =0.526380E-03
3.96 1 F= -.24218612E+02 E0= -.24218739E+02  d E =0.381269E-03
3.98 1 F= -.24226441E+02 E0= -.24226538E+02  d E =0.289445E-03
4.00 1 F= -.24212955E+02 E0= -.24213035E+02  d E =0.240874E-03
4.02 1 F= -.24178788E+02 E0= -.24178881E+02  d E =0.277487E-03
4.04 1 F= -.24125951E+02 E0= -.24126070E+02  d E =0.358309E-03
4.06 1 F= -.24055660E+02 E0= -.24055819E+02  d E =0.477705E-03
###############################################################
```

Figure 8.2 shows the result of all four runs (at 200, 225, 250, and 350 eV), indicating that the convergence is reached at the ENCUT = 250, and the total energy is converged to within 5 meV. The equilibrium lattice parameter is determined to be 3.977 Å, which is in good agreement with the calculated value of 3.975 Å (Bentmann et al. 2008) and is in fair agreement with the experimental value of 3.928 Å (Kittel 2005). Another run with this parameter results in the corresponding energy of –24.2269 eV (–6.057 eV/atom).

The calculation became slower as we increase the cutoff energy (or the *k*-points in the next exercise) as expected. Several important remarks in relation to the ENCUT are as follows:

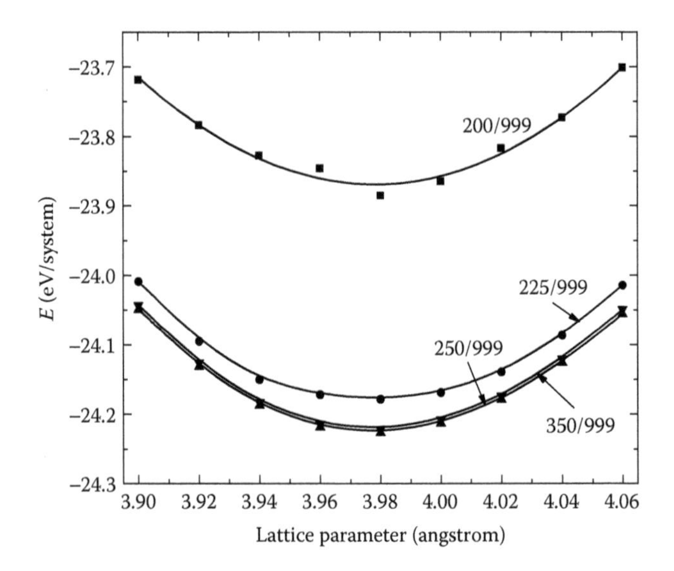

Figure 8.2 Total energy versus lattice parameter of Pt-FCC (four atoms) with various ENCUTs at *k*-point mesh of 9 × 9 × 9.

- Always use the largest cutoff energy as the overall cutoff energy for multicomponent systems.
- When comparing system energies with each other, use the same setup including the ENCUT and the KPOINTS for each system.
- For cell shape and volume relaxations (ISIF = 3), increase the ENCUT by ~30% and use PREC = accurate.

With the changing ENCUT, we can check the corresponding changes in the number of PWs used for the expansion of orbitals on each ***k***-point in the OUTCAR file ($N_{PW} \propto E_{cut}^{3/2}$). For example, when the ENCUT = 250 eV, it will be as follows.

```
############################################################
$ grep "plane waves" OUTCAR
.....
k-point 33: 0.44440.33330.3333 plane waves: 606
k-point 34: 0.44440.44440.3333 plane waves: 611
k-point 35: 0.44440.44440.4444 plane waves: 609
############################################################
```

Note that the NPLWV is the total number of points on FFT grid (NGX*NGY*NGZ) where NGX, NGY, and NGZ are the numbers of grid points in the *x*-, *y*-, and *z*-directions, respectively. If we run the Pt system in the same way but in different crystal structures such as HCP (hexagonal close packed) or BCC (body-centered cubic), we will find out that the FCC structure has the lowest ground-state energy and is the most stable structure at 0 K.

8.4.2 k-*points convergence*

Once we determine the right ENCUT, we do similar calculations in terms of the number of ***k***-points. We remember that the regular grid of ***k***-points in reciprocal space can be reduced by symmetry operations to a few special ***k***-points in the IBZ wedge. These irreducible ***k***-points are crucial for accurately integrating the properties involved. Using the determined ENCUT of 250 eV and the equilibrium lattice parameter of 3.977 Å, we incrementally increase the fineness of the ***k***-points grid (from $2\times2\times2$ to $10\times10\times10$) until a convergence is attained within ~1 meV. Note that the IBZKPT (***k***-points in the IBZ) increases from 1 to 35 in this range of the ***k***-points grid. The rest is the same as in the case of ENCUT convergence, and the results are summarized in Figure 8.3, which indicates that the ***k***-points grid of $9\times9\times9$ is sufficient for the system. However, the IBZKPT files indicate that the even-numbered *k*-points were sampled more evenly under the MP method.

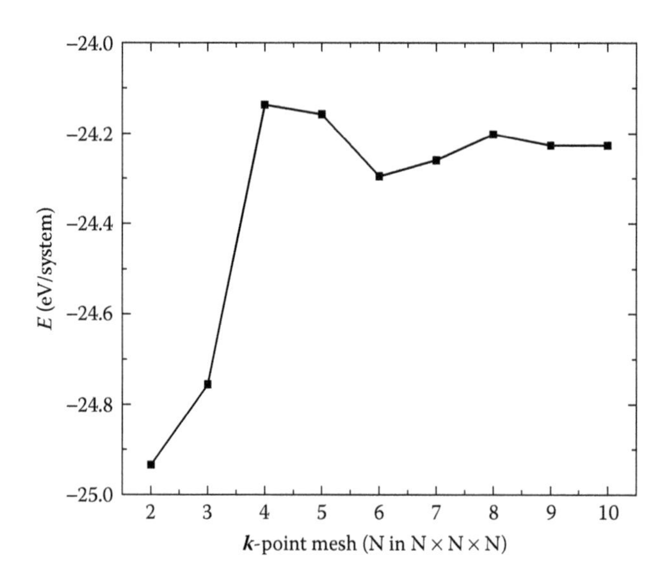

Figure 8.3 Total energy versus **k**-point mesh in a Pt-FCC (four atoms) system at ENCUT = 250 eV and a_0 = 3.977 Å.

8.5 Pt-bulk

We move on to the next run, a bulk Pt of 32 atoms in a 2×2×2 supercell, and calculate some fundamental properties: cohesive energy and vacancy formation energy. From now on, the repeating steps or flags will be largely skipped for the sake of brevity.

8.5.1 Cohesive energy of solid Pt

We first calculate a perfect Pt-bulk to have its cohesive energy. The INCAR file is the same as the Pt-FCC case except that we switch off the writing of CHGCAR and WAVECAR files, which is time-consuming and requires a lot of memory. In addition, we switch on the ISYM = 1 to use symmetry and speed up the calculation.

```
###########################################################
.....
 LCHARG =.FALSE.
 LWAVE =.FALSE.
.....
 ISYM = 1                        # use symmetry
.....
###########################################################
```

As we will start with a perfect POSCAR file, there will be no ionic minimization even though we leave the related flags intact. Since the supercell is eight times bigger than the Pt-FCC, the ***k***-points are reduced to 5×5×5.

```
##############################################################
k-points
0
M
5 5 5          #IBZKPT = 10
0 0 0
##############################################################
```

For the POSCAR file, note that all atoms are in their perfect positions in a 2×2×2 supercell.

```
##############################################################
Pt-bulk-100-222
 3.97700000000000
  2.0000000000000000 0.0000000000000000 0.0000000000000000
  0.0000000000000000 2.0000000000000000 0.0000000000000000
  0.0000000000000000 0.0000000000000000 2.0000000000000000
 32
Selective dynamics
Direct
 0.0000000000000000 0.0000000000000000 0.0000000000000000 F F F
 0.0000000000000000 0.5000000000000000 0.0000000000000000 T T T
 0.2500000000000000 0.2500000000000000 0.0000000000000000 T T T
.....
 0.7500000000000000 0.0000000000000000 0.7500000000000000 T T T
 0.7500000000000000 0.5000000000000000 0.7500000000000000 T T T
##############################################################
```

The run can be monitored by looking at either the nohup.out or OSZICAR file.

```
##############################################################
$ tail nohup.out              # view nohup.out to check progress
.....
reading WAVECAR
entering main loop
  N   E       dE         d eps   ncg  rms              rms(c)
DAV: 1 0.357949543295E+04 0.35795E+04 -0.12273E+05 3864 0.138E+03
DAV: 2 0.564646554651E+02 -0.35230E+04 -0.33960E+04 3840 0.534E+02
$ tail OSZICAR                # view OSZICAR to check progress
.....
RMM: 11 -0.193850977941E+03 0.12127E-02 -0.52223E-03 3840 0.300E-01 0.609E-03
RMM: 12 -0.193851010836E+03 -0.32895E-04 -0.45700E-04 2670 0.308E-02
 1 F= -.19385101E+03 E0= -.19385695E+03 d E =-.193851E+03
##############################################################
```

Here, N is the counter for the electronic iterations, E is the current free energy, dE is the change of the free energy between two iterations, and d eps is the change of the eigenvalues (fixed potential). The column ncg indicates how often the Hamiltonian is operated on orbitals. The column rms gives the initial norm of the residual vector, $\boldsymbol{R} = (\boldsymbol{H} - \varepsilon \boldsymbol{S})|\phi\rangle$, summed over all occupied bands, and is an indication how well the orbitals are converged. The column rms(c) indicates the difference between the input and output charge densities.

After the first five iterations, the update of the charge density starts. The final line shows the free electronic energy F after convergence has been reached. We read the E0 value (the 0 K Energy at $\sigma \rightarrow 0$) on this line, and the energy of a Pt atom in the bulk FCC is then −193.8569/32 = −6.058 eV/atom. Note that this value is slightly lower than −6.056 eV/atom calculated with the Pt-FCC at a temporary lattice parameter of 3.975 Å.

If something goes seriously wrong, VASP will stop itself with a warning message. There are quick ways of stopping a run without saving the WAVECAR and CHGCAR files.

```
############################################################
$ killall -9 vasp-5.2
$ psh node024 "pkill vasp-5.2"
############################################################
```

If a mistake is noticed, one may make a STOPCAR file with a single line.

```
############################################################
LSTOP = .TRUE.
############################################################
```

By copying this file into the current directory, the run stops at the next ionic step after writing the WAVECAR and CHGCAR files so that one can continue the next run with these files after fixing the mistakes.

8.5.1.1 Cohesive energy

Cohesive energy, E_{coh}, is the difference between the average energy of atoms in a bulk solid, E_{bulk}, and that of a free atom, E_{atom}. It is a measure of how strongly atoms bind together when they form a solid and is one of the fundamental properties of materials. In other words, it is the depth of the minimum at the equilibrium bond length minus the energy of an isolated atom. We now have both values and can calculate the cohesive energy:

$$E_{bulk} = -193.85695/32 = -6.058 \text{ eV/atom} \tag{8.3}$$

$$E_{coh} = E_{atom} - E_{bulk} = (-0.528) - (-6.058) = 5.53 \text{ eV/atom} \tag{8.4}$$

This value is in good agreement with the calculated value of 5.53 eV/atom (Bentmann et al. 2008) and in fair agreement with the experimental value of 5.45 eV/atom (Landolt-Börnstein 1991).

8.5.2 Vacancy formation energy of Pt

Metals normally have a vacancy concentration in the order of 10^{-6} at room temperature, and it is obvious that we cannot simulate 10^6 atoms for one vacancy. We rather approximate it in a reasonably sized supercell where the interactions between vacancies in the primary and image supercells can be assumed to be negligibly small. Since we have a vacancy in our system, we should switch off the symmetry flag in the INCAR file.

```
#############################################################
Pt-bulk-1v
.....
ISYM = 0              # use no symmetry
......
#############################################################
```

In the POSCAR file copied from the previous run, the total number of atoms is reduced to 31, and one Pt atom at 0.5 0.5 0.5 is removed to create a vacancy.

```
#############################################################
Pt-bulk-1v
.....
31
.....
(0.5000000000000000 0.5000000000000000 0.5000000000000000
T T T) # removed to create a vacancy
.....
#############################################################
```

The rest will be the same as the case of Pt-bulk. The OSZICAR file shows that the ground-state energy of this 31-atom Pt system with a vacancy is −186.95325 eV/system.

8.5.2.1 Vacancy formation energy

The vacancy formation energy, E_v^f, is

$$E_v^f = E_v - \frac{N-1}{N}E_{\text{bulk}} = -186.95325 - \frac{31}{32}(-193.85695)$$
$$= 0.846\ \text{eV/vacancy} \tag{8.5}$$

where E_v is the total energy of the cell containing a vacancy, N is the number of atoms in the bulk cell, and E_{bulk} is the energy of the perfect bulk (–193.85695 eV) calculated in the previous subsection. The value is not in good agreement with the other calculated value of 0.68 eV/vacancy or the experimental value (from the positron annihilation measurement) of 1.35 eV/vacancy (Schäfer 1987). Mattsson and Mattsson (2002) claimed that the vacancy formation energy of Pt can be improved only by adding a correction for the intrinsic surface error and can result 1.18 eV/vacancy. For other systems such as W (Lee et al. 2009) or SiC (Kim et al. 2009), the defect formation energies by the DFT calculations are generally in good agreement with experimental values. The CONTCAR file has the final atom positions relaxed around the vacancy, and the WAVECAR file is a binary file that contains the last electronic wave functions (orbitals), representing all solutions of the KS equations.

8.5.2.2 CHGCAR plot

The VASP writes out the charge on a regular grid in the CHG and CHGCAR files, and they have almost identical contents: lattice vectors, atom positions, electron charges, and so on. The shape of the grid follows the symmetry of the supercell, and the size of the grid is equal to the dimensions of the FFT grid as written in the OUTCAR file as NGXF, NGYF, and NGZF. By dividing the electron charges from the CHGCAR or CHG files by the supercell volume, one can have a charge density profile. For example, the CHGCAR file resulted from this exercise has the following contents:

```
#############################################################
unknown system                  # CHGCAR-Pt-bulk-1v.txt
 3.97700000000000
  2.000000 0.000000 0.000000
  0.000000 2.000000 0.000000
  0.000000 0.000000 2.000000
 31
Direct
 0.000000 0.000000 0.000000
 .....
 80 80 80
 0.19260317493E+04 0.18748881333E+04 .....
 0.12059793082E+04 0.10387811087E+04 .....
.....
#############################################################
```

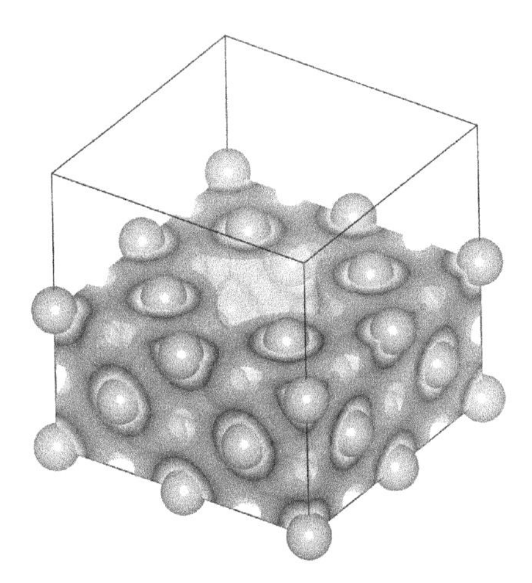

Figure 8.4 Pt-bulk with a vacancy showing isosurfaces of electron density.

Free-downloadable tools such as the "vaspview"* or "VESTA"† can display the file as shown in Figure 8.4. Note that the empty space around the vacancy (Note that the top half of supercell is removed) and more or less uniformly distributed valence electrons around each Pt atom.

Similarly the formation energies of an interstitial, E^f_{inter}, is given by

$$E^f_{\text{inter}} = E^{\text{bulk}}_{\text{inter}} - E^{\text{bulk}} - nE^{\text{atom}} \tag{8.6}$$

where $E^{\text{bulk}}_{\text{inter}}$ and E^{bulk} are the energies of bulk with interstitials and a perfect bulk, n is the number of interstitials, and E^{atom} is the energy of an isolated atom. Various defect formation energies can be calculated from these simplified setups. One should always remember that the primary box should be big enough to minimize any interactions between the defect in the primary box and the image defects in the image boxes.

When two solid phases, for example, A and B, are involved in and form a solid AB, the energy difference is called the formation enthalpy of the solid, which is given by

$$\Delta H_{\text{AB}} = E^{\text{bulk}}_{\text{AB}} - E^{\text{bulk}}_{\text{A}} - E^{\text{bulk}}_{\text{B}} \tag{8.7}$$

where the three terms represent the total energies of the bulk AB, A, and B, respectively.

* VASPview. http://vaspview.sourceforge.net/.

† VESTA (ver. 2.90.1b). 2011. http://www.geocities.jp/kmo_mma/index-en.html.

8.6 Pt(111)-surface

The surface properties of a material are, in fact, as important as the bulk properties. Fundamental properties such as surface energy, work function, adsorption energy, barrier energy for transport of the adatom, and determine the material's usefulness in applications. For example, surface energy is a critical factor in the development of surface morphologies (outward/inward relaxations, reconstruction, buckling, etc.) and for the propagation of a crack to fracture. In addition, work function, adsorption/desorption energies, and barrier energy are deciding factors for oxidation, growth and stability of thin films and nanostructures, corrosion, passivation, and catalytic reactions. Tailoring a better material by design is possible only when these surface phenomena are fully comprehended. In the present section, we will make a slab of Pt atoms and investigate some of the above-mentioned properties.

8.6.1 Pt(111)-slab

We first construct a Pt(111)-slab and calculate its total energy, relaxation energy, and surface energy. These data will serve as references for a further calculation such as adsorption energy and barrier energy.

8.6.1.1 INCAR

```
###########################################################
Starting parameters
 NWRITE = 2                # default, write-flag
 ISTART = 0                # 0-new, 1-cont
 ICHARG = 2                # charge density from: 0-wave,
                           1-file, 2-atom, >10-const
 INIWAV = 1                # default, fills wave-function
                           arrays with random numbers


Electronic Relaxation:
 ENCUT = 400.0
 PREC = Normal             # default, or set as accurate for
                           higher accuracy
 NELM = 100
 EDIFF = 1E-04
 LREAL = Auto              # or.FALSE. (default)
 ALGO = Fast

Ionic Relaxation:
 NSW = 50                  # max number of geometry steps
 IBRION = 1                # ionic relax: 0-MD, 1-quasi-Newton,
                           2-CG, 3-Damped MD
```

```
EDIFFG = -0.02          # force (eV/A)-stopping criterion
                        for geometry steps
ISIF = 1
ISYM = 0
POTIM = 0.20

DOS related values:
ISMEAR = 1
SIGMA = 0.1
############################################################
```

8.6.1.2 KPOINTS

```
############################################################
k-points for Pt-336 (Pt5layers+v7layers)-3.977
0
G
4 4 1                # IBZKPT = 10 including Γ-point
0 0 0
############################################################
```

8.6.1.3 POSCAR

Surface is normally modeled as a slab with layers with a vacuum above periodicity parallel to the slab. There are two convergences to be considered: the slab and the vacuum thicknesses. In this exercise, a surface of Pt(111) as shown in Figure 8.5 is created with a 45-atom slab, and a large vacuum (~16 Å) is included to isolate the surface from the artificial interaction between the neighboring slabs. The Pt(111) surface is chosen because

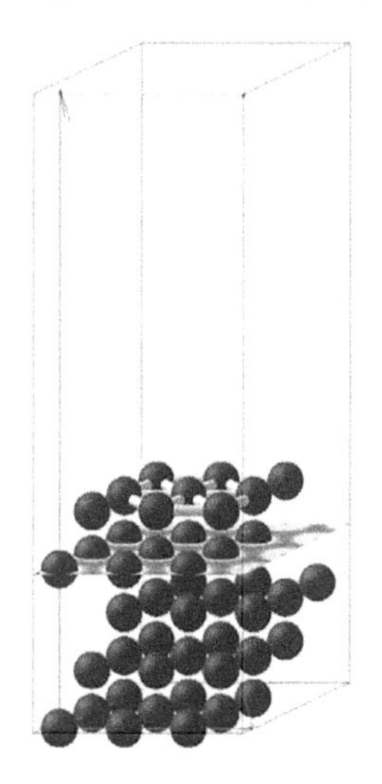

Figure 8.5 Pt(111)-slab system showing the hexagonal arrangement of Pt atoms.

it is the most frequently studied one for the catalytic application of Pt. Note that the Pt atoms on the Pt(111) plane are arranged to form close-packed hexagons, as indicated by the connecting bonds in Figure 8.5.

The relationship between the lattice parameter a for the FCC in a cubic system and the corresponding value for the hexagonal system is

$$a_{\text{HCP}} = \frac{a_{\text{FCC}}}{\sqrt{2}} = \frac{3.977}{\sqrt{2}} = 2.812 \text{ Å} \tag{8.8}$$

Note that this is the distance to the nearest neighbors for each Pt atom. The POSCAR contains a total of 45 Pt atoms arranged in five layers in an a-b-c-a-b sequence of FCC. In fact, up to nine layers may be needed to minimize the interaction between two surfaces through the slab. In this exercise, however, we will use only five layers to be able to finish the run in a reasonable time, accepting some errors.

The supercell of 3 × 3 × 6 (including vacuum layer) is constructed using the following single cell vectors assuming ideal packing (a = 2.812 Å, $c = \sqrt{8/3}a = 1.633a$):

$$(a,0,0), \left(\frac{a}{2}, \frac{\sqrt{3}a}{2}, 0\right), (0,0,c) \tag{8.9}$$

The bottom layer (nine atoms) is fixed in all directions for the stability of the slab, and the bottom+1 layer (nine atoms) is fixed in x- and y-directions only. The remaining three layers are allowed to relax fully. A POSCAR file is newly written for this slab system as the following.

```
#############################################################
Pt(111)-336  (Pt5layers+v7layers)-3.977Å
 2.812
   3.0000000000000000 0.0000000000000000 0.0000000000000000
   1.5000000000000000 2.5980762110000000 0.0000000000000000
   0.0000000000000000 0.0000000000000000 9.7979589711327009
 45
Selective dynamics                     # F: false, T: true
Direct
 0.0000000000000000  0.0000000000000000  0.0000000000000000  F F F
 0.3333333333333333  0.0000000000000000  0.0000000000000000  F F F
 0.6666666666666667  0.0000000000000000  0.0000000000000000  F F F
 0.0000000000000000  0.3333333333333333  0.0000000000000000  F F F
 0.3333333333333333  0.3333333333333333  0.0000000000000000  F F F
 0.6666666666666667  0.3333333333333333  0.0000000000000000  F F F
 0.0000000000000000  0.6666666666666667  0.0000000000000000  F F F
 0.3333333333333333  0.6666666666666667  0.0000000000000000  F F F
 0.6666666666666667  0.6666666666666667  0.0000000000000000  F F F
 0.1111111111111111  0.1111111111111111  0.0833333333333333  F F T
```

```
0.4444444444444444   0.1111111111111111   0.0833333333333333   F F T
0.7777777777777778   0.1111111111111111   0.0833333333333333   F F T
0.1111111111111111   0.4444444444444444   0.0833333333333333   F F T
0.4444444444444444   0.4444444444444444   0.0833333333333333   F F T
0.7777777777777778   0.4444444444444444   0.0833333333333333   F F T
0.1111111111111111   0.7777777777777778   0.0833333333333333   F F T
0.4444444444444444   0.7777777777777778   0.0833333333333333   F F T
0.7777777777777778   0.7777777777777778   0.0833333333333333   F F T
0.2222222222222222   0.2222222222222222   0.1666666666666667   T T T
0.5555555555555556   0.2222222222222222   0.1666666666666667   T T T
0.8888888888888889   0.2222222222222222   0.1666666666666667   T T T
.....
0.1111111111111111   0.7777777777777778   0.3333333333333333   T T T
0.4444444444444444   0.7777777777777778   0.3333333333333333   T T T
0.7777777777777778   0.7777777777777778   0.3333333333333333   T T T
############################################################
```

This Pt(111) surface is the most densely packed plane in the FCC structure and has an atomic density about 15% higher than that of the (001) surface. After the run starts, the initial OUTCAR file will be written, and we can confirm the distances to the nearest neighbors for each Pt atom in the nearest-neighbor table (2.81 Å).

8.6.1.4 Results

The OSZICAR file shows both the unrelaxed system energy (the first line) and the relaxed system energy (the last line).

```
############################################################
$ grep "E0" OSZICAR
 1 F= -.26108942E+03 E0= -.26109064E+03 d E =-.261089E+03
 2 F= -.26111143E+03 E0= -.26111299E+03 d E =-.220102E-01
.....
 6 F= -.26114066E+03 E0= -.26114366E+03 d E =-.160715E-02
 7 F= -.26114082E+03 E0= -.26114380E+03 d E =-.160699E-03
############################################################
```

In Section 8.5, we calculated the energy of a Pt atom in a bulk solid as −6.058 eV/atom. Then, the unrelaxed surface energy, γ_{unrel}, is given by

$$\gamma_s^{\text{unrel}} = \frac{1}{2}\left(E_{\text{slab}}^{\text{unrel}} - NE_{\text{atom}}^{\text{bulk}}\right) = \frac{1}{2}[-261.0906 - 45(-6.058)] \\ = 5.76\ \text{eV/system} \tag{8.10}$$

The surface relaxation energy is given by the difference between the energies after the last (full relaxation) and the first ionic relaxations (no relaxation) written in OSZICAR:

$$\gamma_s^{\text{rel}} = E_{\text{last}} - E_{\text{1st}} = -261.1438 - (-261.0906) = -0.053 \text{ eV/system} \quad (8.11)$$

Note that the surface relaxation (or reconstruction) energy is rather small (<1% of the surface energy). Then, the relaxed surface energy counting the fixed bottom layers is 5.76 − 0.053 = 5.707 eV/system (0.09 eV/Å^2), which is the same as 0.09 eV/Å^2 calculated by a very accurate setup (Getman et al. 2008). Another way of calculating surface energy is fixing the middle layer and relaxing both surfaces. Note that, owing to the difference in planar atomic packing in the FCC solids, the surface energy normally increases in the order of

$$\gamma_{111} < \gamma_{100} < \gamma_{110} \quad (8.12)$$

If we examine the POSCAR and CONTCAR files, we find that the surface layer is expanded by ~1%.

The wave function on this Pt(111) surface decreases rapidly to zero in the vacuum layer. However, the PWs extend into the whole vacuum layer with the same accuracy as the slab. Due to these two facts, the number of required PWs for a slab system increases accordingly, and the computation time takes much longer for a slab (6344 seconds for this 45-atom slab) than for a bulk by a factor of 1–2. The spin polarization is not considered here because the spin, if any, has a tendency to spread out evenly in a supercell. Interface energies between two solids such as the metal–metal and metal–oxide interfaces can be treated in a similar manner.

8.6.2 Adsorption energy

Surface-related processes such as heterogeneous catalysis and chemical vapor deposition are largely determined by the adsorption energy. In metals, the adsorption energy can be interpreted as the interaction of the orbitals of adatom with the orbitals of metallic s-, p-, and d-electrons. This energy plays a critical role in identifying the reaction mechanisms on surfaces. In this exercise, we will study a very simple system, an oxygen atom adsorbed on Pt(111) surface, and calculate the adsorption energy. There are four possible adsorption sites on Pt(111) surface: the hollow-FCC and hollow-HCP sites, the top site above the Pt atom, and the bridge site between two nearest-neighboring Pt atoms (see Figure 8.6). We will take the hollow-FCC site, which is known as the most preferred site for oxygen adatom.

8.6.2.1 POSCAR

For the adsorption energies of gaseous atoms and molecules on a solid surface, we have to make sure to position them close enough to the surface

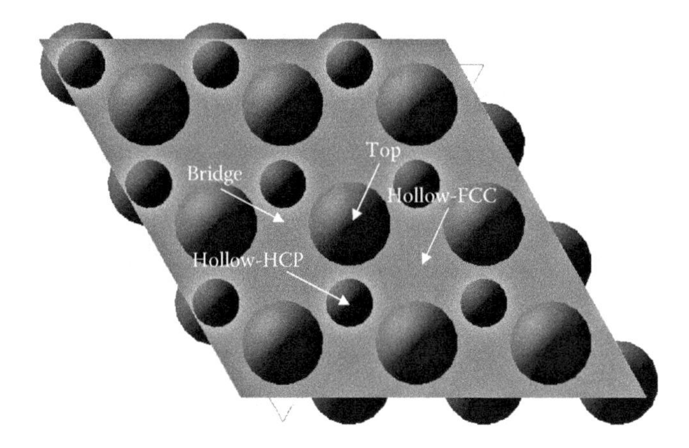

Figure 8.6 Top view of the Pt(111) surface showing two layers (big ball: top layer, small ball: top-1 layer) and adsorption sites of hollow-FCC, hollow-HCP, bridge, and top.

solid atoms. Otherwise, the adsorbates may escape away from the surface. Copy the CONTCAR file from the Pt-slab run, change the name to POSCAR, and edit it.

```
############################################################
Pt(111)-slab-O-FCC          # 3x3x6 (Pt5layers+v7layers)-3.977
 2.81200000000000
  3.0000000000000000 0.0000000000000000  0.0000000000000000
  1.5000000000000000 2.5980762110000000  0.0000000000000000
  0.0000000000000000 0.0000000000000000  9.7979589711327009
 45 1                       # 45 Pt atoms and 1 O atom
Selective dynamics
Direct
 0.0000000000000000  0.0000000000000000  0.0000000000000000 F F F
 0.3333333333333357  0.0000000000000000  0.0000000000000000 F F F
 0.6666666666666643  0.0000000000000000  0.0000000000000000 F F F
 .....
 0.4444408607027651  0.7777756518831712  0.3330202267634754 T T T
 0.7777780929586635  0.7777789561346763  0.3330216995261164 T T T
 0.55555            0.55555            0.37593            T T T
############################################################
```

Here, the total number of atoms changed to 46: 45 for the Pt-slab plus 1 for the adatom O. The atom coordinate of an O atom is added in the last line (0.55555 0.55555 0.37593), which is the hollow-FCC site with three neighboring Pt atoms. The z-coordinate of O adatom is decided based on the reported Pt-O distances. The KPOINTS file is the same as that of Pt-slab.

8.6.2.2 POTCAR

The POTCAR file for both Pt and O is concatenated from two individual POTCARs of Pt and O. We may confirm it by searching the keyword "PBE."

```
#############################################################
$ cat POTCAR-Pt POTCAR-O > POTCAR
$ grep "PBE" POTCAR
 PAW_PBE Pt 05Jan2001
 TITEL = PAW_PBE Pt 05Jan2001
 PAW_PBE O 08Apr2002
 TITEL = PAW_PBE O 08Apr2002
#############################################################
```

8.6.2.3 Results

The energy of the Pt(111) with an O adatom is shown in the following.

```
#############################################################
$ tail nohup.out
 bond charge predicted
   N  E              dE          d eps   ncg    rms          rms(c)
DAV: 1 -0.267250456678E+03  -0.33352E-03  -0.31530E-01 5648   0.911E-01  0.660E-02
RMM: 2 -0.267251312313E+03  -0.85563E-03  -0.44161E-03 5605   0.160E-01  0.182E-01
RMM: 3 -0.267251212787E+03   0.99526E-04  -0.45461E-04 4201   0.503E-02
 10 F= -.26725121E+03 E0= -.26724876E+03 d E =-.112243E-02
.....
#############################################################
```

In the CONTCAR file, we can find that the adsorbed O atom relaxed slightly in the hollow-FCC site (0.55555 0.55555 0.37598).

```
#############################################################
$ cat CONTCAR
.....
0.4384792872038999 0.7897073127914218 0.3345598713973848 T T T
0.7798623700003264 0.7798623896068256 0.3323143963849072 T T T
0.5555539368725361 0.5555528288191630 0.3759815738986362 T T T
.....
#############################################################
```

To calculate the adsorption energy, we have to have the energy of an isolated O atom as the reference state. It is calculated by the same method described in Section 8.2 for a Pt atom.

```
#############################################################
$ tail -5 nohup.out
.....
DAV: 8   -0.158978423065E+01 -0.13379E-03 -0.49997E-05 32 0.412E-02 0.295E-02
DAV: 9   -0.158988686021E+01 -0.10263E-03 -0.47842E-06 32 0.235E-02 0.190E-02
DAV: 10 -0.158991610704E+01 -0.29247E-04 -0.35940E-06 32 0.126E-02
 1 F= -.15899161E+01 E0= -.15513507E+01 d E =-.771308E-01 mag= 2.0000
writing wavefunctions
#############################################################
```

The adsorption energy, E_{ads}, is then obtained as the difference in energy between the slab with O adatom and the clean slab plus an isolated O atom:

$$E_{ads} = \frac{1}{N_O^{atom}}\left(E_{O/Pt(111)}^{slab} - E_{Pt(111)}^{slab} - N_O^{atom}E_O^{atom}\right)$$
$$= -267.2488 - (-261.1438) - (-1.5514) = -4.55 \text{ eV} \tag{8.13}$$

Note that, in another convention, the adsorption energy is defined to be positive for a stable adsorption configuration. The calculated value of −4.55 eV in this exercise is in agreement with another calculated value of −4.68 eV (Pang et al. 2011). Adsorption energies of atoms on other sites such as hollow-HCP, top, bridge, and next to defects or other adsorbates can be calculated in the same manner. Adsorption energies of oxygen are often reported with respect to a free O_2 molecule in the gas phase as the reference. In that case, the adsorption energy in Equation 8.13 will be reduced to −1.42 eV, counting half the binding energy of O_2 (about 3.13 eV/O; Lischka et al. 2007).

8.6.3 *Work function and dipole correction*

This exercise will calculate the work function with the corrected dipole, using the resulted output files from the previous calculation. Work function is defined as the work required for the removal of a Fermi-level electron from the solid surface to infinity (vacuum). And, the dipole, U_{dipole}, perpendicular to the slab's *z*-axis is defined as

$$U_{dipole}(z) = \frac{1}{A}\iint U_{dipole}(\boldsymbol{r})\,dxdy \tag{8.14}$$

Then, the dipole-corrected work function is $-E_F + U_{dipole}$ and depends strongly on the geometry and the nature of the surface.

8.6.3.1 *Work function*

The INCAR file includes five new flags such as the slab's central direct coordinates.

```
###############################################################
Pt(111)-O-hollow(FCC)
.....
Correction for dipole
 IDIPOL = 3
 LDIPOL = .TRUE.       # activate dipole correction
```

```
 LVTOT = .TRUE.        # write averaged local-electrostatic-
                       potential in 3D in LOCPOT
 DIPOL = 0.5 0.5 0.188 # locate the center of slab
 LORBIT = 1
.....
##############################################################
```

The KPOINTS, POSCAR, and POTCAR are the same as the Pt(111)-O case.

8.6.3.2 Results

First read the Fermi level written in the OUTCAR file.

```
##############################################################
$ grep fermi OUTCAR
.....
E-fermi: -0.317 XC(G=0): -6.2420 alpha+bet: -5.3412
##############################################################
```

The new output file, LOCPOT, has the same format as the CHGCAR file and contains the planar-averaged local-electrostatic potential (excluding the XC potential) in three dimensions.

```
##############################################################
unknown system
 2.81200000000000
  3.000000 0.000000 0.000000
  1.500000 2.598076 0.000000
  0.000000 0.000000 9.797959
 45 1
Direct
 0.000000 0.000000 0.000000
 0.333333 0.000000 0.000000
.....
84 84 280
 -.51269861567E+02 -.50785172546E+02 -.49338757533E+02.....
 -.39054609779E+02 -.33762614598E+02 -.27842051812E+02.....
.....
##############################################################
```

Read the potential data along the *z*-direction and convert them to *.dat file with a proper script such as locpot_workfunc.py written by Jinwoo Park and Lanhee Yang (2009, private comm.) at the University of Seoul, Korea.

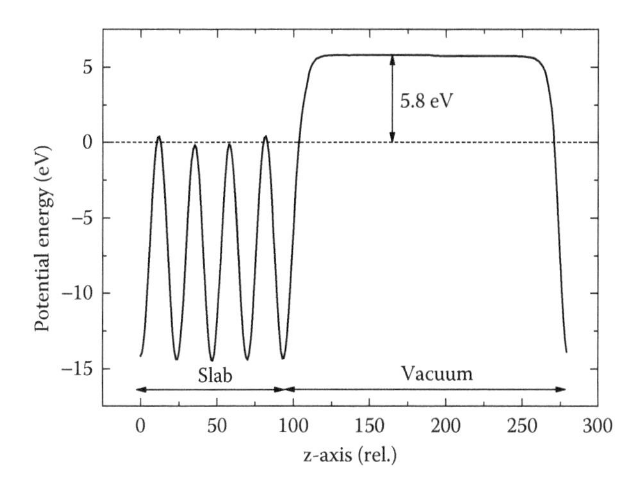

Figure 8.7 Potential energy plot for the Pt(111)-O slab system showing the Fermi energy (= 0 eV) and the work-function (5.8 eV).

```
#############################################################
$ python locpot.py LOCPOT -0.3174 # with numpy and scipy
                                    compiled
#############################################################
```

Here, −0.3174 is the Fermi energy that will be corrected to zero, which is a customary practice. The output.dat file will be generated for the potential energy plot as shown in Figure 8.7. The figure shows the relative positions of Pt(111) layers in the *z*-axis, the Fermi energy, and the work function of 5.8 eV.

8.7 Nudged elastic band method

Human beings may take various paths between two points. One may take the shortest route to be on time, or choose the toughest route to have a good exercise. Unlike human beings, atoms do not care about the shortest or the toughest route. We know very well that atoms (or other matter) under some force always follow the path with the lowest energies under the given conditions, as illustrated in Figure 8.8. Remember that a river always runs through the very bottom of a canyon.

How a system makes any change in configuration from one state to another and identifying the minimum energy path (MEP) that the system takes are an essential aspect in materials study. In this exercise, we will follow the NEB (nudged elastic band) method (Henkelman et al. 2000; Henkelman and Jónsson 2000; Sheppard and Henkelman 2011) to find the MEP and the corresponding barrier energy.

Figure 8.8 Schematic of the human being and atom moving between two points.

8.7.1 Principle of NEB method

Let us examine a potential energy surface as shown in Figure 8.9. Two local minima are identified as the initial and the final states, and two lines are connected between them: one straight band and the other a curved MEP band. As indicated by arrows, we will start the search from the straight line and try to reach the MEP line as efficiently as possible.

8.7.2 Procedure of the NEB method

8.7.2.1 Initial and final states

First, the initial and final states of interest have to be identified by the usual electronic and ionic relaxations. At these two configurations, the energies are minimal, and the forces (first derivatives of the energy) are all zero.

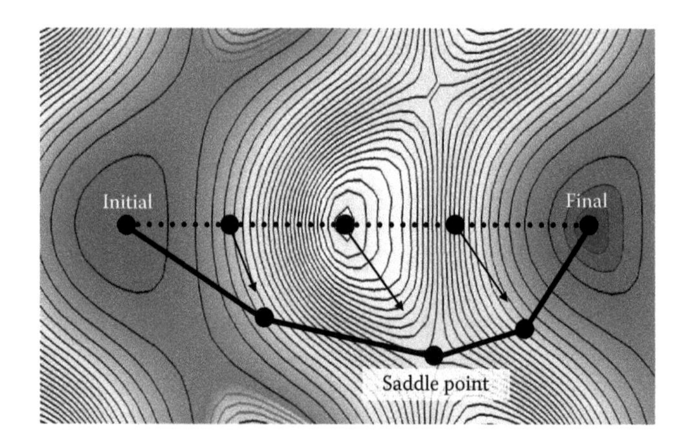

Figure 8.9 Principle of NEB method showing the initial band (dotted line) of three intermediate images, nudging toward the minimum energy path (MEP, solid line) which passes over the saddle point with barrier (activation) energy.

8.7.2.2 Initial band

Since the MEP will be somewhere between the initial and final states, we first draw a straight band between the two states as the trial path and pick several points (images) on that path at equal distances as shown in Figure 8.9. Note that each image represents an intermediate configuration. The number of images depends upon the complexity (curvature) of the path. For most cases in materials, three to seven images may be enough. This simple extrapolation of atomic positions between the initial and the final images may generate atomic positions rather far from the MEP. Then, atoms are under high forces and convergence becomes difficult, which requires some manual adjustments to move the image closer to the MEP.

8.7.2.3 Nudging the band

The search for the MEP will be carried out by moving (nudging) the initial band bit by bit toward the zero-force configurations. To bring the band's nudging under control and to ensure the continuity of the band with equal spacing of the images, a fictitious spring force will be applied on each image along the parallel direction to the band. With this setup, the band will nudge the configuration surface until each image finds the zero-force configuration as illustrated in Figure 8.9. This is analogous to a camel caravan crossing through the dunes: Each camel is connected with a rope to other camels to make sure that everyone is on the route.

8.7.2.4 Force calculation

The NEB method uses a force projection (not the energy) in order to guide the images toward the MEP, including the saddle points. Atoms in each image on the initial band are forced to move by the sum of two forces: spring forces and interatomic forces. In order to direct the band toward the MEP, only the spring forces projected along the band (local tangent) and the interatomic forces projected perpendicular to the band are considered. By ignoring all other components of forces, the images on the straight line will move toward the MEP line. At each nudged configuration, electronic minimization is carried out by the usual DFT calculation, and the forces on each atom are calculated to determine the lower-force directions for the next nudging. After repeating the process, when the forces on each atom go down to the predetermined small number close to zero, the band is assumed to be on the MEP. Since the nudging is based upon forces, the damped MD algorithm (IBRION = 3) is used.

8.7.2.5 NEB method with climb

The NEB/climb method (Henkelman et al. 2000) is designed to make sure one image will be right at the saddle point. To make the highest energy image climb to the saddle point, its true force after some nudging along the tangent is inverted (for example, by applying twice the opposite of the true force). Thus, the climbing image maximizes its energy along the band and minimizes it perpendicular to the band.

8.7.3 Pt(111)-O-NEB

The following example shows how the NEB method finds the MEP and calculates the barrier energy when an O adatom diffuses from the hollow-HCP site to the neighboring hollow-FCC site on the surface of a Pt(111)-slab as shown in Figure 8.10.

8.7.3.1 Pt(111)-slab-O-HCP

For an NEB run, two minimization runs are needed: for initial and final configurations. We already have the final configuration for the Pt(111)-slab-O-FCC as calculated in Section 8.6.3. Here, we will run the same way for the initial configuration, the Pt(111)-slab-O-HCP. The only thing we need is to place an O atom on the hollow-HCP site in the POSCAR file.

```
##############################################################
.....
0.66667 0.33333 0.37598 T T T
##############################################################
```

The run returns the converged energy of –266.87 eV/system and results in the necessary output files for the NEB run, the CONTCAR and OUTCAR files, which will provide atomic positions and the corresponding energy for the initial configuration.

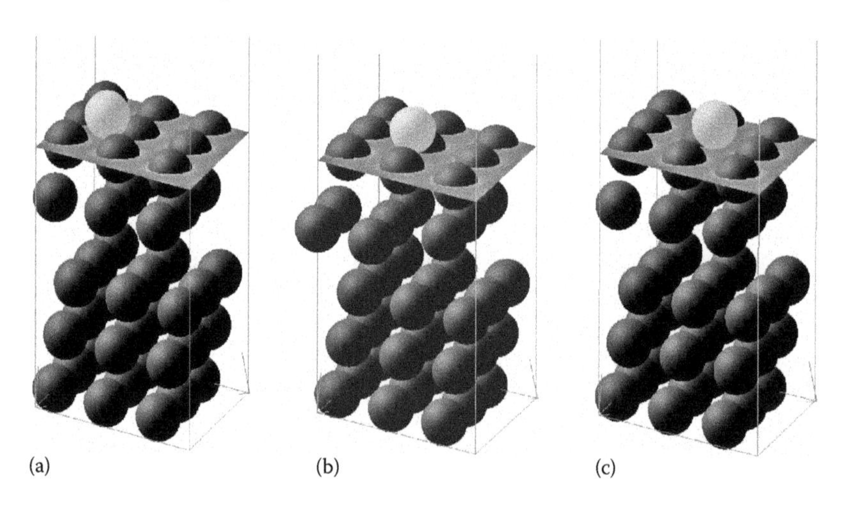

Figure 8.10 Configurations of (a) the initial, (b) the intermediate, and (c) the final images when an O adatom (light grey ball) diffuses from the hollow-HCP site to the next hollow-FCC site on the Pt(111) surface.

8.7.3.2 Run NEB with VTST scripts

The VTST script,* an open source, interfaces with VASP and provides convenient ways to run an NEB calculation. Download and install the package into the VASP source directory and make all *.F files (neb.F, dynmat.F, etc.) into execution files. For an NEB run, three new flags are added, and three flags for the ionic minimizations are modified in the INCAR file.

```
#############################################################
.....
IMAGES = 1      # no. of images excluding two endpoints, set
                a NEB run
                # no. of nodes, must be dividable by no. of
                images.
                # each group of nodes works on one image.
SPRING = -5.0   # spring force (eV/A²) between images
LCLIMB = .TRUE. # turn on the climbing image algorithm
.....
IBRION = 3      # ionic relaxation by Damped MD, best choice
                for a stable NEB run
EDIFFG = -0.05  # force (eV/Å)-stopping criterion for geometry
                steps
                # normally, start with -0.01
POTIM = 0.30    # initial timestep for geo-opt. of each
                iteration
.....
#############################################################
```

The number of images should be in accordance with the number of CPUs since the calculations for each image are allocated evenly to CPUs. Thus, the number of images should be 1, 2, 4, or 8 in our case of 8 CPUs. Then, make a new directory and copy CONTCARs from the above two runs and rename them as CONTCAR1 (O on the hollow-HCP site) and CONTCAR2 (O on the hollow-FCC site). Since this will be a simple straight diffusion, we will use a single image, and the VTST will do the linear interpolation and generate a POSCAR file for the intermediate image.

```
#############################################################
$ ~/vtstscripts/nebmake.pl CONTCAR1 CONTCAR2 1
OK, ALL SETUP HERE
FOR LATER ANALYSIS, PUT OUTCARs IN FOLDERS 00 and 02 !!!
$ ls
00 01 02 CONTCAR1 CONTCAR2 INCAR KPOINTS POTCAR run.vasp
#############################################################
```

* VTST. http://theory.cm.utexas.edu/vtsttools/neb/; http://theory.cm.utexas.edu/vtsttools/downloads/; http://theory.cm.utexas.edu/henkelman/.

Three directories are generated including 00 (initial image), 01 (intermediate image1), and 02 (final image). Any negative number of atomic positions automatically changes into a positive one, for example, −0.001 → 0.999. Copy the OUTCAR file from the Pt(111)-slab-O-HCP into the directory 00, and the OUTCAR file from the Pt(111)-slab-O-FCC into the directory 02. The final E0 data in OUTCAR will be used as the bases to calculate the barrier energies. Also copy the POTCAR and KPOINTS files and start the run. During the run, forces can be monitored from the OUTCAR file.

```
############################################################
$ grep "max atom" OUTCAR
.....
FORCES: max atom, RMS 0.375123 0.072047 # the magnitude of
                                          the forces
############################################################
```

Here, the first number is the maximum force on any atom in the image, and convergence will be reached when the number in every image falls below the prescribed force criteria (<0.05 eV/Å). If the force is too high initially (>10 eV/Å), we may restart the run with a better initial geometry. During the run, the status can be monitored with the vtstscripts/nebef.pl, which writes a summary file, neb*.dat; for example, the data after 19 and 30 ionic relaxations are as in the following.

```
############################################################
$ ~/vtstscripts/nebef.pl > neb19.dat
$ cat neb19.dat
 0 0.01344100 -266.86959100 0.00000000
 1 0.09466700 -266.71675300 0.15283800
 2 0.01829400 -267.24876300 -0.37917200
$ ~/vtstscripts/nebef.pl > neb30.dat
$ cat neb30.dat
 0 0.01344100 -266.86959100 0.00000000
 1 0.04742900 -266.71405400 0.15553700
 2 0.01829400 -267.24876300 -0.37917200
############################################################
```

Here, the data show in order of image number, force, total energy, and difference in total energy between images. All output files (OUTCAR, WAVECAR, CHGCAR, etc.) are written to the directory 01. A typical run normally spends the longest time on the image at the saddle point since the image is forced to move toward the highly unstable state.

8.7.3.3 Results

The nebbarrier.pl script calculates the distances (reaction coordinate) between images along the band and writes the corresponding differences in total energies and the maximum forces on atoms in each image.

```
################################################################
$ ~/vtstscripts/nebbarrier.pl
$ cat neb.dat
 0    0.000000     0.000000     0.001418    0
 1    0.613746     0.155537     0.046151    1
 2    1.738097    -0.379172    -0.002640    2
################################################################
```

The nebresults.pl script writes a series of files for postprocessing.

```
################################################################
$ ~/vtstscripts/nebresults.pl > neb-results.dat
$ ls
 .....movie.vasp neb-results.dat movie.xyz spline.dat exts.
 dat mep.eps nebef.dat.....
################################################################
```

Figure 8.11 shows the difference in total energy versus distance when the O atom moves from the hollow-HCP site to the hollow-FCC site on the Pt(111) surface. The diffusion barrier of the path is calculated as 0.155 eV. The barrier energy for the backward diffusion, from the hollow-FCC site to the hollow-HCP site, is 0.53 eV, which is in good agreement with the other calculated value of 0.5 eV (Pang et al. 2011) and is in fair agreement with the experimental value of 0.43 eV (Wintterlin et al. 1996). It implies that a moderate thermal agitation could move any O adatom on

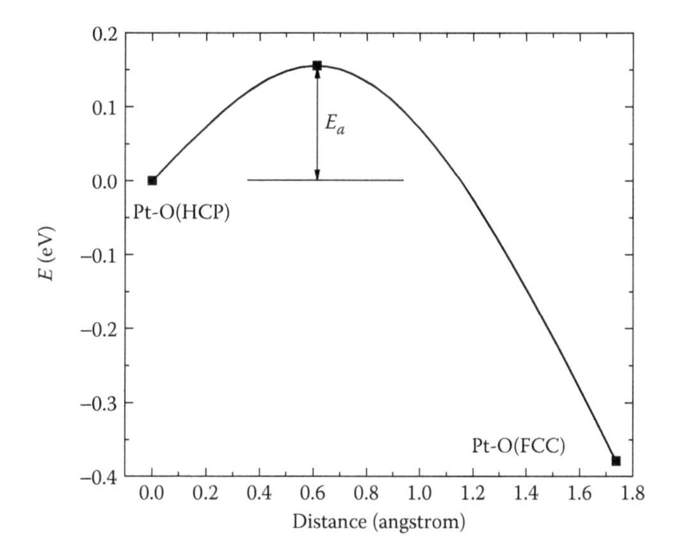

Figure 8.11 Difference in total energy versus distance when an O adatom diffuses from the hollow-HCP site to hollow-FCC site on the Pt(111) surface showing the barrier energy, *Ea*.

the hollow-HCP sites to the more stable hollow-FCC sites. This exercise can be extended to the strained structures of bimetal slab (Mavrikakis et al. 1998) the core-shell nanocluster system, and so on.

8.8 Pt(111)-catalyst

8.8.1 Catalyst

A catalyst is a substance that increases the rate of reaction kinetically without itself being consumed in the reaction. As shown in Figure 8.12, a catalyst provides a lower energy pathway for a given reaction by lowering the activation energy. The thermodynamics of the reaction as indicated by ΔG remains unchanged. The concept of catalysis is as simple as shown in the figure, but the actual process is a very complex one that involves combinations of many structural and electronic factors.

Thus, a detailed knowledge of catalyst surfaces with adsorbates is crucial to understanding the catalytic processes involved. The DFT approach can provide an effective way of finding better catalysts, interpreting experimental results, and providing information that cannot be obtained or is difficult to obtain experimentally. In this section, changes happening on the electrons on a catalytic surface will be studied with respect to the density of states (DOS).

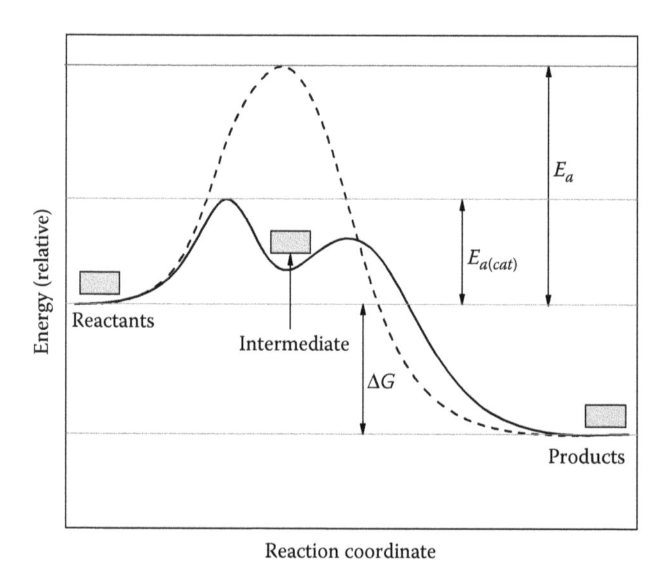

Figure 8.12 Mechanism of a catalyst providing a lower energy pathway for a given reaction by lowering the activation energy.

8.8.2 Density of states

A self-consistent solution to the KS equations results in many KS orbitals and the corresponding KS energies (one-electron eigenvalues) for the set of $\boldsymbol{k}$-points. By grouping them according to their orbitals, we can construct two very important pieces of information: band structure and DOS. A band structure displays the electronic states only along lines of high symmetry in the IBZ, whereas a DOS diagram collects electron states in the whole BZ.

DOS provides the distribution of electronic states in $\boldsymbol{k}$-space in terms of energy and is defined as the number of states at each energy interval that are available to be occupied in a band n:

$$
\begin{aligned}
D(\varepsilon) &= \frac{\text{Number of states between } \varepsilon \text{ and } \partial\varepsilon}{\partial\varepsilon} \\
&= 2\sum_{k} \delta[\varepsilon - \varepsilon_n(\boldsymbol{k})]
\end{aligned}
\tag{8.15}
$$

A high DOS at a specific energy level therefore means that there are many states available for occupation. If we integrate the DOS up to the Fermi energy, it becomes the number of electrons in the system:

$$
n = \int_0^{\varepsilon_F} D(\varepsilon)d\varepsilon \tag{8.16}
$$

Often, we ask for a DOS for a specifically projected way. This PDOS (projected DOS) can be calculated by summing up DOS over orbitals, atoms, species, layers in a slab, and so on. Experimentally, the DOS of a solid is normally measured by photoemission spectroscopy.

8.8.3 Pt(111)-slab-O-DOS

8.8.3.1 Static run

Once a normal minimization is done, the first step is writing a CHGCAR file in a static self-consistent run by using the CONTCAR file and setting NSW = 0 in the INCAR file. Using more $\boldsymbol{k}$-points, that is, $6\times6\times1$, in the KPOINTS file is preferred for a higher accuracy.

8.8.3.2 DOS run

The next step is a continuous run to perform a non-self-consistent calculation using the CHGCAR file from the first step and setting the ICHARG = 11 in the INCAR file.

```
################################################################
ISTART = 1          # continuous run
ICHARG = 11         # keeps the charge fixed
INIWAV = 1          # electr: 0-lowe 1-rand 2-diag
LORBIT = 11         # writes DOSCAR and lm-decomposed PROCAR
                    for PAW
LCHARG = .TRUE.
.....
NEDOS = 1000        # no. of data points
################################################################
```

This will fix the charge density and the potential and find the energy states at each independent *k*-point.

8.8.3.3 Results

The DOSCAR file provides the DOS and the integrated DOS in units of number of states/unit cell, and the PROCAR file provides the s-, p-, d-, and site-projected orbitals of each band.

```
################################################################
$ head DOSCAR
 46 46 1 0
 0.3691449E+02 0.8436000E-09 0.8436000E-09 0.2755186E-08
 0.2000000E-15
 1.000000000000000E-004
 CAR
unknown system
 7.84401694 -21.63116804 1000 -0.31044472 1.00000000
 -21.631 0.0000E+00 0.0000E+00
 -21.602 0.0000E+00 0.0000E+00
.....
################################################################
```

We have to extract the necessary information manually or by using any convenient script available (such as dos.py here), which results in a long list of energy levels and occupancies at each orbital. Here, the atoms selected are two atoms in a bonded pair: a Pt atom and an O atom.

```
################################################################
$ python dos.py DOSCAR tods > pdos-Pt-O-41Pt.dat
$ head pdos-Pt-O-41Pt.dat
 -21.6310000000000 0.0000000000000 0.0000000000000
0.0000000000000
0.0000000000000 0.0000000000000 0.0000000000000
0.0000000000000 0.0000000000000
0.0000000000000 0.0000000000000
.....
```

```
$ python dos.py DOSCAR tods > pdos-Pt-O-46O.dat
$ head pdos-Pt-O-46O.dat
-21.6310000000000 0.0000000000000 0.0000000000000
0.0000000000000 0.0000000000000
0.0000000000000 0.0000000000000 0.0000000000000
0.0000000000000 0.0000000000000
0.0000000000000
.....
############################################################
```

To plot the DOS diagrams from these data files, we have to know the Fermi level.

```
############################################################
$ grep fermi OUTCAR
ISMEAR = 1; SIGMA = 0.20 broadening in eV -4-tet -1-fermi
0-gaus
E-fermi: -0.3104 XC(G=0): -6.2345 alpha+bet: -5.3422
############################################################
```

The above pdos-Pt-O-46O.dat files can be copied into Xmgrace, MS® Excel, or ORIGIN®, and the first column, the energy level, is calibrated so that the Fermi level becomes zero and plotted as shown in Figure 8.13. For comparison, the PDOS for a clean Pt-slab is also shown. Here, the total DOS is projected on each atom as PDOS, and only the d-DOS of Pt and the p-DOS of O are collected. With this diagram, the character and strength of the bond can be qualitatively estimated. For example, it is apparent that the Pt-O bond is mainly due to a hybridization of the Pt-5d orbitals and the O-2p orbitals.

8.9 Band structure of silicon

We have already discussed the band structures of a free electron in Figure 6.10, showing continuous ε-$\boldsymbol{k}$ curves in a quadratic (parabolic) form with a minimum at $\boldsymbol{k} = 0$ (Γ-point). Recall that the energies are periodic in the reciprocal space, and there are many parabolae centered on each of the reciprocal lattice points. It can always be confined to the 1st BZ, and any $\boldsymbol{k}$ outside the 1st BZ can be mapped back into it. The band structure of a real solid shown in Figure 6.10 for Si is very much different from the highly idealized free-electron versions since electrons are now neither completely localized nor completely free, and the band structure thus exhibits features such as band gap formation and nonsymmetry around Γ-point.

In this exercise, we will calculate and construct a band structure for a typical semiconductor, Si, and see how orbitals of isolated atoms form

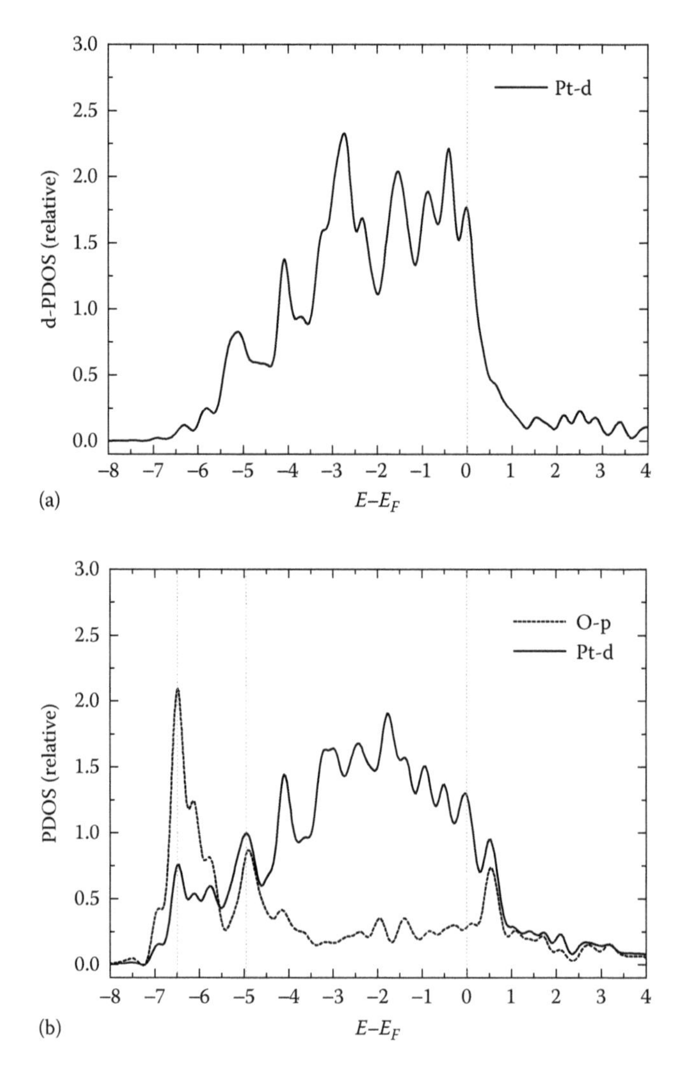

Figure 8.13 PDOS graphs of the Pt(111) system (a) and the Pt(111)-O system (b).

bands and band gap when atoms gather and become a solid. This exercise will demonstrate the calculation of KS energies, ε_{nk}, in terms of band index n and wave-vector index $\boldsymbol{k}$, as we discussed in Section 6.4.3.

8.9.1 Static run for Si

First, the lattice parameter of Si in a diamond structure is determined as $a_0 = 5.47$ Å in the same way already discussed in Section 8.3. To have a band structure of Si, two consecutive runs are needed: a static run to have the converged charge density and a following run to have band energies using the computed charge density along directions of interest non–self-consistently. The first run is carried out by setting ENCUT = 250 in the INCAR file and by using ***k***-points of 7×7×7 in the KPOINTS file. The Si crystals are just two sets of FCC lattice with a displacement of (a/4, a/4, a/4), where a is the length of the conventional cubic lattice. The POSCAR file is written to have an FCC primitive unit cell with two Si atoms.

```
############################################################
Si-diamond
5.470
 0.5000000000000 0.5000000000000 0.0000000000000
 0.0000000000000 0.5000000000000 0.5000000000000
 0.5000000000000 0.0000000000000 0.5000000000000
2
Cartesian
 0.0000000000000 0.0000000000000 0.0000000000000
 0.2500000000000 0.2500000000000 0.2500000000000
############################################################
```

The POTCAR file (potcar.PBE.paw/Si/POTCAR) for Si is copied from the PP library of VASP and placed in the current directory.

```
############################################################
PAW_PBE Si 05Jan2001
 4.00000000000000000
 parameters from PSCTR are:
  VRHFIN =Si: s2p2
  LEXCH = PE
  EATOM = 103.0669 eV, 7.5752 Ry

  TITEL = PAW_PBE Si 05Jan2001
  LULTRA = F use ultrasoft PP ?
  IUNSCR = 1 unscreen: 0-lin 1-nonlin 2-no
......
############################################################
```

After the run, the charge density in CHGCAR file is generated with the following ground-state energy.

```
############################################################
.....
RMM: 9 -0.108208215051E+02 -0.98089E-04 -0.13279E-04 250
0.917E-02
 1 F= -.10820822E+02 E0= -.10820838E+02 d E =0.507108E-04
 writing wavefunctions
############################################################
```

8.9.2 *Run for band structure of Si*

For the continuing run, we need only the CHGCAR file with the usual input files.

8.9.2.1 *INCAR*

In order to have a band structure, we need to do a non-self-consistent run on each desired ***k***-point with one additional flag, ICHARG = 11 in the INCAR file, which reads the provided CHGCAR file from the previous static run and keeps the charge density constant during the run.

```
#############################################################
.....
ISTART = 1
ICHARG = 11
NBANDS = 8                    # number of bands
.....
#############################################################
```

8.9.2.2 *KPOINTS*

The wave vector, ***k***, is a three-dimensional vector spanning within the whole first BZ, but we already know that it is sufficient just to plot the energies along a path connecting points of high-symmetry directions. Table 8.2 shows a list of high-symmetry ***k***-points in the first BZ of an FCC.

Therefore, we only assign enough points in KPOINTS files to make some nice curves, and the orbital-energy calculations will take place only on these predetermined ***k***-points.

Table 8.2 A list of high-symmetry ***k***-points in the first BZ of an FCC[a]

Point	Reciprocal coordinates (units of b_1, b_2, b_3)			Cartesian coordinates (units of $2\pi/a$)		
Γ	0	0	0	0	0	0
X	1/2	0	1/2	0	1	0
W	1/2	1/4	3/4	1/2	1	0
L	1/2	1/2	1/2	1/2	1/2	1/2
Δ	1/4	0	1/4	0	1/2	0
Λ	1/4	1/4	1/4	1/4	1/4	1/4

[a] see Figure 6.11 for the corresponding coordinates.

```
############################################################
k-points        # for band structure
20              # calculates energies for 20 points per
                each line
Line mode
Rec             # positions of high-symmetry k-points in
                the first BZ
0.75 0.5 0.25   # line starting at W to L
0.5 0.5 0.5
0.5 0.5 0.5     # line starting at L to G
0 0 0
0 0 0           # line starting at G to X
0.5 0.5 0
0.5 0.5 0       # line starting at X to W
0.75 0.5 0.25
############################################################
```

The above file instructs VASP to calculate the energies at 80 *k*-points (20 points × 4 lines) along the W-L-Γ-X-W lines (see Figure 6.11).

8.9.2.3 EIGENVAL

There is no need to be concerned about the calculated energy because this run is not a self-consistent run. We are only interested in the eigenvalues, ε_{nk}, in the EIGENVAL file to plot the band structure.

```
############################################################
.....            # a series of 80 data sets starts from here
0.7500000E+00 0.5000000E+00 0.2500000E+00 0.1250000E-01
 1 -1.975362     # ε at band 1 and k-points W
 2 -1.975360
 3 1.856500
 4 1.856511
 5 9.908947
 6 9.908990
 7 10.556935
 8 10.556951
 .....
0.7500000E+00 0.5000000E+00 0.2500000E+00 0.1250000E-01
 1 -1.975362
 2 -1.975360
 3 1.856500
 4 1.856511
 5 9.908947
 6 9.908990
 7 10.556935
 8 10.556951
############################################################
```

The first line is the **k**-point position in reciprocal lattice, and the following eight lines are the corresponding energies of eight bands. The lower four valence bands (numbers 1–4) will be fully occupied at 0 K. Editing the EIGENVAL file, extracting the ε_{nk} values, and setting the Fermi level to zero can be easily done by using a script such as vasp_to_image.py written by Jung-Ho Shin (2010, private comm.; see Appendix C).

After removing all numbers except the ε_{nk} values (i.e., –1.975362, –1.975360, …) and calibrating the Fermi level (5.6980 eV from the OUTCAR file) to be 0 eV, we have a data file, Si-band.dat with **k**-point positions and energies:

```
##########################################################
0.0000 -7.6734 -3.8415 -3.8415 4.2109 4.2110 4.8589 4.8590
0.0186 -7.5070 -3.9561 -3.7059 4.0602 4.2744 4.7916 5.0383
0.0372 -7.3475 -4.0447 -3.5531 3.8956 4.1854 4.9029 5.2583
.....
2.2244 -7.6833 -3.7943 -3.7943 3.5797 3.5797 5.5453 5.5453
2.2430 -7.6779 -3.8205 -3.8205 3.8476 3.8476 5.2460 5.2460
2.2616 -7.6745 -3.8362 -3.8362 4.0895 4.0895 4.9863 4.9864
2.2802 -7.6734 -3.8415 -3.8415 4.2109 4.2110 4.8589 4.8590
##########################################################
```

Figure 8.14 is the final plot, and we can see that Si is an indirect semiconductor with a band gap of $E_g = 0.61$ eV, which represents the smallest energy difference between the two bands at different **k**-points: from the Γ-point to the X-point. As discussed in Section 6.4, this value is only a

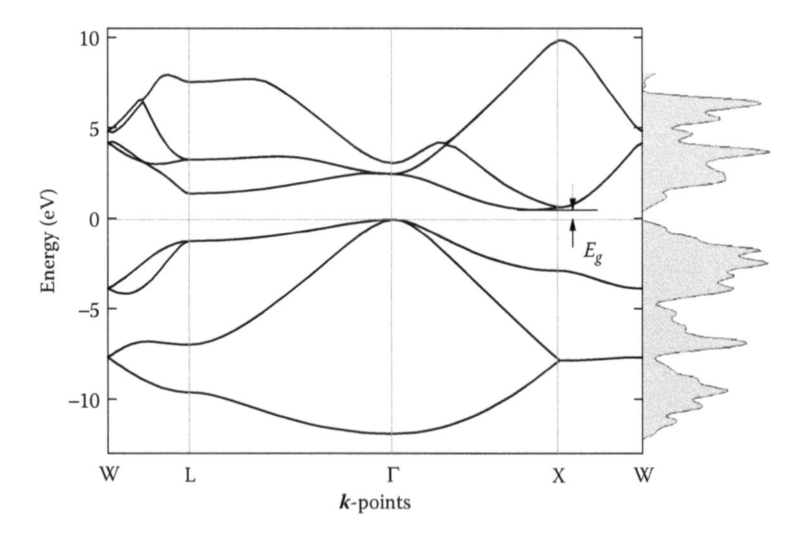

Figure 8.14 Band structure and DOS of Si.

half of the reported experimental band gap, and this underestimation of the band gap is an inherent problem of DFT.

However, the general feature of energy dispersion is well presented, such as four valence bands and four conduction bands, which supports the validity of the DFT scheme. One can also find that the ε_{nk} varies rather smoothly in the **k**-space. In fact, other quantities such as wave functions and electron densities behave similarly, and this confirms that the **k**-points sampling is a reasonable scheme to evaluate the varying quantities in the IBZ.

8.10 Phonon calculation for silicon

This run example of phonon calculation for bulk Si is provided by Prof. Aloysius Soon (2010, private comm.) of Yonsei University in Seoul. Here, we use a convenient phonon-calculation program, "phonopy" (Togo 2010), which is an open source. Phonons are lattice vibrations, propagating through the solid with a wave-vector, **q**, and characteristic frequency $\omega(\mathbf{q})$ in quantized modes. As atoms are interconnected like a series of balls and springs in a lattice, a displacement of one atom (or more) from its equilibrium position will send out such a cascade of atomic movements, resulting data such as force constants, dynamic matrix, and so on. From this information, we can draw a phonon dispersion diagram, $\omega(\mathbf{q})$ versus **k**-point in the first BZ, which can provide rather rich information on that particular solid. Here, we exercise with Si under a very simple setup, the small-displacement method under the frozen-phonon approximation. Note that the phonon dispersion can be obtained experimentally by the neutron scattering.

8.10.1 Input files

Prepare the POSCAR file for a primitive unit cell of Si.

```
#################################################################
Bulk Si
 5.466198843774785
  0.0000000000000000 0.5000000000000000 0.5000000000000000
  0.5000000000000000 0.0000000000000000 0.5000000000000000
  0.5000000000000000 0.5000000000000000 0.0000000000000000
 2
Direct
  0.8750000000000000 0.8750000000000000 0.8750000000000000
  0.1250000000000000 0.1250000000000000 0.1250000000000000
#################################################################
```

First, several supercells are generated (for example, for a 2×2×2 supercell).

```
###############################################################
$ phonopy -d --dim="2 2 2" # generate a 2x2x2 supercell
###############################################################
```

Three files of SPOSCAR, DISP, and POSCAR-{number} will be generated. The SPOSCAR is the perfect supercell structure, the DISP contains the information on displacement directions, and the POSCAR-{number} contains the supercells with atomic displacements. Rename the primitive POSCAR to POSCAR1, and SPOSCAR to POSCAR.

```
###############################################################
$ mv POSCAR POSCAR1
$ mv SPOSCAR POSCAR
###############################################################
```

Prepare the following INCAR and KPOINTS files to run the VASP calculation.

```
###############################################################
PREC=accurate
NSW=1
IBRION=8 # Only available in VASP 5.2
ISMEAR=0
SIGMA=0.01
LWAVE=F
LCHARG=F
###############################################################
```

```
###############################################################
k-points
Automatic mesh
0
Gamma
3 3 3
0.00 0.00 0.00
###############################################################
```

Perform a standard VASP run with the prepared files above.

8.10.2 Phonon calculations

After the VASP run, the DYNMAT file will be generated, and the force constants can be extracted from it as in the following.

```
###########################################################
$ phonopy --fc vasprun.xml
###########################################################
```

This will produce the FORCE_CONSTANTS file that is required for the next step with the INPHON file.

8.10.2.1 INPHON

```
###########################################################
ATOM_NAME=Si
NDIM= 2 2 2   # dimensions of the supercell used with
respect to the unit cell
FORCE_CONSTANTS = READ

ND=5
NPOINTS = 101
QI = 0.00 0.00 0.00 0.00 0.50 0.50 0.25 0.50 0.75 0.375
0.375 0.75 0.00 0.00 0.00
QF = 0.00 0.50 0.50 0.25 0.50 0.75 0.375 0.375 0.75 0.00
0.00 0.00 0.50 0.50 0.50
###########################################################
```

Rename POSCAR to POSCAR2 × 2 × 2 and POSCAR1 to POSCAR, and execute the phonon calculation.

```
###########################################################
$ phonopy -p
###########################################################
```

This produces the required output file for plotting the phonon dispersion, band.yaml. To plot the phonon dispersion, execute the following command.

```
###########################################################
$ bandplot band.yaml
###########################################################
```

The bandplot is saved in the bin directory (same location as phonopy) with a pop-out screen as shown in Figure 8.15. The final output of the phonon dispersion for the 2×2×2 Si supercell is shown in Figure 8.16. To illustrate the importance of convergence, the results for the 3×3×3 and 4×4×4 Si supercells are also plotted along with the 2×2×2 supercell, which shows both the lower acoustic and the upper optical branches.

Based on these basic phonon data, calculations can be extended to the estimations of entropy, free energy, thermal expansion, and so on, which may be beyond the scope of this book.

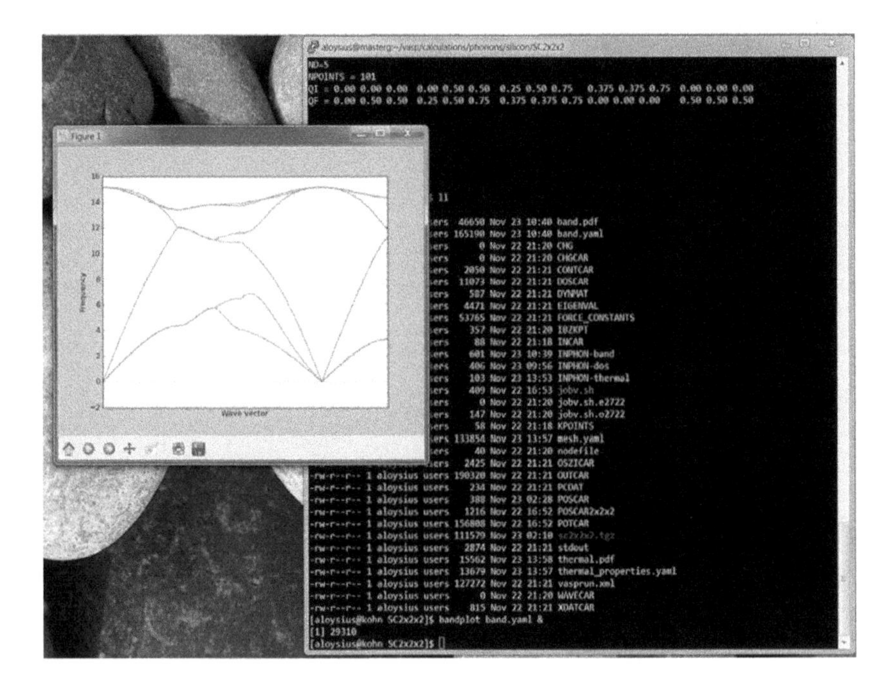

Figure 8.15 A pop-out screen for the calculation of a phonon dispersion diagram of Si.

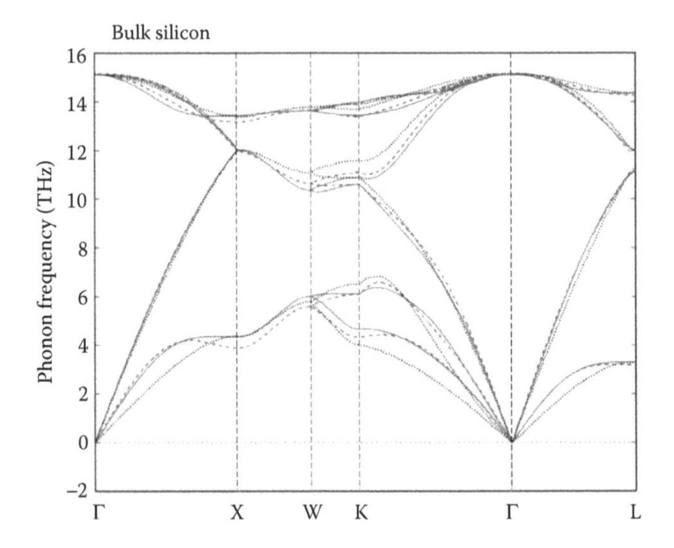

Figure 8.16 The phonon dispersion lines for the 2 × 2 × 2 Si supercell are drawn in dotted lines, while those of the 3 × 3 × 3 Si supercell are drawn in dashed lines and, last, those of the 4 × 4 × 4 Si supercell are shown in solid lines.

Homework

8.1 One of the important roles of simulation in material science is screening tasks before any experimental work. This 'computation-first' strategy is especially effective for a new catalyst development that requires a series of search works in a lab. Let's suppose we are about to find out the strain effect on a Pt-based system. Describe how to model the system in a practically relevant way and list up some important issues to be considered.

8.2 Pt is a well-known catalyst for a long time. Studies have shown that a compound, Pt3Y ($L1_2$ structure), provides promising catalytic properties in terms of catalytic activity, structural stability, lower price, and so on. Let's assume we already obtained the bulk energies of Pt and Y as −6.056 eV/atom and −6.466 eV/atom, respectively. Referring to Section 8.3, construct a FCC lattice (four atoms) for Pt3Y and find:
Equilibrium lattice parameter
Formation enthalpy

8.3 Continuing from the Homework 8.2, construct a Pt3Y(111) slab of 4 × 4 × 2.5 (5 layers, 80 atoms) with vacuum layer of 15 Å and find the adsorption energy of an oxygen atom on the hollow-FCC site surrounded by two Pt and one Y atoms and the Pt-O bond distance. This site is known to be the most preferred site for oxygen adatom in the Pt3Y(111) slab system. Let's assume we already obtained the following energies:
Atomic O energy: −1.5514 eV
Pt3Y-slab system energy: −548.6599 eV
If necessary, an almost-converged CONTCAR file can be provided for students to save their time.

8.4 Considering the Homework 8.3, discuss why we need such a big supercell for the calculation of adsorption energy on a Pt3Y(111) surface.

8.5 Ge is a semiconductor like Si. Construct a band structure diagram for Ge (with the equilibrium lattice parameter of 5.784 Å) by following the steps described in Section 8.9. Can you observe the formation of a band gap on the diagram? If not, list three main causes for that and describe how to remedy it.

References

Bentmann, H., A. A. Demkov, R. Gregory, and S. Zollner. 2008. Electronic, optical, and surface properties of PtSi thin films. *Phys. Rev. B* 78:205302–205310.
Blöchl, P. E. 1994. Projector augmented-wave method. *Phys. Rev. B* 50:17953–17979.
Getman, R. B., Y. Xu, and W. F. Schneider. 2008. Thermodynamics of environment-dependent oxygen chemisorption on Pt(111). *J. Phys. Chem. C* 112:9559–9572.
Hafner, J. 2007. Materials simulations using VASP-a quantum perspective to materials science. *Comp. Phys. Comm.* 177:6–13.

Henkelman, G. and H. Jónsson. 2000. Improved tangent estimate in the nudged elastic band method for finding minimum energy paths and saddle points. *J. Chem. Phys.* 113:9978–9985.
Henkelman, G., B. P. Uberuaga, and H. Jónsson. 2000. A climbing image nudged elastic band method for finding saddle points and minimum energy paths. *J. Chem. Phys.* 113:9901–9904.
Kim, J. H., Y. D. Kwon, P. Yonathan, I. Hidayat, J. G. Lee, J.-H. Choi, and S.-C. Lee. 2009. The energetics of helium and hydrogen atoms in ß-SiC: An ab initio approach. *J. Mat. Sci.* 44:1828–1833.
Kittel, C. 2005. *Introduction to Solid State Physics*, (8th edition). Hoboken, NJ: John Wiley & Sons.
Kresse, G. and J. Furthmüller. 1996. Efficient iterative schemes for ab initio total-energy calculations using a plane-wave basis set. *Phys. Rev. B* 54:11169–11186.
Kresse, G. and J. Hafner. 1994. Ab initio molecular-dynamics simulation of the liquid-metal–amorphous-semiconductor transition in germanium. *Phys. Rev. B* 49:14251–14269.
Kresse, G., M. Marsman, and J. Furthmüller. 2016. VASP the GUIDE. Computational Materials Physics, Universität Wien, Austria.
Landolt-Börnstein. 1991. *Structure Data of Elements and Intermetallic Phase*. Berlin, Germany: Springer.
Lee, S.-C., J.-H. Choi, and J. G. Lee. 2009. Energetics of He and H atoms with vacancies in tungsten: First-principles approach. *J. Nuclear Mat.* 383:244–246.
Lischka, M., C. Mosch, and A. Groß. 2007. Tuning catalytic properties of bimetallic surfaces: Oxygen adsorption on pseudomorphic Pt/Ru overlayers. *Electrochim. Acta.* 52:2219–2228.
Mattsson, T. R. and A. E. Mattsson. 2002. Calculating the vacancy formation energy in metals: Pt, Pd, and Mo. *Phys. Rev. B* 66:214110–214117.
Mavrikakis, M., B. Hammer, and J. K. Nørskov. 1998. Effect of strain on the reactivity of metal surfaces. *Phys. Rev. Lett.* 81:2819–2822.
Pang, Q., Y. Zhang, J.-M. Zhang, and K.-W. Xu. 2011. Structural and electronic properties of atomic oxygen adsorption on Pt(111): A density-functional theory study. *Appl. Surf. Sci.* 257:3047–3054.
Park, J. and L. Yang. 2009. locpot_workfunc.py. Private communication.
Perdew, J. P., K. Burke, and M. Ernzerhof. 1996. Generalized gradient approximation made simple. *Phys. Rev. Lett.* 77:3865–3868.
Perdew, J. P. and Y. Wang. 1992. Accurate and simple analytic representation of the electron-gas correlation energy. *Phys. Rev. B* 45:13244–13249.
Schäfer, H. E. 1987. Investigation of thermal equilibrium vacancies in metals by positron annihilation. *Phys. Status Solidi A* 102:47–65.
Sheppard, D. and G. Henkelman. 2011. Paths to which the nudged elastic band converges. *J. Comput. Chem.* 32:1769–1771.
Togo, A. 2010. Phonopy. http://phonopy.sourceforge.net/
Vanderbilt, D. 1990. Soft self-consistent pseudopotentials in a generalized eigenvalue formalism. *Phys. Rev. B* 41:7892–7895.
Wintterlin, J., R. Schuster, and G. Ertl. 1996. Existence of a "hot" atom mechanism for the dissociation of O_2 on Pt(111). *Phys. Rev. Lett.* 77:123–126.

chapter nine

DFT exercises with MedeA-VASP

We often encounter with runs that require a series of stages, such as calculations for band structure and DOS with hybrid functionals, phonon spectrum and related thermodynamic properties, minimum energy path and barrier energy, interface properties between two phases, *ab initio* MD, and so on. We also have to add dispersion (van der Waals) effect, +U considerations, Bader analysis, and many more, if the system requires it. In addition, their postprocessing based on the calculated data is an important step in computational works. All the above examples are quite time-consuming jobs and, in fact, are out of the range of this introductory book.

However, they could be only clicks away if we employ a proper GUI (graphic user interface) for these nonconventional Density functional theory (DFT) calculations. By using the GUI package loaded with all the essential and handy tools, we can complete jobs conveniently in much less time. Without a doubt, that will be the trend in materials sector in these days and coming future. In this chapter, a materials-oriented GUI, MedeA-VASP,* is introduced along with several examples and the related basics. All exercises in this chapter used an MPI parallel calculation with either 2 nodes/16 cores or 8 nodes/64 cores. Remember that only about 10 years ago, some of these kinds of calculations were considered to be computationally too expensive. Not anymore.

9.1 MedeA-VASP

9.1.1 General features

MedeA-VASP is built on the same approaches as Quantum Espresso and VASP: PBC, plane wave (PW) expansion of the KS orbitals, and pseudopotential (PP) scheme for electronic structural calculations. In fact, it can calculate anything we can imagine practically. However, all operations are carried out not by providing files such as INCAR, KPOINTS, POSCAR, and POTCAR, but via GUI. In addition, a series of tasks, stages, and subroutines are automatically handled in a single job. All you need is clicks, selecting buttons, and typing simple parameters. Referring to Figure 1.3, it is like driving a car equipped with GPS (global positioning system) as shown in Figure 9.1.

* MedeA-VASP: http://www.materialsdesign.com/medea/medea-vasp-52.

Figure 9.1 Driving a car with GPS (≡ GUI) on computational roads.

In addition to conventional DFT calculations, the MedeA-VASP provides:

- *Hybrid functionals*: HSE06 (Heyd et al. 2006), PBE0, B3LYP
- Methods of screened exchange, Hartree-Fock, and GW
- Linear response
- D2/Grimme correction for van der Waals interaction
- Supercell, surface, and interface generation
- Eight main selection buttons and many sub-buttons of selection and fill-up
- All automatic tabulated and plotted results
- Web-based JobServer/TaskServer architecture for each run to save all related files in order
- Many application notes on the website

9.2 Si2-band-HSE06

9.2.1 Introduction

Hybrid functionals are designed to complement the conventional PBE functionals by mixing the HF exchange energy. Note that the HF exchange energy is wave function based and is calculated nonlocally. By incorporating nonlocality into semilocal PBE functionals, we can eliminate most of drawbacks in PBE functionals. Note that electrons move in the exchange potential screened by electron holes of all other electrons, and thus screening is system dependent (normally 25% for metals and semiconductors referencing vacuum as 100% for no screening). The screened hybrid functional, HSE06 (Heyd et al. 2006), is a typical one:

$$E_{xc}^{\mathrm{HSE06}} = \frac{1}{4}E_x^{\mathrm{HF}} + \frac{3}{4}E_x^{\mathrm{PBE,SR}} + E_x^{\mathrm{PBE,LR}} + E_c^{\mathrm{PBE}} \tag{9.1}$$

where the first term is the HF exchange functional, the second and third terms are the short and long range components of the PBE exchange functional, and the last term is the PBE correlation functional. In this section, we calculate the band structure of Si using this hybrid functional and see the difference with the conventional PBE functionals as described in Section 8.9.

9.2.2 *Run steps*

Figure 9.2 shows the main window of MedeA, which is currently set for the Si2-band-HSE06 run in a three-step procedure:

- Single point calculation with the PBE functional and a conventional KPOINTS file (e.g., Gamma 6 × 6 × 6): to save wave functions temporarily for the next step
- Single point calculation with the HSE06 functional and an explicit KPOINTS file (the IBZKPT file plus high-symmetry lines of the BZ with zero fake weights) to determine the nonlocal exchange by KS orbitals
- Band structure plots with PBE and HSE06 functionals

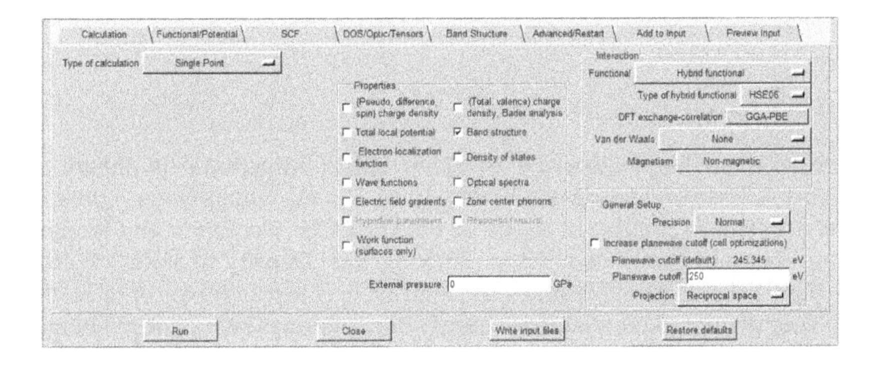

Figure 9.2 Main window of MedeA-VASP for the Si2-band-HSE06 run.

The package automatically writes down all input files including the following INCAR file for the second step:

```
#############################################################
…..
LHFCALC = .TRUE.
HFSCREEN = 0.2                         # AEXX = 0.25 is default
PRECFOCK = Normal
ALGO = Damped
# until orbitals at the new k-points are fully converged
TIME = 0.4
LMAXFOCK = 4
LSORBIT = .TRUE.
INIWAV = 1
ISTART = 1
ICHARG = 12
LCHARG = .TRUE.
…..
#############################################################
```

```
#############################################################
Explicit k-points
  18
Reciprocal
  0.00000000 0.00000000 0.00000000
  0.25000000 0.00000000 0.00000000
  0.50000000 0.00000000 0.00000000
…..
  0.00000000 0.00000000 0.00000000 0.000
  0.00000000 0.05555556 0.05555556 0.000
…..
#############################################################
```

9.2.3 *Results*

The resulted band structure of Si by HSE06 functional is shown in Figure 9.3, which indicates a band gap of 1.15 eV, which is very close to the experimental value of about 1.2 eV. For DOS calculation and its plotting, the only thing we need is checking the "Density of states" button in Figure 9.2. Figure 9.4 shows the total DOS ($s + p + d$) plot of Si by the HSE06 functional. This exercise clearly demonstrates how convenient and time-saving the GUI-equipped VASP package is.

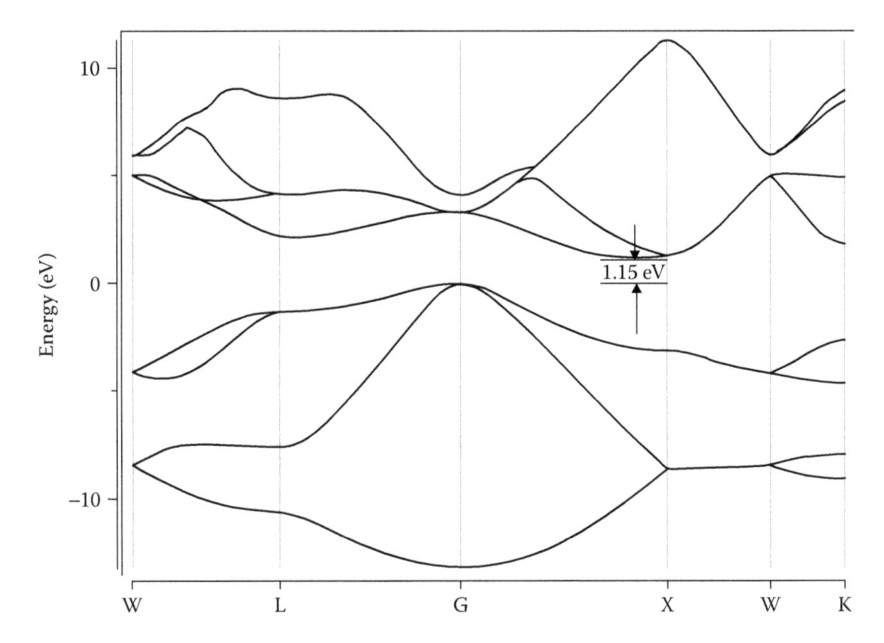

Figure 9.3 Band structure of Si by using the HSE06 functional.

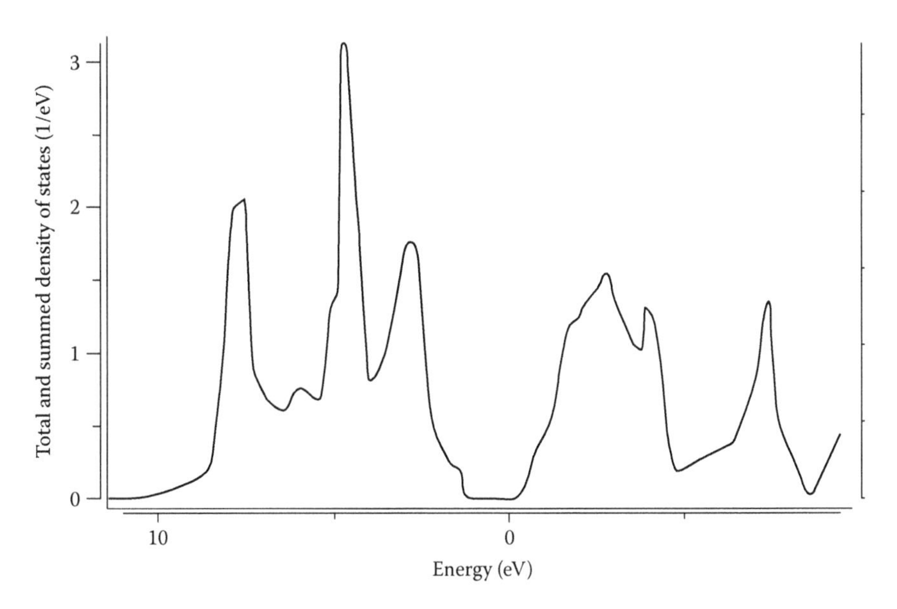

Figure 9.4 Total DOS plot of Si by using the HSE06 functional.

9.3 Si16-phonon

9.3.1 Introduction

We already calculated and plotted the basic phonon spectrum of Si in Section 8.10. In this section, we not only plot the curves but also tabulate thermodynamic properties automatically from the resulted data by using the "Phonon" tool in the MedeA-VASP package.

- *System*: Si supercell (16 atoms)
- *Program*: MedeA-VASP/Phonon
- Asymmetric atom displacements by ±0.02
- *1st step*: Ionic relaxation for supercell
- *2nd step*: Two single point energy calculations on supercells with displaced atoms
- *Goals*: Phonon dispersion, phonon density of states, thermodynamic functions obtained from the phonon density of states within the harmonic approximation

9.3.2 Ionic relaxation for supercell with displacements

The Si supercell after relaxation gives the system energy of −86.7670 eV/16 atoms. With asymmetric atom displacements, the system energy becomes −86.7644 (with displacements of −0.00183, 0.00183, 0.00183) and −86.7644 eV/16 atoms (with displacements of 0.00183, −0.00183, −0.00183).

9.3.3 Results

The corresponding PROCARs for the not-displaced and two displaced cases are as follows:

```
##############################################################
PROCAR new format
# of k-points: 18 # of bands: 41 # of ions: 16
.....
k-point 2: 0.25000000 0.00000000 0.00000000 weight =
0.12500000
band 1 # energy -5.99243678 # occ. 2.00000000
ion s p d tot
1 0.026 0.000 0.000 0.026
2 0.026 0.000 0.000 0.026
.....
##############################################################
```

```
################################################################
PROCAR new format
# of k-points: 18 # of bands: 41 # of ions: 16
.....
k-point 2: 0.25000000 0.00000000 0.00000000 weight =
0.06250000
band 1 # energy -5.99262402 # occ. 2.00000000
ion s p d tot
1 0.025 0.000 0.000 0.026
2 0.026 0.000 0.000 0.026
.....
################################################################
```

```
################################################################
PROCAR new format
# of k-points: 18 # of bands: 41 # of ions: 16
.....
k-point 2: 0.25000000 0.00000000 0.00000000 weight =
0.06250000
band 1 # energy -5.99245565 # occ. 2.00000000
ion s p d tot
1 0.026 0.000 0.000 0.026
2 0.025 0.000 0.000 0.025
.....
################################################################
```

Figures 9.5 and 9.6 show the plot of phonon dispersion curves and total phonon DOS of Si, respectively, which are generated automatically.

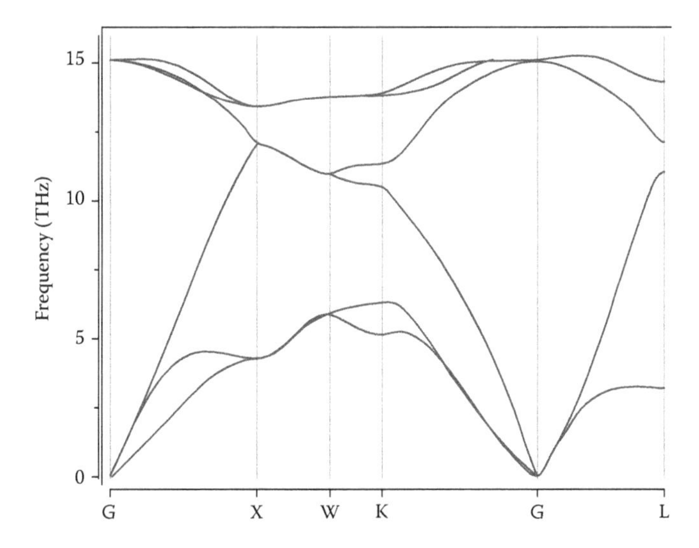

Figure 9.5 Phonon dispersion curves of Si.

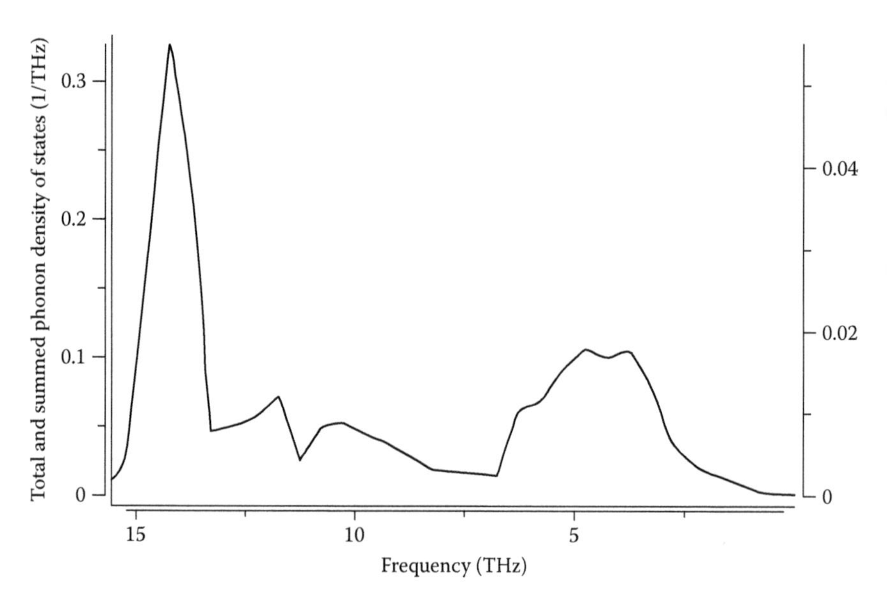

Figure 9.6 Total phonon DOS of Si.

From the data, various thermodynamic functions at different temperatures (T) are also automatically calculated in the job.out file in the MedeA job directory:

```
#############################################################
T    Cv         E(T)-E(0)  S(T)       -(A(T)-E(0))  E(T)      A(T)
K    J/K/mol    kJ/mol     J/K/mol    kJ/mol        kJ/mol    kJ/mol
-------------------------------------------------------------------------
0    0.0000     0.0000     0.0000     -0.0000       5.8946    5.8946
50   2.5703     0.0362     0.9597     0.0118        5.9308    5.8828
100  7.6173     0.2922     4.3088     0.1386        6.1868    5.7559
150  12.2471    0.7921     8.2983     0.4527        6.6866    5.4419
200  15.8058    1.4985     12.3382    0.9692        7.3930    4.9254
250  18.2580    2.3540     16.1463    1.6825        8.2486    4.2121
300  19.9182    3.3111     19.6316    2.5783        9.2057    3.3162
350  21.0612    4.3373     22.7931    3.6403        10.2319   2.2543
400  21.8693    5.4117     25.6612    4.8527        11.3062   1.0418
450  22.4565    6.5205     28.2727    6.2021        12.4151   -0.3075
500  22.8942    7.6548     30.6624    7.6763        13.5494   -1.7817
550  23.2280    8.8082     32.8608    9.2651        14.7028   -3.3705
600  23.4877    9.9764     34.8935    10.9596       15.8709   -5.0650
650  23.6935    11.1561    36.7820    12.7520       17.0507   -6.8575
700  23.8590    12.3450    38.5441    14.6357       18.2396   -8.7411
750  23.9941    13.5415    40.1950    16.6046       19.4360   -10.7100
800  24.1056    14.7440    41.7472    18.6535       20.6386   -12.7589
```

```
850  24.1988 15.9517  43.2115  20.7778      21.8463  -14.8832
900  24.2773 17.1637  44.5969  22.9733      23.0582  -17.0787
950  24.3442 18.3792  45.9114  25.2363      24.2738  -19.3417
1000 24.4015 19.5979  47.1615  27.5633      25.4925  -21.6688
##################################################################
```

Here, each thermodynamic function is represented as follows:

- Cv: Vibrational heat capacity at constant volume
- $E(T) - E(0) = E\_\text{Vib}(T)$: Change in vibrational internal energy from 0 K
- $E(0) = E\_\text{Elec}$ + ZPE (zero-point E, 5.89 kJ/mol): Electronic energy of formation
- $S(T)$: Vibrational entropy at T (note the PV term = 1.49 kJ/mol)
- $-(A(T) - E(0))$: Change in the vibrational Helmholtz free energy from 0 K
- $E(T) = E(0) + E\_\text{Vib}(T)$: Part of enthalpy and free energies, referenced to the elements in their standard state (E_elec = 0.00 kJ/mol)
- $A(T)$: Electronic plus vibrational Helmholtz free energy, $E(T) - \text{TS}(T)$

This exercise clearly demonstrates the usefulness of the package, especially when time is the main concern.

9.4 W12C9-Co28-interface

9.4.1 Introduction

Among many useful tools in MedeA, the Interfaces tool for generating a two-phase system from two different phases is one of the best. This section utilizes the tool to explore the interfacial property of WC–Co tool bits. Note that the composite has the necessary hardness and ductility coming from the covalent-bonded WC particles and the metallic-bonded Co matrix, respectively. The latter (5–20 volume %) also provides densification ability to the system via the liquid-phase sintering route at around 1450°C. Referring to the MedeA's Application Note (Materials Design, Inc. 2008a), we simply compare energies of the composite system with two surface systems in this exercise.

- *Programs*: MedeA-VASP/Interfaces
- Need three separate calculations with the WC(001) and the Co(001) surfaces of HCP, and the WC–Co interface
- *Goal*: To compare the WC–Co interface with the WC(111) and the Co(111) surfaces, and to confirm the energy difference and its usefulness in a very naïve fashion

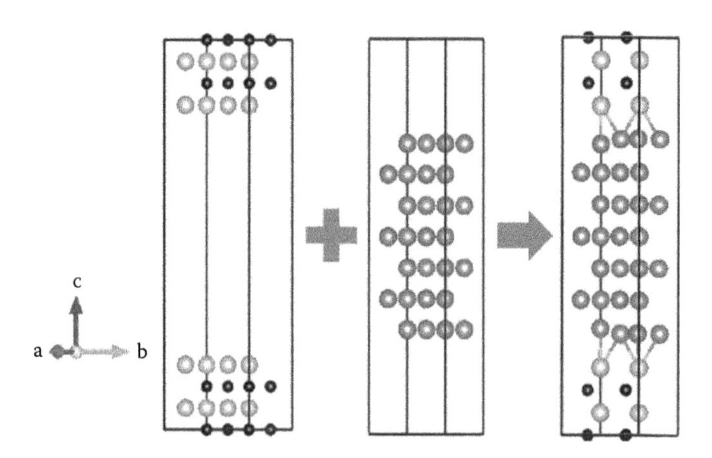

Figure 9.7 Surface and interface models for the WC–Co composite (gray big ball: W, black small ball: C, dark gray big ball: Co) drawn by VESTA.

9.4.2 *Surface models for WC and Co*

Surface models for WC(001) (total seven layers, four layers of W and three layers of C) and Co(001) (seven layers) are prepared as shown in Figure 9.7. To join the two phases together in the next step, vacuum layers are located in the middle for WC and on top and bottom for Co. Note that the WC(001) has W-terminated surfaces, which is known to be more stable.

9.4.3 *Interface model for WC–Co*

The prepared two surface models are joined together to have a supercell with two identical interfaces (total 49 atoms: 12 W, 9 C, and 28 Co), as shown in Figure 9.7. Here, the equilibrium lattice parameters of WC are as the reference because its elastic modulus is much higher than that of Co. MedeA tries various fitting possibilities and provides the best WC(111)-Co(111) interface with a mismatch less than 5%. The composite model is ready for full and static structural relaxations under the properly maintained PBC with the following INCAR file:

```
############################################################
SYSTEM = W12 Co28 C9
PREC = Accurate
ENCUT = 400.000
IBRION = 2
NSW = 100
ISIF = 2
NELMIN = 2
EDIFF = 1.0e-05
EDIFFG = -0.02
```

```
VOSKOWN = 1
NBLOCK = 1
NELM = 60
ALGO = Fast (Davidson and RMM-DIIS)
ISPIN = 1
.....
############################################################
```

Note that KPOINTS of Gamma $3 \times 3 \times 1$ is enough because the system has a long dimension in the *z*-axis.

9.4.4 Results

The relaxed CONTCAR file is as follows:

```
############################################################
W12 Co28 C9 (P1) ~ 1.1 (VASP)
1.00000000000000
5.0045590624301859 -0.000000184897273  0.0000000012054240
-2.5022792631875470 4.3340754360842615 0.0000000012067625
0.0000004532963527 0.0000007857136124 25.0454974798881373
Co W C
28 12 9
Direct
0.1466606700554536 0.3305760427232585 0.7444883704057190
0.1466606700517303 0.3305760427195352 0.2555116295942383
.....
0.6664171845356677 0.6659547489741584 0.8855410871859775
0.6665898938150008 0.6659897612422512 0.0000000000000071
############################################################
```

The CHGCAR file on VESTA (see website in references) shows the isosurface of electron density, which indicates strong bonding between W and Co atoms with bonding lengths of 2.41–2.59 Å (Figure 9.8). The results in terms of the net effect can be summarized as follows:

- W12C9: E0 = –232.4549 eV/system
- Co28: E0 = –193.3571 eV/system
- W12C9–Co28: E0 = –437.1215 eV/system

$$\begin{aligned} \Delta E &= E_{\mathrm{WC-Co}} - E_{\mathrm{WC}} - E_{\mathrm{Co}} \\ &= -11.3095/\mathrm{system} = -261\ \mathrm{meV}/\mathrm{A}^2 \end{aligned} \tag{9.2}$$

The negative energy change implies that the interface system has more stable structure compared to the two surface systems together. This is why the composite is so successful in application such as cutting tools.

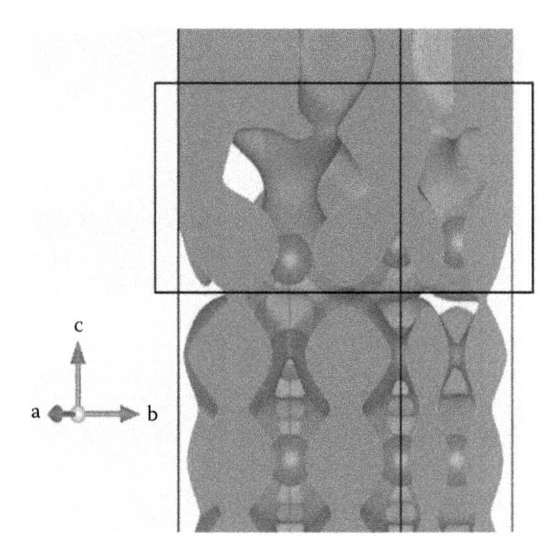

Figure 9.8 Isosurface of electron density at the WC–Co interface drawn by VESTA.

Co is a transition metal with localized electron densities on site, and correlates with other electrons strongly. If we consider this +U consideration for Co and WC–Co with the effective +U value of 5 for Co, the results are a little different, but the conclusion is the same:

```
#############################################################
….  .
# U parameter
LMAXMIX = 4
# CHGCAR file contains information up to LMAXMIX for the
on-site PAW occupancy, band drawing: 2 (default), 4:
d-orbital, 6: f-orbital
# LDAU = .TRUE.            # +U on
LDAUTYPE = 2               # Dudarev
LDAUL = 2 2 2
# l-quantum number on which U acts ((1._q,0._q) for each
type
LDAUU = 5 0 0              # U coefficient (coulomb
                           interaction) for each species
LDAUJ = 0 0 0              # J coefficient (exchange) for each
                           species
….  .
#############################################################
```

Note that W is also a transition metal, but has no +U effect as WC.

- Co28(+U): E0 = −92.3344 eV/system
- W12C9: E0 = −232.4549 eV/system
- W12C9–Co28(+U): E0 = −340.4258 eV/system

$$\begin{aligned} \Delta E &= E_{\mathrm{WC-Co}} - E_{\mathrm{WC}} - E_{\mathrm{Co}} \\ &= -15.6365/\mathrm{system} = -360\ \mathrm{meV}/\mathrm{A}^2 \end{aligned} \tag{9.3}$$

Note that a more rigorous calculation requires the effect of carbon chemical potential.

The same approach can be applied to systems with interface such as the NiSi2–Si(100) contact for the CMOS (complementary MOS) device for calculations of the SBH (Schottky barrier height) in terms of a doping element at the metal/semiconductor interface (MedeA 2015).

9.5 *$Mg_4(Mo_6S_8)_3$-barrier energy*

9.5.1 *Introduction*

In this section, we perform DFT calculations on the $(Mo_6S_8)_3$ chevrel phase with the insertion of four Mg atoms to evaluate their transport properties and to have better understanding of ionic moves in this battery system by using the MedeA-VASP/Transition State Search tool. The basics are the same as described in Section 8.7 (Henkelman et al. 2000).

- *Programs*: MedeA-VASP/Transition State Search
- *Pseudopotentials*: Mo_sv (4p6 4d5 5s1), S (3s2 3p4), and Mg_sv (2p6 3s2) as valence electrons
- *Energy cutoff*: 520 eV, gamma-centered 4 × 4 × 4 *k*-point mesh (*k*-spacing = 0.2/Å)
- Total energy convergence < 10^{-5} eV/system, force convergence ≤ 0.01 eV/Å
- Barrier energy by the nudged elastic band (NEB) method with 10 images including initial and final configurations within the structural convergence of −0.05 eV/Å
- *Goal*: To demonstrate a calculation of barrier energy for transport of Mg ion in the system

The $M_{36}(Mo_6S_8)_3$ supercell is shown in Figure 9.9a along with four representative paths for a 3D diffusion of M ions in Figure 9.9b. Here, M represents the 36 sites available for insertion of the foreign atom. However, it is known that only 6 of 36 sites are actually available for the Mg case to maintain local electroneutrality within the host crystal.

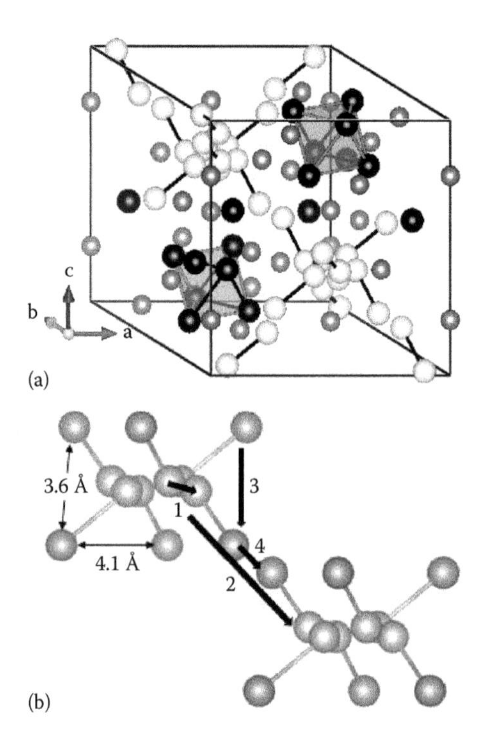

Figure 9.9 Computational model. (a) $M_{36}(Mo_6S_8)_3$ supercell (black ball; Mo, small gray ball; S, white ball; M insertion sites). (b) Four paths for 3D diffusion on M insertion sites in $M_{36}(Mo_6S_8)_3$ drawn by VESTA.

Let us first find out the electron transfer between ions by Bader analysis by simply checking "Bader analysis" on MedeA/Calculation/Properties. The job.out provides the following table and we can tell how each atom's charge is transferred and the system maintains the overall electroneutrality.

```
#############################################################
      Atom      Valence      Charge       Volume      Distance
                charge       transfer     (Ang^3)     (Ang)
      -----     -------     ---------    ---------    --------
      Mo        13.1605      0.8395      13.5367        1.0712
      Mo        13.1264      0.8736      13.9169        1.0901
.....
      S         6.8661      -0.8661      24.1203        0.0798
      S         6.9813      -0.9813      24.1775        1.1789
.....
      Mg        0.3931       1.6069      7.1454         0.8423
      Mg        0.3876       1.6124      6.9421         0.8656
.....
#############################################################
```

Referring to Figure 9.9b, usual interpolation between initial and final images will be fine as the initial NEB calculation for paths 1, 2, and 4, because these paths are more or less straight. Path 3, however, has an S atom very close to the path of moving Mg ion in a distance of less than 1 Å. We have to provide an intermediate image so that the Mg ion moves around the S atom. Otherwise, the Mg ion will be kicked off by strong repulsion and the run will be out of order.

In MedeA-VASP, initial images can be selected from specified configurations to define a specific pathway by checking "Initial images from specified systems," as shown in Figure 9.10. The following is the modified POSCAR file for the intermediate image supplied as a specified system into the Transition State Search/Calculation window:

```
##########################################################
Mg4-in-23 site
1.00000000000000
 9.7018033199999998  0.0000000000000000  0.0000000000000000
-4.9894916599999997  8.3710029600000002  0.0000000000000000
 0.0882629000000000  -0.0988088000000000 10.5050624900000003
Mo S Mg
18 24 4
Direct
0.3478538327547664 0.8311947565722873 0.0603182966748119
…..
0.4500000593567948 0.6006165445012214 0.7006471312412932
##########################################################
```

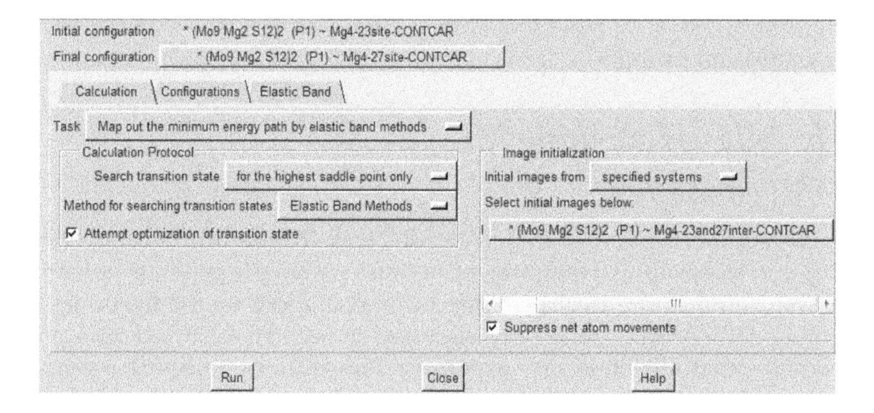

Figure 9.10 Transition State Search/Calculation window in MedeA-VASP to select a specified configuration to define a specific pathway (e.g., Number of images = 1 excluding the initial and final images).

The following is the nearest-neighbor distance between the Mg ion and the S atom in the OUTCAR file generated at the start of the run with the above POSCAR file:

```
##############################################################
…. .
46  0.450  0.601  0.701  -32  2.09  34  2.28  41  2.41
…. .
##############################################################
```

It shows that #46 Mg (in total numbering) in the 0.450, 0.601, 0.701 position has a distance of 2.09 Å with #32 S (in total numbering), which can make the NEB calculation run smoothly.

9.5.2 *NEB run*

The actual run used 10 images based on the structures from linear interpolation from the initial to the specified intermediate image, and from that to the final image.

9.5.3 *Results*

Figure 9.11 shows (a) barrier energies of Path3 for the $Mg_4(Mo_6S_8)_3$ system and (b) the corresponding trajectories of the Mg ion (Mg#4) drawn by VESTA. Note that the Path3 goes around the S14 atom, which is near to the path, and the resulted barrier energy is about 0.77 eV. Further calculations indicate that the Path1 and 4 are almost barrierless and Path2 has a barrier energy of about 0.4–0.5 eV. This concludes that Path2 is operative as the major transporting route for the Mg ion in the $Mg_4(Mo_6S_8)_3$ system. Note that barrier energy is effective exponentially in a diffusion process.

9.6 *Si14-H2*-ab initio *MD*

9.6.1 *Introduction*

In this section, we run an MD in *ab initio* way (Kresse and Hafner 1994). This is a rather serious commitment, because we have to generate potential between atoms from electronic structure calculations on the fly. Basics of *ab initio* MD (AIMD) are already discussed in Sections 6.6.3 and 6.6.4, and we know that the job is very expensive. However, by selecting a simple system of Si14 and by referring to steps and tips provided by MedeA's Application Note (Materials Design, Inc. 2008b), we can handle the job without any difficulty and learn the way AIMD progress.

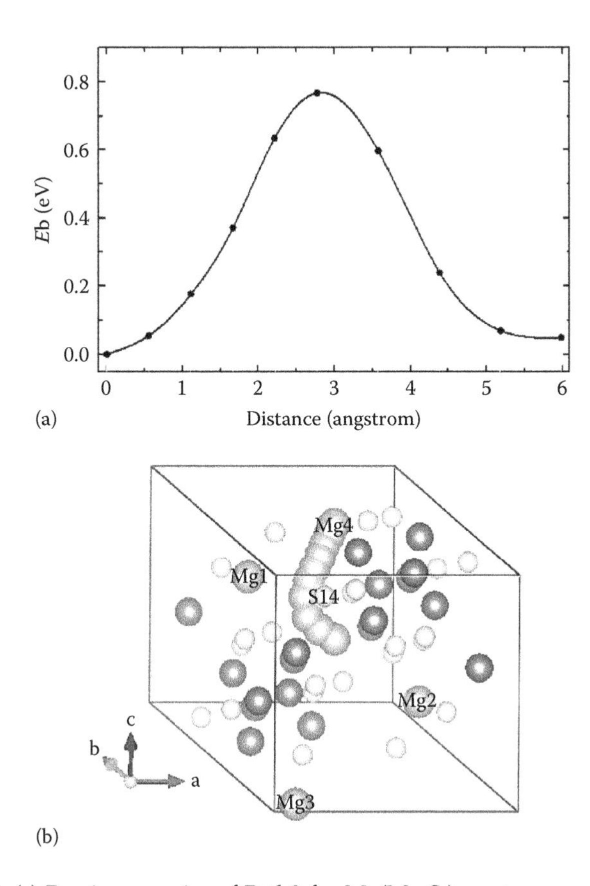

Figure 9.11 (a) Barrier energies of Path3 for $Mg(Mo_6S_8)_3$ systems and (b) the corresponding trajectories of the Mg ion (Mg#4) drawn by VESTA.

- *Programs*: MedeA-VASP/Molecular Dynamics/temperature scaling (nVE ensemble) as shown in Figure 9.12
- *System*: A slab of Si14(001) with seven layers
- Hydrogenated bottom layer with two H atoms
- Total energy convergence < 10^{-5} eV/system, force convergence ≤ 0.02 eV/Å for the surface model of Si14-H2
- Barrier energy by the NEB method with 10 images including initial and final configurations within structural convergence of −0.05 eV/Å
- *Goal*: To observe the reconstruction event of surface under the stimulated annealing from 600 to 300 K and to understand how the broken surface atoms reduce their surface energy

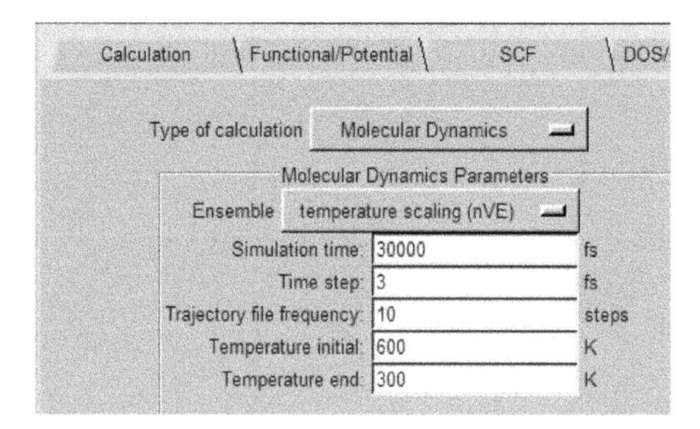

Figure 9.12 MedeA-VASP/Molecular Dynamics window set for an AIMD run under the temperature scaling (nVE ensemble).

9.6.2 Run steps

We start with the optimized bulk Si (lattice parameter = 5.469643 Å). We load this structure and create a surface structure:

```
############################################################
menu > Edit > Build surfaces > Miller indices 0 0 1 > Search
############################################################
```

The newly generated slab system has the (001) plane parallel to the *xy*-plane, and we select seven layers by trimming both surfaces and the gap (10 Å) between the slabs, and finally select centered P1. Now we generate a new structure of a 2 × 2 × 1 supercell by using the Edit > Build supercell. We make the bottom surface inert by attaching H atoms (hydrogenation).

Now we apply a smart tip suggested by MedeA, selecting one surface atom and moving it by 0.075 in the *x*-direction. This breaks the symmetry of the surface and initiates the reconstruction process easily. Figure 9.13 shows the resulting structure at the left.

```
#############################################################
# Si12(100)-H2
......
IBRION = 0
NSW = 10000
EDIFF = 1.0e-04
EDIFFG = -0.02
ISYM = 0
SMASS = -1
POTIM = 3.0
```

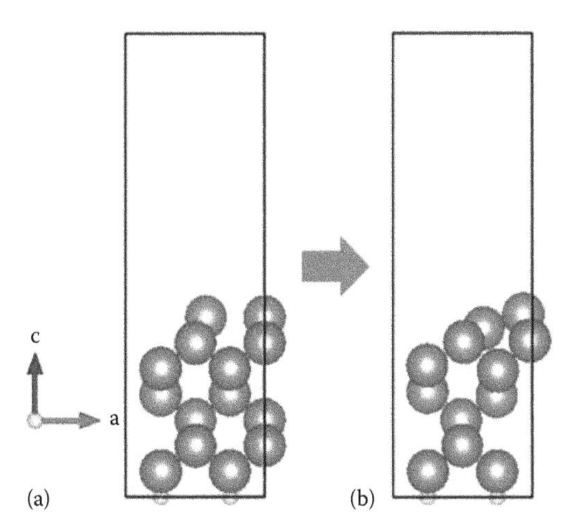

Figure 9.13 Reconstruction of surface atoms in the Si14H2 system by AIMD drawn by VESTA. (a) initial and (b) final.

```
TEBEG = 600
TEEND = 300
…. .
############################################################
```

We perform structural optimization again. Note that convergence criteria are relaxed for this AIMD to reduce run time.

9.6.3 Results

The relaxed CONTCAR file is as follows:

```
############################################################
Si14(001)-H2 # Bottom Si atoms and H2 fixed
1.0
  7.6801700592 0.0000000000 0.0000000000
  0.0000000000 3.8801701069 0.0000000000
  0.0000000000 0.0000000000 24.5037250519
  Si H
  14 2
Direct
    0.491842002 0.995242000 0.166754007
    0.652965009 0.993247986 0.363896012
    0.250000000 0.494834989 0.055406999
    0.245006993 0.494141996 0.284094006
    0.750000000 0.494834989 0.055406999
    0.756410003 0.494129002 0.272792995
    0.004245000 0.995324016 0.166981995
```

```
    0.937753022 0.993686020 0.393552989
    0.001730000 0.495469004 0.111175001
    0.984290004 0.493571013 0.337480009
    0.747614026 0.994158983 0.218651995
    0.497878999 0.495492011 0.111051999
    0.511260986 0.493333995 0.334583998
    0.248586997 0.994419992 0.227045000
    0.250000000 0.989669025 0.000000000
    0.750000000 0.989669025 0.000000000
#############################################################
```

Even for this rather simple system, calculation takes about 2 hours on one node/eight processors. Figure 9.13 (at the right) shows that Si atoms form surface dimers (Si–Si distance of 2.305 Å) with the bond axis tilted with respect to (001) in accordance with the reported experiments. This demonstrates quite clearly that the combination of experiment and computation can be the most reliable way of reaching the right results. Remember that the (001) surface of Si is the typical substrate surface in the manufacturing of semiconducting devices.

References

Henkelman, G., B. P. Uberuaga, and H. Jonsson. 2000. A climbing image nudged elastic band method for finding saddle points and minimum energy paths. *J. Chem. Phys.* 113:9901–9904.

Heyd, J., G. E. Scuseria, and M. Ernzerhof. 2006. Erratum: Hybrid functionals based on a screened Coulomb potential. *J. Chem. Phys.* 118:8207 (2003); *J. Chem. Phys.* 124:219906(E).

Kresse, G. and J. Hafner. 1994. Ab initio molecular-dynamics simulation of the liquid-metal–amorphous-semiconductor transition in germanium. *Phys. Rev. B* 49:14251–14269.

Materials Design, Inc. 2008a. Application note, interface energy of metal-ceramic interface Co-WC using ab initio thermodynamics. http://www.materialsdesign.com/appnote/.

Materials Design, Inc. 2008b. Application note, surface reconstruction of Si(001). http://www.materialsdesign.com/appnote/.

VESTA (ver. 2.90.1b). 2011. http://www.geocities.jp/kmo_mma/index-en.html.

Appendix A

List of symbols and abbreviations

acceleration	$\boldsymbol{a}$
all electron	AE
atomic mass	m_I
band index	n
body-centered cubic	BCC
Bohr magneton	μ_B
Boltzmann constant	k_B
bond order	B_{ij}
Born–Oppenheimer molecular dynamics	BOMD
Brillouin zone	BZ
bulk modulus	B
Car–Parrinello MD	CPMD
charges of nucleus	Z, Z_1, Z_2
constants	a, A, C, C_1, C_2, C_i
coordination number	N_{coor}
density functional theory	DFT
density of states	DOS
diffusion coefficient	D
effective potential, KS	U_{eff}
electron density	ρ
electronic mass	m
embedded atom method	EAM
energy, cutoff	E_{cut}
energy, KS orbital or electron gas	ε
equilibrium interatomic distance	r_0
exchange correlation	XC
exchange-correlation energy	E_{xc}
external energy	E_{ext}
face-centered cubic	FCC
fast Fourier transformation	FFT

fictitious mass	μ
force	$\boldsymbol{F}$
Fourier coefficients	c_i, c_{nk}
fractional occupancy	f_i
frequency	v
generalized gradient approximation	GGA
gradient	$d/dx, \nabla$
Hamiltonian matrix	H
Hamiltonian operator	$\hat{H}$
Hamiltonian operator, KS one-electron	$\hat{H}_{KS}$
Hartree energy	E_H
Hartree–Fock	HF
hexagonal close packed	HCP
integer numbers	n
irreducible Brillouin zone	IBZ
k-point	$\boldsymbol{k}$
kinetic energy	E_{kin}
kinetic energy of electron	E_i^{kin}
kinetic energy of nucleus	E_I^{kin}
Kohn–Sham orbital	ϕ
Kronecker delta	δ_{ij}
Lagrange multiplier	λ, Λ_{ij}
Laplacian operator	∇^2
local density approximation	LDA
local spin-polarized density approximation	LSDA
mean-square displacement	MSD
minimum energy path	MEP
molecular dynamics	MD
momentum	$\boldsymbol{p}$
Monte Carlo	MC
number of atoms	N
number of electrons, orbital number	n
number of PWs	N_{pw}
number of timesteps	$N_{\Delta t}$
order of N	$O(N)$
periodic boundary conditions	PBC
periodic function, lattice	u
Planck's constant	h
Planck's constant, reduced	$\hbar$
plane wave	PW, ϕ_{PW}
position vector, electronic coordinate	$\boldsymbol{r}$
potential	U
potential (attractive and repulsive)	U_A, U_R

potential cutoff distance	r_{cut}
potential energy by electron–electron interaction	U_{ij}
potential energy by nucleus–electron interaction	U_{Ii}
potential energy by nucleus–nucleus interaction	U_{IJ}
potential, potential energy	U
pressure	P
primitive real lattice vector	$\boldsymbol{a}_i$
primitive reciprocal lattice vector	$\boldsymbol{b}_i$
projector augmented wave	PAW
pseudopotential	PP
real lattice vector, residuum	$\boldsymbol{R}$
reciprocal lattice vector	$\boldsymbol{G}$
simple cubic	SC
spin-up, spin-down	$\uparrow, \downarrow$
stress	σ
surface tension, surface energy	γ
time	t
timestep	Δt
temperature, absolute	T
total energy	E
ultrasoft pseudopotential	USPP
velocity	$\boldsymbol{v}$
volume	V, Ω (reciprocal lattice)
strain	ε
Young's modulus	E
wave function (of a system)	Ψ
wave function (of an electron)	ψ
wave vector	$\boldsymbol{k}$
Wigner–Seitz radius	r_s
X-ray diffraction	XRD

Appendix B

Linux basic commands

$ 1 (after top)	# show status of all CPUs
$ adduser aaa	# add a user aaa
$ cat file1	# display all of file1
$ cd ~	# move to home directory (~: tilde)
$ cd ..	# move to upper directory
$ cd aaa	# move to directory aaa
$ chmod 755	# change all files to execution mode
$ chmod +x aaa	# change aaa file to execution mode
$ cp filename1 filename2	# copy filename1 as filename2
$ ↵ d	# exit from Linux
$ dir	# show directories and files
$ echo Hello, world! > file1	# create directory file1 and save the text in file1
$ free	# show status of memory only
$ grep "F=" OSZICAR	# show lines with "F=" in OSZICAR file
$ head -n aaa	# show the first n lines of file aaa
$ history 5	# show last five commands
$ jobs -1	# list process IDs
$ kill -9 2316	# kill run number 2316
$ killall -9 vasp-5.2.2	# kill all CPUs running on VASP
$ less POTCAR	# show the first part of POTCAR file (↵ for more)
$ ls	# list directories and files
$ ls –la	# list directories and files in long/all formats
$ man passwd	# show manual for passwd
$ mkdir aaa	# make new directory aaa
$ more file1	# display first part of file1 (↵ for more)
$ mv file1 dir1	# move file1 under dir1
$ mv file1 file2	# change file name from file1 to file2
$ nohub vasp &	# run VASP in a background mode
$ passwd	# change password

$ passwd aaa	# changing password for user aaa
$ pestat	# show load for each node
$ poweroff	# turn off supercom
$ ps -ef	# show process status for all
$ ps -ef | grep jglee	# show process status for jglee
$ psh node1-node2 "killall -9 vasp"	# kill VASP on multiple nodes
$ pwd	# show the present working directory
$:q	# exit from editor without save
$:q!	# exit from editor
$ qsub filename	# submit filename to be calculated
$ qdel job ID	# delete job
$ qstat –u username	# show the status of username's jobs
$ qstat –f	# show status of nodes and jobs
$ rm file2	# remove file2
$ rmdir aaa	# remove directory aaa
$ scp POTCAR IP1:IP2	# copy POTCAR from IP1to IP2
$ ssh node193	# move to node193
$ ssh ready aaa	# set remote access for user aaa
$ su -	# move to root
$ su – jglee	# to jglee
$ tail -f aaa	# show the update of the last 10 lines of file aaa
$ tail POTCAR	# show the last parts of POTCAR
$ top	# show status of CPU/Memory/Swap
$ touch file1	# create an empty file1
$ userdel aaa	# remove account for aaa
$ userdel -r aaa	# remove account/directory for aaa
$ vi file1	# show file1 on editor

Appendix C
Convenient scripts

The following scripts are two of many scripts available for the postprocessing of VASP data. They are given here simply to demonstrate how convenient they are in the course of various VASP runs.

C.1 Generation of out.dat file for work function graph

This script, locpot_workfunc.py, is written by Jinwoo Park and Lanhee Yang (2009, private comm.) at the University of Seoul to read the potential data from the LOCPOT along the z-direction and convert them to *.dat file for plotting.

```
###############################################################
#!/usr/bin/env python

## Original algorithm: Byungdeok Yu <ybd@uos.ac.kr>
## PYTHON VERSION reduck and yangd5d5

import sys, math, string, re, numpy
usage = 'Usage: %s inputfile(LOCPOT) FermiEnergy(option)
\n' % sys.argv[0]
usage = usage + 'If you do not insert the FermiEnergy,
write just average values. \n'
usage = usage + 'If you input the FermiEnergy, values
consider Fermi Energy \n'

foottext = '\n## Thank you \n## Jinwoo Park <reduck96@
physics.uos.ac.kr> \n## Lanhee Yang <yangd5d5@physics.uos.
ac.kr>'

version = 20090224
```

```
## PRINT COPYRIGHT
print "## LOCPOT - Workfunction x-y plot convert for VASP"
print "## Version %s \n" % version

# Check option

FermiE = 0.0 # Initialize Fermi Energy
if len(sys.argv) == 1:
  print usage
  print foottext
  sys.exit(1)

if len(sys.argv) == 2:
  infilename = sys.argv[1]
  print 'Do not consider Fermi Energy\n'

if len(sys.argv) == 3:
  infilename = sys.argv[1]
  FermiE = float(sys.argv[2])
  print 'Inserted Fermi Energy: %f \n' % FermiE

#try:
# infilename = sys.argv[1]
# except:

# print usage
# print foottext
# sys.exit(1)

# Done Check option

# FILE CHECK --
try:
  ifile = open(infilename, 'r') # open file for reading
except:
  print '---- WARNING! WRONG FILE NAME ---- \n---- CHECK
  YOUR LOCPOT FILE ----\n'
  print foottext
  sys.exit(1)

# read and print comment line
print 'Subject: '+ ifile.readline()

# Jump primitive vector configuration
for i in range(4):
  ifile.readline()
# Read Number of kind of atoms and read numbers each other
```

```
ReadNumberOfKindOfAtoms = ifile.readline().split()
NumberOfKindOfAtoms = len(ReadNumberOfKindOfAtoms)

# Sum total number of atoms
TotalNumberOfAtoms = 0
for i in range(NumberOfKindOfAtoms):
  TotalNumberOfAtoms = TotalNumberOfAtoms +
int(ReadNumberOfKindOfAtoms[i])
# debug print TotalNumberOfAtoms
# Jump comment(dummy - Cartesian? or Direct?)
ifile.readline()

for i in range(TotalNumberOfAtoms):
  ifile.readline()

#Jump Comment(dummy blank line)
ifile.readline()

#Read number of X/Y/Z elements
NumberOfElements = ifile.readline().split()

#NumberOfElements[0]: number of X-axis elements
#NumberOfElements[1]: number of Y-axis elements
#NumberOfElements[2]: number of Z-axis elements
NX = int(NumberOfElements[0])
NY = int(NumberOfElements[1])
NZ = int(NumberOfElements[2])

# debug -- print NX, NY, NZ

print "Matrix: %d x %d x %d (X, Y, Z)" % (NX, NY, NZ)
# empty array for DATA[NX, NY, NZ]
DATA=numpy.zeros((NX, NY, NZ),float)

## WHILE LOOP
## Outmost NZ / Middle NY / INNER NX
x=0; y=0; z=0; i=0;

line = ifile.readline().split() # read first line
count = len(line)

while z<NZ:
  while y<NY:
    while x<NX:
      if count == 0:
        line = ifile.readline().split()
        count = len(line)
```

```
      i=0
      DATA[x,y,z]=float(line[i])
# debug print 'DATA[%d,%d,%d] = %f' % (x,y,z, DATA[x, y,z])
     count=count-1
     i = i+1
     x = x+1
     y=y+1
     x=0
    z=z+1
    y=0

## DONE READ DATA

## take average of x-axis values
x=0; y=0; z=0;
XAVR=numpy.zeros((NY, NZ),float) # 2D array
while z<NZ:
  while y<NY:
   while x<NX:
     XAVR[y,z] = XAVR[y,z] + DATA[x,y,z]
      x=x+1
     XAVR[y,z] = XAVR[y,z]/float(NX)
     y=y+1
     x=0
   z=z+1
   y=0

## take average of y-axis values
y=0; z=0;
AVR=numpy.zeros((NZ),float) # 1D array
while z<NZ:
  while y<NY:
   AVR[z] = AVR[z] + XAVR[y,z]
   y=y+1
# AVR[z] = AVR[z]/float(NZ) 20090224> fix NZ->NY, Thanks
Prof. Yu
  AVR[z] = AVR[z]/float(NY)
  z=z+1
  y=0

### CHECK
#z=0;
#while z<NZ:
# print '%d %f' % (z, AVR[z])
# z=z+1
#
```

```
# FILE WRITE
outfile = open("output.dat","w")
for i in range(0,NZ):
  inputtext ='%d %f \n' % (i, AVR[i]-float(FermiE))
  outfile.write(inputtext)
outfile.close()

print '--> RESULT FILE: output.dat\n'

############################################################
# print foot-text
print foottext

# file close
ifile.close()

### Copyright - see http://spcc.uos.ac.kr/copyright
### SPCC, University of Seoul
############################################################
```

C.2 Generation of *.dat file for band structure

The following script is written by Jung-Ho Shin (2010, private comm.) in the computational science center at KIST, Seoul, for the automatic generation of a *.dat file from the EIGENVAL and OUTCAR files to draw a band-structure diagram (the drawing function may be switched on, if one wants to see the drawing on Linux with this script). Note that this script also calculates the distances for each **k**-point. First install numpy module.

```
############################################################
$ yum install numpy
############################################################
```

The script reads the EIGENVAL and OUTCAR files.

C.3 EIGENVAL

Number of bands: From the third number (see the small box in the following figure) in the line 6.

*Distances between **k**-points*: From the numbers (see the big box in the following figure) in the line 8 and those of the next **k**-points.

```
161[backup] 119> head -20 EIGENVAL
   2    2    1    1
  0.2001288E+02  0.3839590E-09  0.3839590E-09  0.3839590E-09  0.2000000E-15
   1.000000000000000E-004
  CAR
 Si
     8    140    20

  0.5000000E+00  0.0000000E+00  0.5000000E+00  0.7142857E-02
   1        -1.9417
   2        -1.9417
   3         3.0162
   4         3.0162
   5         6.5598
   6         6.5598
   7        15.9975
   8        15.9975
   9        16.8838
  10        16.8838
  11        18.4171
  12        18.4171
```

```
############################################################
# !/usr/bin/python
# vasp_to_image.py: #
# from EIGENVAL, OUTCAR #
# cf. './vasp_to_image.py -h' #
import os
import sys
import getopt
from numpy import *
# default variables to be controlled #
inEigenvalName = 'EIGENVAL'          # the file name of
                                     "EIGENVAL" format
inOutcarName = 'OUTCAR'              # the file name of
                                     "OUTCAR" format
outDataName = 'band.dat'             # the name of a data(2D
                                     coordinate) file
optPersist = 1                       # (1 or 0) if 1, the plot
                                     remains on the screen
gpTitle = '<Energy Band Structure>'  # graph's title
gpSize = '0.7, 1.0'                  # size of the plot
gpXrange = '[*:*]'                   # X axis range
gpXlabel = 'K-points'                # X axis label
gpXticLabel = 0                      # X tic's label, if 0,
                                     expression by x value
gpYrange = '[*:*]'                   # Y axis range
gpYlabel = 'Energy(eV)'              # Y axis label
gpTerm = 'x11'                       # a type of terminal
                                     and command
```

```
gpPlotMethod = 'with line'  # plot method
gpPicForm = 'pos'           # image format
gpPosDefault = 'landscape enhanced color solid "Times-
Roman" 15'
                            # postscript's defaults
gpPicName = 'band.eps'      # image file's name
gpKey = 'nokey'             # 'key' or 'nokey', if 'nokey',
                            don't display the label of lines
# default variables to be fixed #
blnEigenval = 7             # the number of comment lines
                            in "EIGENVAL" file
# options #
optlist, args = getopt.getopt(sys.argv[1:],
       'o:p:k:h', ['kpoints=', 'eigenval=',
          'outcar=', 'blnKpoint=',
          'title=', 'size=',
          'xRange=', 'xLabel=',
          'xTicLabel=', 'yRange=',
          'yLabel=', 'term=',
          'plotMethod=', 'form=',
          'epsDefault=', 'picName='])

for op, p in optlist:
  if op == '-o': outDataName = p
  if op == '-p': optPersist = int(p)
  if op == '-k': gpKey = p
  if op == '-h': help = 1

  if op == '--eigenval': inEigenvalName = p
  if op == '--outcar': inOutcarName = p
  if op == '--title': gpTitle = p
  if op == '--size': gpSize = p
  if op == '--xRange': gpXrange = p
  if op == '--xLabel': gpXlabel = p
  if op == '--xTicLabel': gpXticLabel = p.split()
  if op == '--yRange': gpYrange = p
  if op == '--yLabel': gpYlabel = p
  if op == '--term': gpTerm = p
  if op == '--plotMethod': gpPlotMethod = p
  if op == '--form': gpPicForm = p
  if op == '--epsDefault': gpPosDefault = p
  if op == '--picName': gpPicName = p
# print default variables #
print ""
print ""
print "** Printing Variables **"
print ""
```

```
print "-o: the file name of output data"
print "-p: if 1, the plot remains on the screen (0 or 1)"
print "-k: if 'nokey', don't display the label of lines",
print "('key' or 'nokey')"
print ""
print "--eigenval=: the file name of EIGENVAL format"
print "--outcar=: the file name of OUTCAR format"
print "--title=: graph's title"
print "--size=: size of the plot, ratio of x and y axis"
print "--xRange=: X axis range"
print "--xLabel=: X axis label"
print "--xTicLabel=: X tic label, about greek letter, add
'gk' at the front of alphabet."
print ": usage) --xTicLabel=' X W gkG L '"
print "--yRange=: Y axis range"
print "--yLabel=: Y axis label"
print "--term=: a type of terminal"
print "--plotMethod=: plotting method by Gnuplot
command(plot)"
print "--form=: image format"
print "--epsDefault=: postscript's defaults"
print "--picName=: image file's name"
print ""

print "-o '%-s'" % outDataName
print "-p %-d" % optPersist
print "-k '%-s'" % gpKey
print "--eigenval='%-s'" % inEigenvalName
print "--outcar='%-s'" % inOutcarName
print "--title='%-s'" % gpTitle
print "--size='%-s'" % gpSize
print "--xRange='%-s'" % gpXrange
print "--xLabel='%-s'" % gpXlabel

if gpXticLabel==0: print "--xTicLabel=%d" % gpXticLabel
else:
  print "--xTicLabel='",
  for string in gpXticLabel: print "%s" % string,
  print "'"

print "--yRange='%-s'" % gpYrange
print "--yLabel='%-s'" % gpYlabel
print "--term='%-s'" % gpTerm
print "--plotMethod='%s'" % gpPlotMethod
print "--form='%-s'" % gpPicForm
print "--epsDefault='%-s'"% gpPosDefault
print "--picName='%-s'" % gpPicName
print ""
print ""
```

```
if help==1: sys.exit()
# open file #
inEigenval = open(inEigenvalName, 'r')
outData = open(outDataName, 'wb')
outGpInput = open('gnuinput.in', 'wb')
# extract 3D coordinates from the input line #
def extract(inLine):
  list = inLine.split()
  x = float(list[0])
  y = float(list[1])
  z = float(list[2])
  return x, y, z
# calculate distance between two points #
def distCrd(x1, y1, z1, x2, y2, z2):
  x21 = abs(x2 - x1)
  y21 = abs(y2 - y1)
  z21 = abs(z2 - z1)
  dist21 = sqrt(x21*x21 + y21*y21 + z21*z21)
  return dist21
# obtain y values(energy) from the "EIGENVAL" file #
def obtainy():
  xMinorCrd = inEigenval.readline()          # xyz
  x, y, z = extract(xMinorCrd)
  i=0; yValueList=[]
  while i < nplotLine:
    list = inEigenval.readline().split()
    yValue = float(list[1])
    yValueList.append(yValue)
    i = i+1
  next = inEigenval.readline()               # blank line
  return x, y, z, yValueList, next
# append x tic label #
def appendXticLabel(order):
  if gpXticLabel==0: xMajorList.append('%s, '% xMinor)
    else:
      if len(gpXticLabel) < order+1: print 'ERROR =>
      Needed more tic labels of x axis!!'; sys.exit()
      if gpXticLabel[order][0:2]=='gk': gpXticLabel[order]
      = "{/Symbol "+gpXticLabel[order][2]+"}"
      xMajorList.append('"%s" %s, '% (gpXticLabel[order],
      xMinor))
      order = order+1
    return order
# read comment(blank) lines #
i=0
while i < blnEigenval:
  if i==5:
    list = inEigenval.readline().split()
    nplotLine = int(list[2])
```

```
  else:
    line = inEigenval.readline()
  i = i+1
# transfer format and write into the output file #
fermiE = (os.popen('grep E-fermi %s | cut -d":" -f 2' %
inOutcarName)).read().split()[0]
print 'Fermi energy is %.4f(eV)\n' % float(fermiE)

j=1; l=0
xMajorList=[]
xMinor, xMinorXb, xMinorYb, xMinorZb = 0, 0, 0, 0

l = appendXticLabel(l)

while 1:
  xMinorXa, xMinorYa, xMinorZa, yValueList, next =
  obtainy()
  dist = distCrd(xMinorXb, xMinorYb, xMinorZb, xMinorXa,
  xMinorYa, xMinorZa)

  if (xMinorXa == xMinorXb and xMinorYa == xMinorYb and
  xMinorZa == xMinorZb): k=0
  else: k=1

  xMinorXb, xMinorYb, xMinorZb = xMinorXa, xMinorYa,
  xMinorZa

  if j==1: dist=0
  xMinor = xMinor + dist

  if not (j>1 and k==0):
    outData.write('%9.4f' % xMinor)
  else:
    l = appendXticLabel(l)

  for yValue in yValueList:
    yValue = yValue-float(fermiE)
    if not (j>1 and k==0): outData.write('%9.4f' % yValue)

  if not (j>1 and k==0): outData.write('\n')
  if not next: break
  j=2

l = appendXticLabel(l)
# use Gnuplot module #
xticsStr = ''.join(xMajorList)

if optPersist==1: optPersist='persist'
```

```
else: optPersist='nopersist'
outGpInput.write("set term %s %s\n" % (gpTerm, optPersist))

outGpInput.write("set title '%s'\n" % gpTitle)
outGpInput.write("set size %s\n" % gpSize)

if gpKey=='key': outGpInput.write("set key\n")
else: outGpInput.write("unset key\n")

outGpInput.write("set xlabel '%s'\n" % gpXlabel)
if gpXrange=='[*:*]': gpXrange = '[*:'+str(xMinor+0.001*xMi
nor)+']'
outGpInput.write("set xrange %s\n" % gpXrange)
outGpInput.write("set xtics (%s)\n" % xticsStr[0:-2])
outGpInput.write("set ylabel '%s'\n" % gpYlabel)
outGpInput.write("set yrange %s\n" % gpYrange)

outGpInput.write("plot '%s' using 1:2 %s\n" % (outDataName,
gpPlotMethod))

i=3
while i < nplotLine+2:
  outGpInput.write("replot '%s' using 1:%d %s\n" %
  (outDataName, i, gpPlotMethod))
  i = i+1

if gpPicForm == 'pos': outGpInput.write("set term %s %s\n" %
(gpPicForm, gpPosDefault))
else: outGpInput.write('set term %s\n' % gpPicForm)

outGpInput.write("set output '%s'\n" % gpPicName)
outGpInput.write("replot '%s' using 1:%d %s\n" %
(outDataName, nplotLine+1, gpPlotMethod))
outGpInput.write("set term %s %s\n" % (gpTerm, optPersist))

inEigenval.close()
outData.close()
outGpInput.close()

#os.popen('gnuplot gnuinput.in') # open windows for
plotting graph on LINUX
#############################################################
```

Appendix D

The Greek alphabet

Α, α alpha	Β, β beta	Γ, γ gamma	Δ, δ delta
Ε, ε epsilon	Ζ, ζ zeta	Κ, η eta	Θ, θ theta
Ι, ι iota	Κ, κ kappa	Λ, λ lambda	Μ, μ mu
Ν, ν nu	Ξ, ξ ksi	Ο, ο omicron	Π, π pi
Ρ, ρ rho	Σ, σ sigma	Τ, τ tau	Υ, υ upsilon
Φ, φ phi	Χ, χ chi	Ψ, ψ psi	Ω, ω omega

Appendix E
SI prefixes

10^{24}	Yotta, Y	10^{21}	Zetta, Z	10^{18}	Exa, E
10^{15}	Peta, P	10^{12}	Tera, T	10^{9}	Giga, G
10^{6}	Mega, M	10^{3}	Kilo, k	10^{2}	Hecto, h
10^{1}	Deka, da	10^{-1}	Deci, d	10^{-2}	Centi, c
10^{-3}	Milli, m	10^{-6}	Micro, μ	10^{-9}	Nano, n
10^{-12}	Pico, p	10^{-15}	Femto, f	10^{-18}	Atto, a
10^{-21}	Zepto, z	10^{-24}	Yocto, y		

Appendix F
Atomic units

Atomic units	Values	Remarks
1 electron charge (e)	1.60218×10^{-19} C	1 eV = 1.6×10^{-19} J
1 electron mass (m_e)	9.11×10^{-31} kg	
1 Bohr (length)	0.529 Å (1s radius in H)	Unit of Ψ: bohr$^{-3/2}$
1 Hartree (energy)	27.21 eV 4.36×10^{-18} J = 627.51 kcal/mole	el.-el. energy at 1 Bohr distance * 1s energy in H = −0.5 Hartree
1 $\hbar$ (h "dash")	1.05457×10^{-34} J·s $h/2\pi$	Reduced Planck's constant
1 Coulomb const.	8.98755×10^{9} kg·m^3·s^{-2}·C^{-2} $1/4\pi\varepsilon_0$	

Index

Note: Page numbers followed by f and t refer to figures and tables, respectively.

E

F

G

H

O

P

T

U

V

For Product Safety Concerns and Information please contact our EU representative GPSR@taylorandfrancis.com
Taylor & Francis Verlag GmbH, Kaufingerstraße 24, 80331 München, Germany

www.ingramcontent.com/pod-product-compliance
Ingram Content Group UK Ltd.
Pitfield, Milton Keynes, MK11 3LW, UK
UKHW021116180425
457613UK00005B/103

* 9 7 8 1 4 9 8 7 4 9 7 3 2 *